METHODS IN MOLECULAR BIOLOGY

Series Editor
John M. Walker
School of Life and Medical Sciences
University of Hertfordshire
Hatfield, Hertfordshire, UK

For further volumes:
http://www.springer.com/series/7651

For over 35 years, biological scientists have come to rely on the research protocols and methodologies in the critically acclaimed *Methods in Molecular Biology* series. The series was the first to introduce the step-by-step protocols approach that has become the standard in all biomedical protocol publishing. Each protocol is provided in readily-reproducible step-by-step fashion, opening with an introductory overview, a list of the materials and reagents needed to complete the experiment, and followed by a detailed procedure that is supported with a helpful notes section offering tips and tricks of the trade as well as troubleshooting advice. These hallmark features were introduced by series editor Dr. John Walker and constitute the key ingredient in each and every volume of the *Methods in Molecular Biology* series. Tested and trusted, comprehensive and reliable, all protocols from the series are indexed in PubMed.

Next Generation Culture Platforms for Reliable In Vitro Models

Methods and Protocols

Edited by

Tiziana A.L. Brevini

Laboratory of Biomedical Embryology, Università degli Studi di Milano, Milan, Italy

Alireza Fazeli

Department of Pathophysiology, University of Tartu, Tartu, Estonia

Kursad Turksen

Ottawa, Ontario, Canada

 Humana Press

Editors
Tiziana A.L. Brevini
Laboratory of Biomedical Embryology
Università degli Studi di Milano
Milan, Italy

Alireza Fazeli
Department of Pathophysiology
University of Tartu
Tartu, Estonia

Kursad Turksen
Ottawa, Ontario, Canada

ISSN 1064-3745   ISSN 1940-6029   (electronic)
Methods in Molecular Biology
ISBN 978-1-0716-1245-3   ISBN 978-1-0716-1246-0   (eBook)
https://doi.org/10.1007/978-1-0716-1246-0

Preface

In Quest of the Perfect In Vitro Model

Almost a quarter of a century ago, the pioneering studies carried out by Mina Bissell, a chemist at the Lawrence Berkeley National Laboratory's Division of Cell and Molecular Biology in California, demonstrated that culturing murine mammary cells in a solution of extracellular matrix, much like what surrounds them in vivo, induced cells to assemble themselves into 3D structures and, most importantly, to regain their ability to produce milk in vitro. These experiments first demonstrated that the way the cells were organized was making them either preserve their physiological function or misbehave and that, in 2D culture systems, cells lost their architecture as well as the hallmark of their tissue of origin. Bissell's work on mammary cells is presently considered to be the beginning of the 3D culture and organoid era and, since these first observations, a great number of tridimensional cell culture platforms have been developed to create models of organs from intestines, liver, pancreas, ovary, and brain that may help in drug testing, organ development, and disease mechanism studies. Indeed, we now have a clear notion that traditional in vitro 2D culture systems fail to imitate the physiological and biochemical features of cells in the original tissue. Growing evidence suggests, as a matter of fact, that the differences between the microenvironment provided by in vitro cell culture models and those of in vivo tissues are significant and can cause deviations in cell response and behavior. This is, in turn, heavily affecting the translation of results from bench to bedside, while, inevitably, causing the use of large numbers of animals in experimental testing. All these aspects eventually contribute to the high attrition rate that the pharmaceutical industry has been struggling with over the past years when trying to introduce a new prescription medicine into the market, largely due to unsatisfactory, misleading testing models during the early preclinical phase.

It is clear that in vitro cell models are one of the milestones of present scientific investigation due to their flexibility, relatively small size, and ease, which clearly points to the need of a huge effort addressed to a continuous improvement of our in vitro models in an attempt to ensure their predictive capability.

In this volume, we have attempted to compile examples of a number of well-characterized, reproducible in vitro platforms in which these designs are being met. Investigators have contributed original protocol chapters that will encourage and stimulate the scientific community to design and produce models for the laboratory that mimic cell guidance conditions as they occur in vivo. The protocols collected describe powerful strategies to exploit chemical cues involved in cell differentiation processes. Special emphasis is given to the use of methods for purification and characterization of exosomes and other secreted vesicles, as well as micro- and noncoding RNAs, that have been demonstrated to control the tuning of the in vivo micro- and macroenvironment, in order to ensure the optimal soluble environment in vitro.

Various methodologies are described to improve/encourage a specific cell phenotype, generated, for example, by modulating substrate stiffness, or creating functional scaffolds and biological matrix, and/or enriching the culture setup with the addition of epigenetic modifiers that boost cell differentiation.

It is evident that the mechanical cues represented by 3D architecture-related information, specific matrix supports, and 3D scaffold systems are still to be optimized. However, the protocols presented here clearly demonstrate that their use is particularly advantageous, to produce organized arrangements of cells in a controllable and reproducible manner, which lead to the derivation of well-defined, naturally organized tissue structure, displaying an architecture consisting of discrete layers of alternative cell types and functions. An even more sophisticated tuning may be envisaged, with the inclusion of nanoparticles that are the same size as many biological entities (such as Coronaviridae). Nanoparticles are the rising stars, the newcomers of modeling in vitro. It is important to keep in mind that they are able to interact with biological matter at subcellular scale and can represent a powerful vehicle for useful factors that may be used to empower or selectively block specific cell pathways. We are convinced that they represent one of the missing pieces of the "perfect model" puzzle and are fully convinced that further research is mandatory in their direction.

The protocols gathered here are faithful to the mission statement of the *Methods in Molecular Biology* series. They are well established and described in an easy to follow step-by-step fashion so as to be valuable for not only experts but also novices in the field. That goal is achieved because of the generosity of the contributors who have shared their protocols in this volume, and we are very grateful for their efforts.

In particular, we would like to thank the COST organization and COST Action CellFit members for their contributions.

Our great appreciation goes to Dr. John Walker, the Editor-in-Chief of the *Methods in Molecular Biology* series, for giving us the opportunity to create this volume and for supporting us along the way.

We are also grateful to Patrick Marton, the Executive Editor of *Methods in Molecular Biology* and the Springer Protocols collection, and Anna Rakovsky, Assistant Editor for *Methods in Molecular Biology*, for their continuous support from idea to completion of this project.

We would also like to thank David C. Casey, the Editor for *Methods in Molecular Biology*, for his outstanding editorial work during the production of this volume.

Finally, we acknowledge Sarumathi Hemachandrane, Anand Ventakachalam, and the rest of the production crew for their work in putting together an outstanding volume.

Milan, Italy   *Tiziana A.L. Brevini*
Tartu, Estonia   *Alireza Fazeli*
Ottawa, ON, Canada   *Kursad Turksen*

Contents

Preface . . . . *v*

Contributors . . . . *ix*

1   Recreating the Follicular Environment: A Customized Approach
for In Vitro Culture of Bovine Oocytes Based on the Origin
and Differentiation State 1
Alberto Maria Luciano, Rodrigo Garcia Barros,
Ana Caroline Silva Soares, Jose Buratini, Valentina Lodde,
and Federica Franciosi

2   Clinostat 3D Cell Culture: Protocols for the Preparation
and Functional Analysis of Highly Reproducible, Large, Uniform
Spheroids and Organoids 17
Krzysztof Wrzesinski, Helle Sedighi Frandsen, Carlemi Calitz,
Chrisna Gouws, Barbara Korzeniowska, and Stephen J. Fey

3   Protocol to Study the Role of Extracellular Vesicles During
Induced Stem Cell Differentiation 63
Kelly C. S. Roballo, Carlos E. Ambrosio, and Juliano C. da Silveira

4   A Simplified Method for Three-Dimensional (3D) Porcine
Preantral Follicles Culture Utilizing Hydrophobic Microbioreactors 75
Malgorzata Duda, Lucia Gizler, and Gabriela Gorczyca

5   Use of Transparent Liquid Marble: Microbioreactor
to Culture Cardiospheres 85
Jeffrey Aalders, Laurens Léger, Davide Piras,
Jolanda van Hengel, and Sergio Ledda

6   Isolation, Culture, and Characterization of Primary
Bovine Endometrial, Epithelial, and Stromal Cells for 3D
In Vitro Tissue Models 103
Antonio Murillo and Marta Muñoz

7   Using Decellularization/Recellularization Processes
to Prepare Liver and Cardiac Engineered Tissues 111
Matteo Ghiringhelli, Yousef Abboud, Snizhanna V. Chorna,
Irit Huber, Gil Arbel, Amira Gepstein, Georgia Pennarossa,
Tiziana A.L. Brevini, and Lior Gepstein

8   Use of Virus-Mimicking Nanoparticles to Investigate Early
Infection Events in Upper Airway 3D Models 131
Georgia Pennarossa, Alireza Fazeli, Sergio Ledda,
Fulvio Gandolfi, and Tiziana A.L. Brevini

9   Creation of a Bioengineered Ovary: Isolation of Female
Germline Stem Cells for the Repopulation of a Decellularized
Ovarian Bioscaffold 139
Georgia Pennarossa, Matteo Ghiringhelli, Fulvio Gandolfi,
and Tiziana A.L. Brevini

10   A Two-Step Protocol to Erase Human Skin Fibroblasts
     and Convert Them into Trophoblast-like Cells 151
     Sharon Arcuri, Fulvio Gandolfi, Edgardo Somigliana,
     and Tiziana A.L. Brevini

11   3D-ViaFlow: A Quantitative Viability Assay for Multicellular Spheroids 159
     Joel Mario Vej-Nielsen and Adelina Rogowska-Wrzesinska

12   Method to Disassemble Spheroids into Core and Rim
     for Downstream Applications Such as Flow Cytometry,
     Comet Assay, Transcriptomics, Proteomics, and Lipidomics................. 173
     Helle Sedighi Frandsen, Martina Štampar, Joel Mario Vej-Nielsen,
     Bojana Žegura, and Adelina Rogowska-Wrzesinska

13   Isolation of Extracellular Vesicles (EVs) Using Size-Exclusion
     High-Performance Liquid Chromatography (SE-HPLC) 189
     Keerthie Dissanayake, Kasun Godakumara, and Alireza Fazeli

14   Isolation of Extracellular Vesicles (EVs) Using Benchtop Size
     Exclusion Chromatography (SEC) Columns 201
     Qurat Ul Ain Reshi, Mohammad Mehedi Hasan,
     Keerthie Dissanayake, and Alireza Fazeli

15   Measurement of the Size and Concentration and Zeta Potential
     of Extracellular Vesicles Using Nanoparticle Tracking Analyzer 207
     Keerthie Dissanayake, Getnet Midekessa, Freddy Lättekivi,
     and Alireza Fazeli

16   Isolation, Characterization, and MicroRNA Analysis
     of Extracellular Vesicles from Bovine Oviduct and Uterine Fluids 219
     Karina Cañón-Beltrán, Meriem Hamdi, Rosane Mazzarella,
     Yulia N. Cajas, Claudia L. V. Leal, Alfonso Gutiérrez-Adán,
     Encina M. González, Juliano C. da Silveira, and Dimitrios Rizos

17   Biomimetic 3D-Bone Tissue Model 239
     Mahmut Parmaksiz, Ayşe Eser Elçin, and Yaşar Murat Elçin

18   Using the Air–Liquid Interface Approach to Foster Apical–Basal
     Polarization of Mammalian Female Reproductive Tract Epithelia In Vitro 251
     Shuai Chen and Jennifer Schoen

19   Preparation of Biological Scaffolds and Primary Intestinal Epithelial
     Cells to Efficiently 3D Model the Fish Intestinal Mucosa 263
     Nicole Verdile, Anna Szabó, Rolando Pasquariello,
     Tiziana A.L. Brevini, Sandra Van Vlierberghe,
     and Fulvio Gandolfi

20   Use of Porous Polystyrene Scaffolds to Bioengineer Human
     Epithelial Tissues In Vitro 279
     Lydia Costello, Nicole Darling, Matthew Freer, Steven Bradbury,
     Claire Mobbs, and Stefan Przyborski

Index *297*

Contributors

JEFFREY AALDERS • *Medical Cell Biology Research Group, Department of Human Structure and Repair, Faculty of Medicine and Health Sciences, Ghent University, Ghent, Belgium*

YOUSEF ABBOUD • *Sohnis Research Laboratory for Cardiac Electrophysiology and Regenerative Medicine, The Rappaport Faculty of Medicine and Research Institute, Technion-Israel Institute of Technology, Haifa, Israel*

CARLOS E. AMBROSIO • *Department of Veterinary Medicine, Faculty of Animal Science and Food Engineering, University of Sao Paulo, São Paulo, Brazil*

GIL ARBEL • *Sohnis Research Laboratory for Cardiac Electrophysiology and Regenerative Medicine, The Rappaport Faculty of Medicine and Research Institute, Technion-Israel Institute of Technology, Haifa, Israel*

SHARON ARCURI • *Laboratory of Biomedical Embryology, Department of Health, Animal Science and Food Safety and Centre for Stem Cell Research, Università degli Studi di Milano, Milan, Italy*

RODRIGO GARCIA BARROS • *Reproductive and Developmental Biology Laboratory, Department of Health, Animal Science and Food Safety, University of Milan, Milan, Italy*

STEVEN BRADBURY • *Department of Biosciences, Durham University, Durham, UK*

TIZIANA A.L. BREVINI • *Laboratory of Biomedical Embryology, Department of Health, Animal Science and Food Safety and Center for Stem Cell Research, Università degli Studi di Milano, Milan, Italy*

JOSE BURATINI • *Department of Structural and Functional Biology, Institute of Biosciences, Sao Paulo State University, Botucatu, Brazil; Biogenesi, Reproductive Medicine Centre, Monza, Italy*

YULIA N. CAJAS • *Department of Animal Reproduction, National Institute for Agriculture and Food Research and Technology (INIA), Madrid, Spain*

CARLEMI CALITZ • *Department of Medical Cell Biology, Uppsala University, Uppsala, Sweden*

KARINA CAÑÓN-BELTRÁN • *Department of Animal Reproduction, National Institute for Agriculture and Food Research and Technology (INIA), Madrid, Spain*

SHUAI CHEN • *Institute of Reproductive Biology, Leibniz Institute for Farm Animal Biology (FBN), Dummerstorf, Germany*

SNIZHANNA V. CHORNA • *Sohnis Research Laboratory for Cardiac Electrophysiology and Regenerative Medicine, The Rappaport Faculty of Medicine and Research Institute, Technion-Israel Institute of Technology, Haifa, Israel*

LYDIA COSTELLO • *Department of Biosciences, Durham University, Durham, UK*

JULIANO C. DA SILVEIRA • *Department of Veterinary Medicine, Faculty of Animal Science and Food Engineering, University of Sao Paulo, São Paulo, Brazil*

NICOLE DARLING • *Department of Biosciences, Durham University, Durham, UK*

KEERTHIE DISSANAYAKE • *Institute of Biomedicine and Translational Medicine, University of Tartu, Tartu, Estonia*

MALGORZATA DUDA • *Department of Endocrinology, Institute of Zoology and Biomedical Research, Faculty of Biology, Jagiellonian University in Krakow, Kraków, Poland*

AYŞE ESER ELÇIN • *Tissue Engineering, Biomaterials and Nanobiotechnology Laboratory, Ankara University Faculty of Science, and Ankara University Stem Cell Institute, Ankara, Turkey*

YAŞAR MURAT ELÇIN • *Tissue Engineering, Biomaterials and Nanobiotechnology Laboratory, Ankara University Faculty of Science, and Ankara University Stem Cell Institute, Ankara, Turkey; Biovalda Health Technologies, Inc., Ankara, Turkey*

ALIREZA FAZELI • *Academic Unit of Reproductive and Developmental Medicine, Department of Oncology and Metabolism, University of Sheffield, Sheffield, UK; Department of Pathophysiology, Institute of Biomedicine and Translational Medicine, University of Tartu, Tartu, Estonia*

STEPHEN J. FEY • *CelVivo ApS, Blommenslyst, Denmark*

FEDERICA FRANCIOSI • *Reproductive and Developmental Biology Laboratory, Department of Health, Animal Science and Food Safety, University of Milan, Milan, Italy*

HELLE SEDIGHI FRANDSEN • *Institute for Biochemistry and Molecular Biology, University of Southern Denmark, Odense, Denmark*

MATTHEW FREER • *Department of Biosciences, Durham University, Durham, UK*

FULVIO GANDOLFI • *Laboratory of Biomedical Embryology, Department of Agricultural and Environmental Sciences—Production, Landscape, Agroenergy and Center for Stem Cell Research, Università degli Studi di Milano, Milan, Italy*

AMIRA GEPSTEIN • *Sohnis Research Laboratory for Cardiac Electrophysiology and Regenerative Medicine, The Rappaport Faculty of Medicine and Research Institute, Technion-Israel Institute of Technology, Haifa, Israel*

LIOR GEPSTEIN • *Sohnis Research Laboratory for Cardiac Electrophysiology and Regenerative Medicine, The Rappaport Faculty of Medicine and Research Institute, Technion-Israel Institute of Technology, Haifa, Israel; Cardiology Department, Rambam Health Care Campus, Haifa, Israel*

MATTEO GHIRINGHELLI • *Sohnis Research Laboratory for Cardiac Electrophysiology and Regenerative Medicine, The Rappaport Faculty of Medicine and Research Institute, Technion-Israel Institute of Technology, Haifa, Israel*

LUCIA GIZLER • *Department of Endocrinology, Institute of Zoology and Biomedical Research, Faculty of Biology, Jagiellonian University in Krakow, Kraków, Poland*

KASUN GODAKUMARA • *Institute of Biomedicine and Translational Medicine, University of Tartu, Tartu, Estonia*

ENCINA M. GONZÁLEZ • *Department of Anatomy and Embryology, Veterinary Faculty, Complutense University of Madrid (UCM), Madrid, Spain*

GABRIELA GORCZYCA • *Department of Endocrinology, Institute of Zoology and Biomedical Research, Faculty of Biology, Jagiellonian University in Krakow, Kraków, Poland*

CHRISNA GOUWS • *Pharmacen™, Centre of Excellence for Pharmaceutical Sciences, North-West University, Potchefstroom, South Africa*

ALFONSO GUTIÉRREZ-ADÁN • *Department of Animal Reproduction, National Institute for Agriculture and Food Research and Technology (INIA), Madrid, Spain*

MERIEM HAMDI • *Department of Animal Reproduction, National Institute for Agriculture and Food Research and Technology (INIA), Madrid, Spain*

MOHAMMAD MEHEDI HASAN • *Institute of Biomedicine and Translational Medicine, University of Tartu, Tartu, Estonia*

IRIT HUBER • *Sohnis Research Laboratory for Cardiac Electrophysiology and Regenerative Medicine, The Rappaport Faculty of Medicine and Research Institute, Technion-Israel Institute of Technology, Haifa, Israel*

BARBARA KORZENIOWSKA • *CelVivo ApS, Blommenslyst, Denmark*

FREDDY LÄTTEKIVI • *Institute of Biomedicine and Translational Medicine, University of Tartu, Tartu, Estonia*

CLAUDIA L. V. LEAL • *Department of Animal Reproduction, National Institute for Agriculture and Food Research and Technology (INIA), Madrid, Spain; Department of Veterinary Medicine, College of Animal Science and Food Engineering, University of São Paulo, Pirassununga, São Paulo, Brazil*

SERGIO LEDDA • *Department of Veterinary Medicine, University of Sassari, Sassari, Italy*

LAURENS LÉGER • *Medical Cell Biology Research Group, Department of Human Structure and Repair, Faculty of Medicine and Health Sciences, Ghent University, Ghent, Belgium*

VALENTINA LODDE • *Reproductive and Developmental Biology Laboratory, Department of Health, Animal Science and Food Safety, University of Milan, Milan, Italy*

ALBERTO MARIA LUCIANO • *Reproductive and Developmental Biology Laboratory, Department of Health, Animal Science and Food Safety, University of Milan, Milan, Italy*

ROSANE MAZZARELLA • *Department of Veterinary Medicine, College of Animal Science and Food Engineering, University of São Paulo, Pirassununga, São Paulo, Brazil*

GETNET MIDEKESSA • *Institute of Biomedicine and Translational Medicine, University of Tartu, Tartu, Estonia*

CLAIRE MOBBS • *Department of Biosciences, Durham University, Durham, UK*

MARTA MUÑOZ • *Servicio Regional de Investigación y Desarrollo Agroalimentario (SERIDA), Asturias, Spain*

ANTONIO MURILLO • *Producción Animal e Industrialización, Facultad de Ingeniería, Universidad Nacional de Chimborazo (UNACH), Riobamba, Ecuador*

MAHMUT PARMAKSIZ • *Tissue Engineering, Biomaterials and Nanobiotechnology Laboratory, Ankara University Faculty of Science, and Ankara University Stem Cell Institute, Ankara, Turkey*

ROLANDO PASQUARIELLO • *Department of Agricultural and Environmental Sciences— Production, Landscape, Agroenergy, Center for Stem Cell Research, Università degli Studi di Milano, Milan, Italy*

GEORGIA PENNAROSSA • *Laboratory of Biomedical Embryology, Department of Health, Animal Science and Food Safety and Center for Stem Cell Research, Università degli Studi di Milano, Milan, Italy*

DAVIDE PIRAS • *Department of Veterinary Medicine, University of Sassari, Sassari, Italy*

STEFAN PRZYBORSKI • *Department of Biosciences, Durham University, Durham, UK; Reprocell Europe, Sedgefield, UK*

QURAT UL AIN RESHI • *Institute of Biomedicine and Translational Medicine, University of Tartu, Tartu, Estonia*

DIMITRIOS RIZOS • *Department of Animal Reproduction, National Institute for Agriculture and Food Research and Technology (INIA), Madrid, Spain*

KELLY C. S. ROBALLO • *Department of Veterinary Medicine, Faculty of Animal Science and Food Engineering, University of Sao Paulo, São Paulo, Brazil; College of Health Sciences, School of Pharmacy, University of Wyoming, Laramie, WY, USA*

ADELINA ROGOWSKA-WRZESINSKA • *Institute for Biochemistry and Molecular Biology, University of Southern Denmark, Odense, Denmark*

JENNIFER SCHOEN • *Institute of Reproductive Biology, Leibniz Institute for Farm Animal Biology (FBN), Dummerstorf, Germany*

ANA CAROLINE SILVA SOARES • *Department of Structural and Functional Biology, Institute of Biosciences, Sao Paulo State University, Botucatu, Brazil*

EDGARDO SOMIGLIANA • *Ospedale Maggiore Policlinico, Università degli Studi di Milano, Milan, Italy*

MARTINA ŠTAMPAR • *Department of Genetic Toxicology and Cancer Biology, National Institute of Biology, Ljubljana, Slovenia*

ANNA SZABÓ • *Polymer Chemistry and Biomaterials Group, Centre of Macromolecular Chemistry, Ghent University, Ghent, Belgium*

JOLANDA VAN HENGEL • *Medical Cell Biology Research Group, Department of Human Structure and Repair, Faculty of Medicine and Health Sciences, Ghent University, Ghent, Belgium*

SANDRA VAN VLIERBERGHE • *Polymer Chemistry and Biomaterials Group, Centre of Macromolecular Chemistry, Ghent University, Ghent, Belgium*

JOEL MARIO VEJ-NIELSEN • *Institute for Biochemistry and Molecular Biology, University of Southern Denmark, Odense, Denmark*

NICOLE VERDILE • *Department of Agricultural and Environmental Sciences—Production, Landscape, Agroenergy, Center for Stem Cell Research, Università degli Studi di Milano, Milan, Italy*

KRZYSZTOF WRZESINSKI • *CelVivo ApS, Blommenslyst, Denmark; Pharmacen™, Centre of Excellence for Pharmaceutical Sciences, North-West University, Potchefstroom, South Africa*

BOJANA ŽEGURA • *Department of Genetic Toxicology and Cancer Biology, National Institute of Biology, Ljubljana, Slovenia*

Chapter 1

Recreating the Follicular Environment: A Customized Approach for In Vitro Culture of Bovine Oocytes Based on the Origin and Differentiation State

Alberto Maria Luciano, Rodrigo Garcia Barros, Ana Caroline Silva Soares, Jose Buratini, Valentina Lodde, and Federica Franciosi

Abstract

The mammalian ovary is a large source of oocytes organized into follicles at various stages of folliculogenesis. However, only a limited number of them can be used for in vitro embryo production (IVEP), while most have yet to complete growth and development to attain full meiotic and embryonic developmental competence. While the in vitro growth of primordial follicles in the ovarian cortex has the potential to produce mature oocytes, it is still at an experimental stage. The population of early antral follicles (EAFs), instead, may represent a reserve of oocytes close to completing the growth phase, which might be more easily exploited in vitro and could increase the number of female gametes dedicated to IVEP.

Here we present in vitro culture strategies that have been developed utilizing physiological parameters to support the specific needs of oocytes at distinct stages of differentiation, in order to expand the source of female gametes for IVEP by maximizing the attainment of fertilizable oocytes. Furthermore, these culture systems provide powerful tools to dissect the molecular processes that direct the final differentiation of the mammalian oocyte.

Key words Oocyte, Cumulus cells, Gap junction, Chromatin, Germinal vesicle, Intercellular communication, Meiotic arrest, Prematuration, Oogenesis, In vitro oocyte growth

1 Introduction

In vitro embryo production (IVEP) technologies have been developed to allow the transfer of embryos obtained through in vitro maturation (IVM) of immature oocytes, followed by in vitro fertilization (IVF) and in vitro culture (IVC) of early embryos. However, the efficiency is still limited, and the small supply of mature, fertilizable oocytes is one of the main limitations to the success rate of assisted reproduction, in cattle as well as in other mammals.

Tiziana A.L. Brevini et al. (eds.), *Next Generation Culture Platforms for Reliable In Vitro Models: Methods and Protocols*, Methods in Molecular Biology, vol. 2273, https://doi.org/10.1007/978-1-0716-1246-0_1,

The mammalian ovary is a potential large source of oocytes that are enclosed into follicles at various stages of development, from the primordial to the ovulatory ones. Nonetheless, only a limited number of such oocytes can be readily submitted to routine IVEP protocols and generate viable embryos. These oocytes are the ones that have reached the so-called fully grown stage and are enclosed in middle and large antral follicles (MAFs and LAFs, respectively). Yet not all of them reach the blastocyst stage of embryonic development when subjected to standard IVEP. Furthermore, most of the oocytes in an ovary have yet to reach the fully grown stage and ultimately acquire full meiotic and embryonic developmental competence [1]. Specifically, these oocytes span from the ones enclosed in primordial follicles to the ones enclosed in early antral follicles (EAFs), that are still in the growing phase.

While in vitro growth of primordial follicles enclosing the resting oocyte pool in the ovarian cortex has the potential to produce mature oocytes [2], this technique is still experimental, and an efficient exploitation of this reserve seems far from being applicable on a large scale. By contrast, the population of oocytes enclosed in EAFs may represent a more readily exploitable source that would increase the number of female gametes that can successfully undergo IVEP [3] (Table 1). Finally, optimization of culture systems for fully grown oocytes isolated from MAFs may increase the overall efficiency.

Oocytes enclosed in EAFs and MAFs represent an extremely heterogeneous cell population, with very distinct morphological

Table 1
Extent of follicle reserve and follicle categories in 4–8-year-old bovine ovaries (data were extrapolated from [44, 74–77]). The bottom part of the table are the proposed optimal exploitation strategies

Follicle category	Primordial (<0.1 mm)	Primary and secondary (>0.1 to <0.5)	Early antral (≥0.5 to <2 mm)	Antral (2–8 mm)
Number/ovary	65,000	25,000	120	25
Atresia	<5%	5%	30%	60%
Chromatin	–	–	GV0	GV1, GV2, GV3
Optimal exploitation strategies				
Ex-vivo (following culling) ovaries	Freezing/ culture [2]	Freezing/ culture [2]	L-IVCO [3, 17]	Pre-IVM (GV1) and direct IVM (GV2 and GV3) [9, 70]
In vivo	–	–	–	Synchronization followed by OPU and direct IVM [15]

and functional characteristics [4, 5]. In bovine, nearly all of the oocytes isolated from EAFs (0.5–2 mm) are still growing, the chromatin appears mostly uncondensed and dispersed throughout the nucleoplasm, in the so-called GV0 configuration [6]. At this stage, oocytes are still transcriptionally active and are functionally fully coupled through gap junctions (GJs) to the surrounding cumulus cells [6–8]. On the other hand, oocytes from MAFs (2–8 mm), which are the most commonly used for IVEP, are characterized by various degrees of progressive chromatin compaction, namely GV1, GV2, and GV3 configurations. The 3 classes are equally represented within the populations of oocytes collected from MAFs [6, 9]. The transition from GV0 to the higher classes of chromatin compaction, up to GV3, is accompanied by progressive transcriptional silencing [7], changes in the epigenetic signatures such as global DNA methylation [10] and histone modifications [11, 12] and changes in cytoplasmic organelle redistribution and nuclear architecture [7]. More importantly, these changes are also accompanied by a gradual acquisition of meiotic and developmental competence [3, 6, 7, 13, 14].

In the present work, we describe cultural strategies customized to better fulfill the physiological needs of oocytes at the distinct stages of differentiation. The described procedures induced an increased chromatin compaction, that was accompanied by a progressive and significant increase in developmental capacity (Fig. 1). Notably, the transition to higher degrees of chromatin compaction is associated with an increase in developmental competence and has

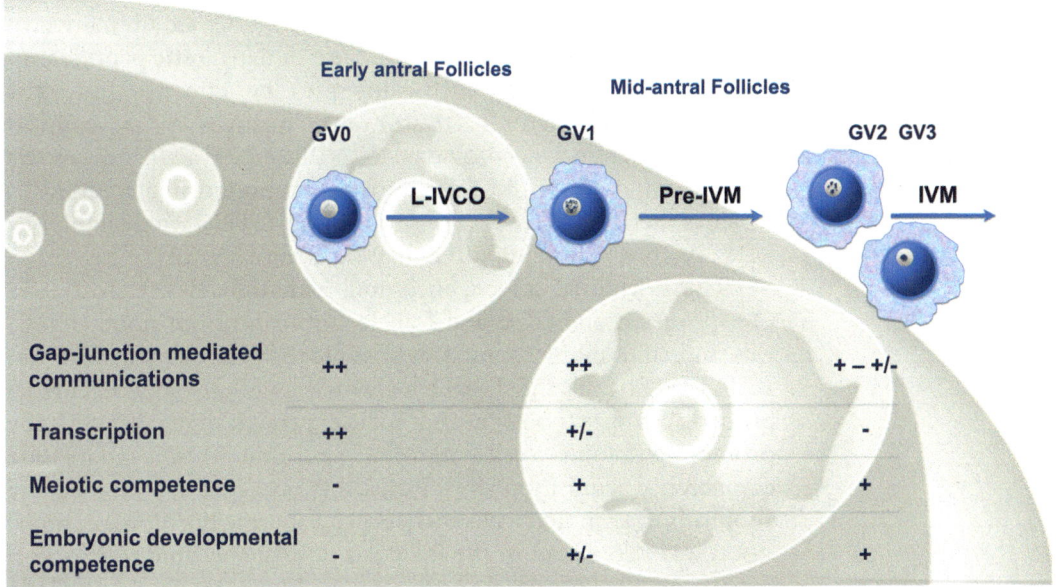

Fig. 1 Multiple approach: a customized culture system for each stage specific need

recently been observed in vivo [15, 16]. Specifically, oocytes that were "synchronized" at the GV2 stage using a mild FSH stimulation before ovum pick up (OPU), reached significantly higher blastocyst rates when submitted to IVM/IVF/IVC [15].

Precisely, in this method chapter we will describe the following:

1. The long in vitro culture of oocytes (L-IVCO), targeted to growing oocytes isolated from EAFs [17];

2. The prematuration (pre-IVM), mostly beneficial to the GV1-enriched population of oocytes isolated from MAFs [9];

3. The IVM, mostly beneficial to the GV2 and GV3-enriched population of oocytes isolated from MAFs [9].

The rationale adopted in designing and optimizing the above-listed treatments stems from the idea of recreating the follicular environment in which a developmentally competent oocyte grows and differentiates until the first meiotic division is completed.

Each of the proposed protocols has been optimized considering the following main physiological characteristics and experimental evidences.

Once the oocytes have acquired the molecular machinery necessary to resume meiosis, the follicular environment acts to keep the cell cycle arrested by supporting the intraoocyte content of cyclic nucleotides [18]. In vivo, the protracted arrest ensures that the oocyte undergoes the final steps of differentiation that confer developmental competence, by avoiding untimely meiotic resumption. This condition was recreated in vitro by supplementing the L-IVCO and pre-IVM media with cilostamide [3, 14, 19–21], an inhibitor of phosphodiesterase-3 (PDE3) which is specific to the oocyte in bovine as in most mammals [22]. As an alternative to pharmacological treatment, also the use of natriuretic peptide precursor C (NPPC) has been shown to be effective [14, 23–28]. NPPC is the natural activator of the guanylyl cyclase–coupled natriuretic peptide receptor type-2 (NPR2) [29] that induces the production of $3'$–$5'$ cyclic guanosine monophosphate (cGMP), which is then transferred via GJs to the oocyte [30] where it inhibits PDE3A, thus maintaining the meiotic arrest [31, 32].

The hormonal milieu, including estradiol (E_2), testosterone (T), progesterone (P_4), and follicle stimulating hormone (FSH), was studied to mimic as much as possible the follicular conditions in EAFs [33–36]. E_2 and T (which also acts as an estrogen precursor) primarily support oocyte and follicle growth [21, 37, 38], by promoting granulosa cells intercellular communication and meiotic competence acquisition [39]. P_4 has instead a role in inhibiting the apoptosis in granulosa cells [40]. Since a specific hormonal composition was also found in the follicular fluid at the time of selection for dominance [28, 41–44], the pre-IVM medium can be similarly supplemented with E_2, T, and P_4 [28].

Studies conducted in several mammalian species show that supplementation of the IVM medium with FSH improves oocyte quality [45–50], prompting the widespread use of FSH as hormonal supplementation in IVM protocols. Even though there is not consensus on the exact signaling mode, FSH at high dosage seems to exert its effects through the activation of the epidermal growth factor (EGF) network [45, 50–53], which is physiologically triggered in vivo by the surge in luteinizing hormone (LH) [54]. However, FSH levels similar to the ones used in IVM proved to be harmful for the culture of growing oocytes, that underwent early closure of the GJ-mediated communications with the cumulus cells, precocious meiotic resumption, and insufficient growth [3]. The preservation of cumulus oocyte communication functionality during in vitro growth and prematuration protocols represents a key point, and specifically the intracellular cAMP concentration evoked by FSH stimulation plays a crucial role [3, 13, 19, 20, 55–65]. By conducting dose–response curves, it was established that one thousand times less FSH was instead beneficial during the growing phase, sustained the functionality of the GJs, and promoted transcriptional activity and oocyte growth and differentiation [3].

During pre-IVM, FSH concentrations similar to the ones used in L-IVCO also improved oocyte quality by sustaining GJ coupling with the cumulus cells [9, 13, 14, 28].

Other aspects that are instead peculiar to the L-IVCO aimed at promoting a 3D structure and supporting the transcriptional activity. Specifically, the first was achieved by increasing the viscosity of the medium to mimic the physiological viscosity of the follicular fluid, and by culturing the cumulus oocyte complexes (COCs) on a collagen coated surface (Fig. 2) [17, 66].

Finally the inclusion of zinc sulfate is supported by recent results showing the role of this trace element in promoting the differentiation and transcriptional activity of bovine growing oocytes [67].

The procedures described herein represent multiple culture strategies that have been developed to support the specific oocyte needs in relation to the developmental step, with the aim of expanding the source of oocytes usable for IVEP. Nonetheless, these protocols will provide useful tools to dissect the cellular and molecular process that control the final oocyte development.

2   Materials

All media are prepared with embryo tested ultrapure water. Disposable, sterile plasticware is from BD Falcon by Corning, NUNC IVF Line, and Sterilin™ by Thermo Scientific, where duly specified. Final filtration of all stock solutions, as well as preparation of

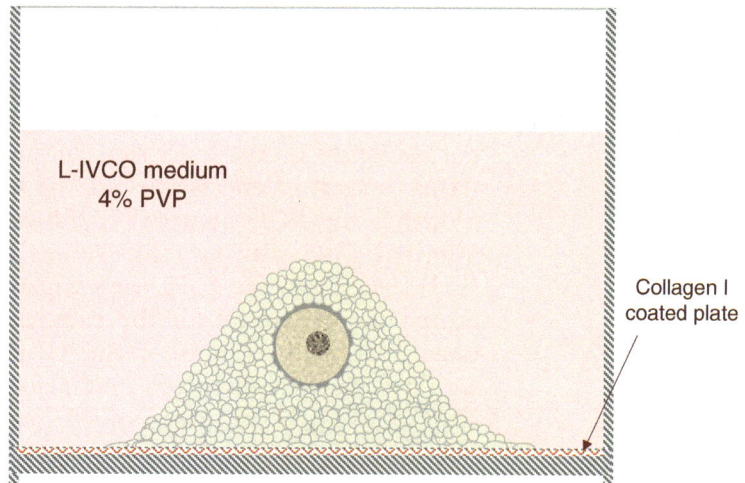

Fig. 2 Schematic representation of the COCs growing in the BioCoat™ Collagen I 96-well plate during L-IVCO. To promote the three-dimensional organization of cells, the COCs are cultured in L-IVCO medium where viscosity is increased by adding 4% PVP to mimic the physiological viscosity of the follicular fluid and by culturing the COCs on a collagen coated surface that promote the formation of a 3D-like culture [17, 66].The stem of granulosa cells adheres to the coated bottom while the cumulus oophorous protrudes into the medium in the center of the well

working solutions, is performed using sterile techniques under a biohazard laminar flow cabinet or laminar flow hood to keep sterility. All glassware is exclusively used for embryo culture media preparation and is high pressure steam-sterilized by autoclaving at 121 °C for 20 min. After use, glassware is immediately washed and rinsed with running tap water for 30 min, rinsed three times with 18.2 mΩ water, then dried completely and covered with aluminum foil until sterilization.

All the procedures are conducted at 26–28 °C unless otherwise specified.

2.1   Collection Media

1. Prepare HEPES-based manipulation medium (HM199) from Medium 199 with Earle's salts supplemented with 0.68 mM L-glutamine, 25 mM HEPES, 0.4% bovine serum albumin (BSA) Fraction V, 0.164 mM penicillin, 0.048 mM streptomycin, and 1790 u/L heparin. HM199 are aliquoted in 20 mL in Sterilin tube (Thermo Scientific) and stored at 4 °C for 6 months.

2. Prepare meiotic arrest holding medium (HM199-Cilo) with HM199 supplemented with 5 μM of the selective PDE3 inhibitor cilostamide [14] (*see* **Note 1**).

3. Warm sealed aliquots of HM199 and HM199-Cilo at 38.5 °C in a dry thermostatic chamber before use.

2.2 Long In Vitro Oocyte Culture (L-IVCO) Medium

1. Prepare the basic oocyte culture medium (bM199) from Medium 199 with Earle's salts supplemented with 25 mM sodium bicarbonate, 21.3 µg/mL phenol red, 75 µg/mL kanamycin, and 4% polyvinylpyrrolidone (PVP; 360k molecular weight) (*see* **Note 2**).

2. Five hours before culture (*see* **Note 3**), prepare the L-IVCO medium by supplementing the above bM199 with 2 mM GlutaMAX™, 0.4% BSA fatty acid free, 0.2 mM sodium pyruvate, 0.1 mM cysteamine, 0.15 µg/mL zinc sulfate, 10^{-4} IU/mL recombinant human FSH (r-hFSH), 10 ng/mL E_2, 50 ng/mL T, 50 ng/mL P_4, and 5 µM Cilostamide [17].

3. Fill a BioCoat™ Collagen I 96-well plate (Corning) with 200 µL of L-IVCO medium and equilibrate for at least 4 h before use for oocyte culture at 38.5 °C and 5% CO_2 in air, maximum humidity. For all additives' storage, *see* **Note 4**.

2.3 Pre-IVM Medium

1. Prepare the oocyte pre-IVMmedium from Medium 199 with Earle's salts supplemented with 25 mM sodium bicarbonate, 2 mM GlutaMAX™, 0.4% BSA fatty acid free, 0.2 mM sodium pyruvate, 0.1 mM cysteamine, 50 µg/mL of kanamycin, 10^{-4} IU/mL r-hFSH, and 10 µM Cilostamide [9, 14] (*see* **Note 5**). Equilibrate in the incubator at 38.5 °C and 5% CO_2 in air, maximum humidity (*see* **Note 6**).

2. Fill each well of a NUNC IVF four-well plate (Thermo Scientific) with 500 µL of pre-IVM medium and equilibrate for at least 4 h before use at 38.5 °C and 5% CO_2 in air, maximum humidity.

2.4 IVM Medium

1. Prepare IVM medium from Medium 199 with Earle's salts supplemented with 25 mM sodium bicarbonate, 2 mM GlutaMAX™, 0.4% BSA fatty acid free, 0.2 mM sodium pyruvate, 0.1 mM cysteamine, 50 µg/mL of kanamycin, and 0.1 IU/mL of r-hFSH.

2. Fill each well of a NUNC IVF four-well plate with 500 µL of IVM medium and equilibrate for at least 4 h before use for oocyte culture at 38.5 °C and 5% CO_2 in air, maximum humidity.

3 Methods

The recommended operative procedure is to first proceed with the isolation of COCs from MAFs (2–8 mm) by aspiration through a needle connected to a vacuum pump (Subheading 3.1). These COCs will be divided in subgroups and either subjected directly to IVM (Subheading 3.2) or to pre-IVM (Subheading 3.3) followed by IVM as detailed below. COCs from EAFs (0.5–2 mm) are isolated by individual follicle dissection from slices of ovarian cortex (Subheading 3.4) taken from the ovaries that already underwent MAFs aspiration. These oocytes are submitted to L-IVCO (Subheading 3.5) and, at the end of L-IVCO, can be transferred to IVM.

All the procedures are conducted at 36–38 °C, unless otherwise indicated.

3.1 Isolation of COCs from MAFs

1. Collect bovine ovaries from 4 to 8-year-old Holstein dairy cows subjected to routine veterinary inspection and in accordance to the specific health requirements stated in Council Directive 89/556/ECC and subsequent modifications, at the abattoir. Transport to the laboratory, within 3 h, in sterile saline maintained at 28 °C.

2. Retrieve COCs from 2 to 8 mm follicles with an 18-gauge needle mounted on an aspiration pump (COOK-IVF, Australia) with a vacuum pressure of 28 mm/Hg (see **Note 7**). Once the content of the MAFs has been aspirated, keep the ovaries in warm saline for procedures described in Subheading 3.4.

3. Examine the morphology of the COCs under a stereomicroscope and select the complexes medium brown in color, with five or more complete compact layers of cumulus cells [68, 69]. These COCs will be further divided in 3 subclasses, according to previously described morphological features [9, 70, 71]: Class 1, with homogeneous ooplasm and compact cumulus cells; Class 2, with minor granulation of the ooplasm and compact cumulus cells; Class 3, with highly granulated ooplasm and slight expansion of cumulus cell layers (Fig. 3). As previously demonstrated, Class 1 is GV1-enriched and Class 2/3 are GV2/3-enriched [9].

3.2 IVM

Group the Class 2 and Class 3 COCs, wash twice in HM199, then culture in groups of 30–34 in 500 μL of the IVM medium for 24 h, in four-well dishes at 38.5 °C under 5% CO_2 in humidified air (see **Note 8**).

3.3 Pre-IVM

Wash twice the Class 1 COCs in HM199, then culture in groups of approximately 30 COCs in 500 μL of the pre-IVM medium for 6 h, in four-well dishes (NUNC) at 38.5 °C under 5% CO_2 in humidified air. After pre-IVM culture, wash COCs of Class 1 twice in HM199, then transfer them in 500 μL of IVM medium for 24 h, in four-well dishes (NUNC) at 38.5 °C under 5% CO_2 in humidified air (see **Note 8**).

3.4 Isolation of COCs from EAFs

1. Prepare ovarian cortex slices as previously described [17]. Under a horizontal laminar flow hood, with warm plate and a stereomicroscope equipped with a heating stage, place

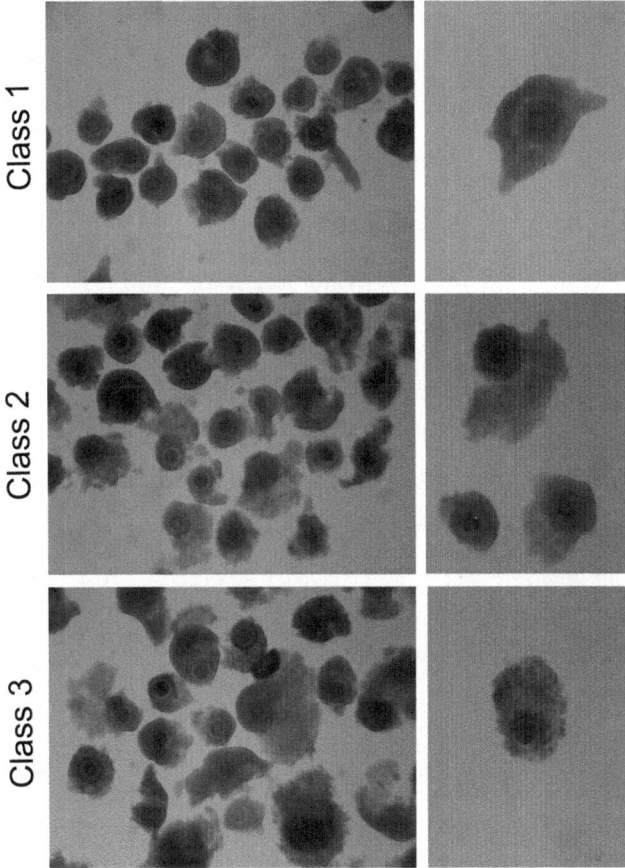

Fig. 3 Representative pictures of COCs isolated from MAFs. After collection, COCs are separated according to morphological criteria. Class 1: homogeneous ooplasm and absence of expansion of outer layer of cumulus cells; Class 2: minor granulation of the ooplasm and/or beginning of expansion of outer layer of cumulus cells; Class 3: highly granulated ooplasm and few cumulus cells layers showing expansion

one ovary at the time on a sterile Teflon cutting board. Using a surgical blade n. 22 mounted on a scalpel handle, cut slices 1.5–2 mm thick of ovarian cortex and parallel to the major axis of the organ. Maintain the slices of ovarian cortex in a sterile glass petri dish covered with HM199 on a warm plate at 38.5 °C.

2. Place the ovarian cortex slice in a 60 mm glass petri dish with 2–3 mL of HM199. Under the stereomicroscope select the follicles between 0.5 and 2 mm by using a micrometer-equipped eyepiece. Remove the ovarian stroma surrounding the follicle to expose it on the edge using a surgical blade no. 22 mounted on a scalpel handle.

Carefully, holding the slice with tweezers, make a slit on the follicle wall using a 26G needle mounted on a syringe to release the follicular content, including the COC, follicular fluid and clumps of cells.

3. Under the stereomicroscope select COCs medium brown in color, with five or more complete compact layers of cumulus cells, intact zona pellucida and homogeneous cytoplasm. The diameter of the oocyte, excluding the zona pellucida, must be within a range of 100–110 μm (*see* **Note 9**). Place the isolated COCs in HM199-Cilo.

Limit each round of COCs isolation to 30 min, then incubate the selected COCs in a 35 mm petri dish with 2 mL of L-IVCO at 38.5 °C under 5% CO_2 in humidified air, until the collection procedure is completed and the COCs can be transferred individually into a well of the L-IVCO 96-well plate.

3.5   Long In Vitro Oocyte Culture

1. Place each selected COC singularly in the center of a well of the 96-well plate containing the previously prepared L-IVCO medium (*see* Subheading 2.2) and incubate the plate for 5 days at 38.5 °C and 5% CO_2 in air, maximum humidity.

2. On day 2 and day 4 renew half of the medium by removing 100 μL of medium and replacing with 100 μL of freshly prepared and incubator-equilibrated L-IVCO medium. Perform the medium renewal gently under the stereomicroscope to avoid detaching of the COCs from the bottom of the well.

3. On day 5, select all COCs with compact cumulus cell investment, with no sign of cumulus expansion and cell degeneration while discard COCs showing abundant loss of cumulus cells extending for more than 50% of the oocyte surface, and signs of cell degeneration and cell debris, as previously described in [17].

4. Wash the selected in vitro grown COCs twice in HM199 then transfer them in 500 μL of IVM medium for 24 h, in four-well dishes (NUNC) at 38.5 °C under 5% CO_2 in humidified air (*see* **Note 8**).

4   Notes

1. Meiotic arrest during the collection procedure can also be maintained by using the nonselective PDE inhibitor 3-isobutyl-1-methyl-xanthine (IBMX) at the final concentration 0.5 mM [13].

2. To prepare the 4% PVP basic L-IVCO, first prepare an 8% PVP solution by dissolving 40 g of PVP in 500 mL of embryo tested

cell culture water in a sterile bottle. Seal the cap tightly and keep in a water bath at 100 °C for 40 min. Leave to cool. Prepare basic L-IVCO 2× by dissolving the Medium 199 with Earle's salts powder in 500 mL of embryo tested cell culture water and supplement with 25 mM sodium bicarbonate, 21.3 μg/mL of phenol red, 75 μg/mL of kanamycin. Adjust the pH to 7.4 and filter-sterilize. Mix basic L-IVCO 2× and 8% PVP solution 1:1 to obtain basic L-IVCO medium.

3. All $NaHCO_3$-buffered media must be equilibrated in the incubator for a minimum of 4 h before use. This step allows the gas exchange with the correct CO_2 pressure to obtain a pH of approximately 7.4.

4. Additives such as sodium pyruvate and cysteamine are prepared as stock solutions 100× and stored at −20 °C for up to 3 months. The BSA stock solution is 50× and is stored at +4 °C for up to 3 months. The r-hFSH is prepared as 100 IU/mL stock in PBS containing 0.1% of BSA and stored at −20 °C up to 3 months. Steroids are prepared as stock solutions 1000× in absolute ethanol and stored at −20 °C for up to 1 year.

5. For pre-IVM, depending on the experimental design, cilostamide can be replaced with 100 nM natriuretic peptide precursor C (NPPC) [14, 28, 72].

6. In order to further reproduce the follicle environment during the phase of selection for dominance, pre-IVM medium can additionally be supplemented with 500 ng/mL E_2, 50 ng/mL P_4, and 50 ng/mL T [16, 28].

7. During follicle aspiration, aspirate aliquots of about 0.5 mL of HM199 through the needle into the tube circuit every 2 mL of aspirated liquid to prevent formation of clots. Avoid excessive dilution of follicular fluid with HM199 to effectively maintain oocytes in meiotic arrest [73]. The whole procedure (follicle aspiration and COCs selection) must be performed in approximately 60 min to avoid meiotic resumption of Class 1 oocytes. Alternatively, use the HM199-Cilo as holding medium to pool all Class 1 COCs during selection.

8. After IVM, oocytes can be submitted to IVF and presumptive zygotes cultured for 7 days as previously described [3, 9, 14, 28].

9. Measure the diameter of the oocytes using a micrometer-equipped eyepiece or other appropriate tools such as a microscope-mounted camera and software that allows measurements. The diameter of the oocyte excluding the zona pellucida must be within a range of 100–110 μm. Discard COCs with larger or smaller oocytes and with irregular and nonround shaped oocyte.

Acknowledgments

This work was supported by Regione Lombardia PSR 2014/2020 - Operazione 10.2.01 "Conservazione della Biodiversità Animale e Vegetale"—no. 201801061529—INNOVA to AML and by MIUR, PRIN 2017 Linea b n. 20172N2WL3_002 to FF.

References

1. Conti M, Franciosi F (2018) Acquisition of oocyte competence to develop as an embryo: integrated nuclear and cytoplasmic events. Hum Reprod Update 24(3):245–266

2. Telfer EE (2019) Future developments: in vitro growth (IVG) of human ovarian follicles. Acta Obstet Gynecol Scand 98 (5):653–658

3. Luciano AM, Franciosi F, Modina SC, Lodde V (2011) Gap junction-mediated communications regulate chromatin remodeling during bovine oocyte growth and differentiation through cAMP-dependent mechanism(s). Biol Reprod 85(6):1252–1259

4. Fair T, Hyttel P, Greve T (1995) Bovine oocyte diameter in relation to maturational competence and transcriptional activity. Mol Reprod Dev 42(4):437–442

5. Luciano AM, Lodde V (2013) Changes of large-scale chromatin configuration during mammalian oocyte differentiation. In: Coticchio G, Albertini DF, De Santis L (eds) Oogenesis. Springer, London, pp 93–108

6. Lodde V, Modina S, Galbusera C, Franciosi F, Luciano AM (2007) Large-scale chromatin remodeling in germinal vesicle bovine oocytes: interplay with gap junction functionality and developmental competence. Mol Reprod Dev 74(6):740–749

7. Lodde V, Modina S, Maddox-Hyttel P, Franciosi F, Lauria A, Luciano AM (2008) Oocyte morphology and transcriptional silencing in relation to chromatin remodeling during the final phases of bovine oocyte growth. Mol Reprod Dev 75(5):915–924

8. Luciano AM, Franciosi F, Dieci C, Lodde V (2014) Changes in large-scale chromatin structure and function during oogenesis: a journey in company with follicular cells. Anim Reprod Sci 149(1-2):3–10

9. Dieci C, Lodde V, Labreque R, Dufort I, Tessaro I, Sirard MA, Luciano AM (2016) Differences in cumulus cell gene expression indicate the benefit of a pre-maturation step to improve in-vitro bovine embryo production. Mol Hum Reprod 22(12):882–897

10. Lodde V, Modina SC, Franciosi F, Zuccari E, Tessaro I, Luciano AM (2009) Localization of DNA methyltransferase-1 during oocyte differentiation, in vitro maturation and early embryonic development in cow. Eur J Histochem 53 (4):e24

11. Labrecque R, Lodde V, Dieci C, Tessaro I, Luciano AM, Sirard MA (2015) Chromatin remodelling and histone m RNA accumulation in bovine germinal vesicle oocytes. Mol Reprod Dev 82(6):450–462

12. Lodde V, Luciano AM, Franciosi F, Labrecque R, Sirard MA (2017) Accumulation of chromatin remodelling enzyme and histone transcripts in bovine oocytes. Results Probl Cell Differ 63:223–255

13. Lodde V, Franciosi F, Tessaro I, Modina SC, Luciano AM (2013) Role of gap junction-mediated communications in regulating large-scale chromatin configuration remodeling and embryonic developmental competence acquisition in fully grown bovine oocyte. J Assist Reprod Genet 30(9):1219–1226

14. Franciosi F, Coticchio G, Lodde V, Tessaro I, Modina SC, Fadini R, Dal Canto M, Renzini MM, Albertini DF, Luciano AM (2014) Natriuretic peptide precursor C delays meiotic resumption and sustains gap junction-mediated communication in bovine cumulus-enclosed oocytes. Biol Reprod 91(3):61

15. Soares ACS, Marques KNG, Braganca LGM, Lodde V, Luciano AM, Buratini J (2020) Synchronization of germinal vesicle maturity improves efficacy of in vitro embryo production in Holstein cows. Theriogenology 154:53–58

16. Soares ACS, Sakoda JN, Gama IL, Bayeux BM, Lodde V, Luciano AM, Buratini J (2020) Characterization and control of oocyte large-scale chromatin configuration in different cattle breeds. Theriogenology 141:146–152

17. Barros RG, Lodde V, Franciosi F, Luciano AM (2020) In vitro culture strategy for oocytes from early antral follicle in cattle. J Vis Exp 161:e61625

18. Zhang M, Su YQ, Sugiura K, Xia G, Eppig JJ (2010) Granulosa cell ligand NPPC and its receptor NPR2 maintain meiotic arrest in mouse oocytes. Science 330(6002):366–369

19. Dieci C, Lodde V, Franciosi F, Lagutina I, Tessaro I, Modina SC, Albertini DF, Lazzari G, Galli C, Luciano AM (2013) The effect of cilostamide on gap junction communication dynamics, chromatin remodeling, and competence acquisition in pig oocytes following parthenogenetic activation and nuclear transfer. Biol Reprod 89(3):68

20. Shu YM, Zeng HT, Ren Z, Zhuang GL, Liang XY, Shen HW, Yao SZ, Ke PQ, Wang NN (2008) Effects of cilostamide and forskolin on the meiotic resumption and embryonic development of immature human oocytes. Hum Reprod 23(3):504–513

21. Alam MH, Lee J, Miyano T (2018) Inhibition of PDE3A sustains meiotic arrest and gap junction of bovine growing oocytes in in vitro growth culture. Theriogenology 118:110–118

22. Tsafriri A, Chun SY, Zhang R, Hsueh AJ, Conti M (1996) Oocyte maturation involves compartmentalization and opposing changes of cAMP levels in follicular somatic and germ cells: studies using selective phosphodiesterase inhibitors. Dev Biol 178(2):393–402

23. Romero S, Sanchez F, Lolicato F, Van Ranst H, Smitz J (2016) Immature oocytes from unprimed juvenile mice become a valuable source for embryo production when using C-type natriuretic peptide as essential component of culture medium. Biol Reprod 95(3):64

24. Sanchez F, Lolicato F, Romero S, De Vos M, Van Ranst H, Verheyen G, Anckaert E, Smitz JEJ (2017) An improved IVM method for cumulus-oocyte complexes from small follicles in polycystic ovary syndrome patients enhances oocyte competence and embryo yield. Hum Reprod 32(10):2056–2068

25. Zhang T, Fan X, Li R, Zhang C, Zhang J (2018) Effects of pre-incubation with C-type natriuretic peptide on nuclear maturation, mitochondrial behavior, and developmental competence of sheep oocytes. Biochem Biophys Res Commun 497(1):200–206

26. Soto-Heras S, Menendez-Blanco I, Catala MG, Izquierdo D, Thompson JG, Paramio MT (2019) Biphasic in vitro maturation with C-type natriuretic peptide enhances the developmental competence of juvenile-goat oocytes. PLoS One 14(8):e0221663

27. Caixeta FM, Sousa RV, Guimaraes AL, Leme LO, Spricigo JF, Netto SB, Pivato I, Dode MA (2017) Meiotic arrest as an alternative to increase the production of bovine embryos by somatic cell nuclear transfer. Zygote 25 (1):32–40

28. Soares ACS, Lodde V, Barros RG, Price CA, Luciano AM, Buratini J (2017) Steroid hormones interact with natriuretic peptide C to delay nuclear maturation, to maintain oocyte-cumulus communication and to improve the quality of in vitro-produced embryos in cattle. Reprod Fertil Dev 29(11):2217–2224

29. Zhang M, Su YQ, Sugiura K, Wigglesworth K, Xia G, Eppig JJ (2011) Estradiol promotes and maintains cumulus cell expression of natriuretic peptide receptor 2 (NPR2) and meiotic arrest in mouse oocytes in vitro. Endocrinology 152 (11):4377–4385

30. Richard S, Baltz JM (2014) Prophase I arrest of mouse oocytes mediated by natriuretic peptide precursor C requires GJA1 (connexin-43) and GJA4 (connexin-37) gap junctions in the antral follicle and cumulus-oocyte complex. Biol Reprod 90(6):137

31. Norris RP, Ratzan WJ, Freudzon M, Mehlmann LM, Krall J, Movsesian MA, Wang H, Ke H, Nikolaev VO, Jaffe LA (2009) Cyclic GMP from the surrounding somatic cells regulates cyclic AMP and meiosis in the mouse oocyte. Development 136(11):1869–1878

32. Vaccari S, Weeks JL 2nd, Hsieh M, Menniti FS, Conti M (2009) Cyclic GMP signaling is involved in the luteinizing hormone-dependent meiotic maturation of mouse oocytes. Biol Reprod 81(3):595–604

33. Henderson KM, McNeilly AS, Swanston IA (1982) Gonadotrophin and steroid concentrations in bovine follicular fluid and their relationship to follicle size. J Reprod Fertil 65 (2):467–473

34. Dieleman SJ, Bevers MM, Poortman J, van Tol HT (1983) Steroid and pituitary hormone concentrations in the fluid of preovulatory bovine follicles relative to the peak of LH in the peripheral blood. J Reprod Fertil 69 (2):641–649

35. Kruip TA, Dieleman SJ (1985) Steroid hormone concentrations in the fluid of bovine follicles relative to size, quality and stage of the oestrus cycle. Theriogenology 24 (4):395–408

36. Sakaguchi K, Yanagawa Y, Yoshioka K, Suda T, Katagiri S, Nagano M (2019) Relationships between the antral follicle count, steroidogenesis, and secretion of follicle-stimulating hormone and anti-Mullerian hormone during follicular growth in cattle. Reprod Biol Endocrinol 17(1):88

37. Endo M, Kawahara-Miki R, Cao F, Kimura K, Kuwayama T, Monji Y, Iwata H (2013)

Estradiol supports in vitro development of bovine early antral follicles. Reproduction 145 (1):85–96

38. Walters KA, Allan CM, Handelsman DJ (2008) Androgen actions and the ovary. Biol Reprod 78(3):380–389

39. Makita M, Miyano T (2015) Androgens promote the acquisition of maturation competence in bovine oocytes. J Reprod Dev 61 (3):211–217

40. Luciano AM, Pappalardo A, Ray C, Peluso JJ (1994) Epidermal growth factor inhibits large granulosa cell apoptosis by stimulating progesterone synthesis and regulating the distribution of intracellular free calcium. Biol Reprod 51 (4):646–654

41. Ireland JJ, Roche JF (1983) Growth and differentiation of large antral follicles after spontaneous luteolysis in heifers: changes in concentration of hormones in follicular fluid and specific binding of gonadotropins to follicles. J Anim Sci 57(1):157–167

42. Beg MA, Bergfelt DR, Kot K, Ginther OJ (2002) Follicle selection in cattle: dynamics of follicular fluid factors during development of follicle dominance. Biol Reprod 66 (1):120–126

43. Tessaro I, Luciano AM, Franciosi F, Lodde V, Corbani D, Modina SC (2011) The endothelial nitric oxide synthase/nitric oxide system is involved in the defective quality of bovine oocytes from low mid-antral follicle count ovaries. J Anim Sci 89(8):2389–2396

44. Modina SC, Tessaro I, Lodde V, Franciosi F, Corbani D, Luciano AM (2014) Reductions in the number of mid-sized antral follicles are associated with markers of premature ovarian senescence in dairy cows. Reprod Fertil Dev 26 (2):235–244

45. Wang W, Niwa K (1995) Synergetic effects of epidermal growth factor and gonadotropins on the cytoplasmic maturation of pig oocytes in a serum-free medium. Zygote 3(4):345–350

46. Izadyar F, Zeinstra E, Bevers MM (1998) Follicle-stimulating hormone and growth hormone act differently on nuclear maturation while both enhance developmental competence of in vitro matured bovine oocytes. Mol Reprod Dev 51(3):339–345

47. Merriman JA, Whittingham DG, Carroll J (1998) The effect of follicle stimulating hormone and epidermal growth factor on the developmental capacity of in-vitro matured mouse oocytes. Hum Reprod 13(3):690–695

48. Luvoni GC, Chigioni S, Perego L, Lodde V, Modina S, Luciano AM (2006) Effect of gonadotropins during in vitro maturation of feline oocytes on oocyte-cumulus cells functional coupling and intracellular concentration of glutathione. Anim Reprod Sci 96(1-2):66–78

49. Modina S, Abbate F, Germana GP, Lauria A, Luciano AM (2007) Beta-catenin localization and timing of early development of bovine embryos obtained from oocytes matured in the presence of follicle stimulating hormone. Anim Reprod Sci 100(3-4):264–279

50. Franciosi F, Manandhar S, Conti M (2016) FSH regulates mRNA translation in mouse oocytes and promotes developmental competence. Endocrinology 157(2):872–882

51. Sirard MA, Desrosier S, Assidi M (2007) In vivo and in vitro effects of FSH on oocyte maturation and developmental competence. Theriogenology 68(Suppl 1):S71–S76

52. Caixeta ES, Machado MF, Ripamonte P, Price C, Buratini J (2013) Effects of FSH on the expression of receptors for oocyte-secreted factors and members of the EGF-like family during in vitro maturation in cattle. Reprod Fertil Dev 25(6):890–899

53. Khan DR, Guillemette C, Sirard MA, Richard FJ (2015) Characterization of FSH signalling networks in bovine cumulus cells: a perspective on oocyte competence acquisition. Mol Hum Reprod 21(9):688–701

54. Park JY, Su YQ, Ariga M, Law E, Jin SL, Conti M (2004) EGF-like growth factors as mediators of LH action in the ovulatory follicle. Science 303(5658):682–684

55. Luciano AM, Pocar P, Milanesi E, Modina S, Rieger D, Lauria A, Gandolfi F (1999) Effect of different levels of intracellular cAMP on the in vitro maturation of cattle oocytes and their subsequent development following in vitro fertilization. Mol Reprod Dev 54(1):86–91

56. Guixue Z, Luciano AM, Coenen K, Gandolfi F, Sirard MA (2001) The influence of cAMP before or during bovine oocyte maturation on embryonic developmental competence. Theriogenology 55(8):1733–1743

57. Luciano AM, Modina S, Vassena R, Milanesi E, Lauria A, Gandolfi F (2004) Role of intracellular cyclic adenosine 3′,5′-monophosphate concentration and oocyte-cumulus cells communications on the acquisition of the developmental competence during in vitro maturation of bovine oocyte. Biol Reprod 70 (2):465–472

58. Modina S, Luciano AM, Vassena R, Baraldi-Scesi L, Lauria A, Gandolfi F (2001) Oocyte developmental competence after in vitro maturation depends on the persistence of cumulus-oocyte comunications which are linked to the

intracellular concentration of cAMP. Ital J Anat Embryol 106(2 Suppl 2):241–248

59. Atef A, Paradis F, Vigneault C, Sirard MA (2005) The potential role of gap junction communication between cumulus cells and bovine oocytes during in vitro maturation. Mol Reprod Dev 71(3):358–367

60. Albuz FK, Sasseville M, Lane M, Armstrong DT, Thompson JG, Gilchrist RB (2010) Simulated physiological oocyte maturation (SPOM): a novel in vitro maturation system that substantially improves embryo yield and pregnancy outcomes. Hum Reprod 25 (12):2999–3011

61. Nogueira D, Ron-El R, Friedler S, Schachter M, Raziel A, Cortvrindt R, Smitz J (2006) Meiotic arrest in vitro by phosphodiesterase 3-inhibitor enhances maturation capacity of human oocytes and allows subsequent embryonic development. Biol Reprod 74 (1):177–184

62. Ozawa M, Nagai T, Somfai T, Nakai M, Maedomari N, Fahrudin M, Karja NW, Kaneko H, Noguchi J, Ohnuma K, Yoshimi N, Miyazaki H, Kikuchi K (2008) Comparison between effects of 3-isobutyl-1-methylxanthine and FSH on gap junctional communication, LH-receptor expression, and meiotic maturation of cumulus-oocyte complexes in pigs. Mol Reprod Dev 75 (5):857–866

63. Vanhoutte L, Nogueira D, Dumortier F, De Sutter P (2009) Assessment of a new in vitro maturation system for mouse and human cumulus-enclosed oocytes: three-dimensional prematuration culture in the presence of a phosphodiesterase 3-inhibitor. Hum Reprod 24(8):1946–1959

64. Zeng HT, Ren Z, Guzman L, Wang X, Sutton-McDowall ML, Ritter LJ, De Vos M, Smitz J, Thompson JG, Gilchrist RB (2013) Heparin and cAMP modulators interact during pre-in vitro maturation to affect mouse and human oocyte meiosis and developmental competence. Hum Reprod 28(6):1536–1545

65. Sanchez F, Romero S, De Vos M, Verheyen G, Smitz J (2015) Human cumulus-enclosed germinal vesicle oocytes from early antral follicles reveal heterogeneous cellular and molecular features associated with in vitro maturation capacity. Hum Reprod 30(6):1396–1409

66. Hirao Y, Itoh T, Shimizu M, Iga K, Aoyagi K, Kobayashi M, Kacchi M, Hoshi H, Takenouchi N (2004) In vitro growth and development of bovine oocyte-granulosa cell complexes on the flat substratum: effects of high polyvinylpyrrolidone concentration in culture medium. Biol Reprod 70(1):83–91

67. Lodde V, Garcia Barros R, Dall'Acqua PC, Dieci C, Robert C, Bastien A, Sirard MA, Franciosi F, Luciano AM (2020) Zinc supports transcription and improves meiotic competence of growing bovine oocytes. Reproduction 159(6):679–691

68. Gordon I (2003) Laboratory Production of Cattle Embryos. In: Biotechnology in Agriculture No. 27. CABI Publishing, Cambridge MA (USA)

69. Stringfellow D, Givens MD (2010) Manual of the International Embryo Transfer Society (IETS). A procedural guide and general information for the use of embryo transfer technology emphasizing sanitary procedures, 4th edn. International Embryo Transfer Society (IETS), Champaign, IL (USA)

70. Blondin P, Sirard MA (1995) Oocyte and follicular morphology as determining characteristics for developmental competence in bovine oocytes. Mol Reprod Dev 41(1):54–62

71. Hazeleger NL, Hill DJ, Stubbing RB, Walton JS (1995) Relationship of morphology and follicular fluid environment of bovine oocytes to their developmental potential in vitro. Theriogenology 43(2):509–522

72. Sanchez F, Le AH, Ho VNA, Romero S, Van Ranst H, De Vos M, Gilchrist RB, Ho TM, Vuong LN, Smitz J (2019) Biphasic in vitro maturation (CAPA-IVM) specifically improves the developmental capacity of oocytes from small antral follicles. J Assist Reprod Genet 36 (10):2135–2144

73. Gilchrist RB, Luciano AM, Richani D, Zeng HT, Wang X, Vos MD, Sugimura S, Smitz J, Richard FJ, Thompson JG (2016) Oocyte maturation and quality: role of cyclic nucleotides. Reproduction 152(5):R143–R157

74. Katska L, Smorag Z (1984) Number and quality of oocytes in relation to age of cattle. Anim Reprod Sci 7(5):451–460

75. Lussier JG, Matton P, Dufour JJ (1987) Growth rates of follicles in the ovary of the cow. J Reprod Fertil 81(2):301–307

76. Scaramuzzi RJ, Baird DT, Campbell BK, Driancourt MA, Dupont J, Fortune JE, Gilchrist RB, Martin GB, McNatty KP, McNeilly AS, Monget P, Monniaux D, Vinoles C, Webb R (2011) Regulation of folliculogenesis and the determination of ovulation rate in ruminants. Reprod Fertil Dev 23(3):444–467

77. Silva-Santos KC, Santos GM, Siloto LS, Hertel MF, Andrade ER, Rubin MI, Sturion L, Melo-Sterza FA, Seneda MM (2011) Estimate of the population of preantral follicles in the ovaries of Bos taurus indicus and Bos taurus taurus cattle. Theriogenology 76(6):1051–1057

Chapter 2

Clinostat 3D Cell Culture: Protocols for the Preparation and Functional Analysis of Highly Reproducible, Large, Uniform Spheroids and Organoids

Krzysztof Wrzesinski, Helle Sedighi Frandsen, Carlemi Calitz, Chrisna Gouws, Barbara Korzeniowska, and Stephen J. Fey

Abstract

Growing cells as 3D structures need not be difficult. Often, it is not necessary to change cell type, additives or growth media used. All that needs to be changed is the geometry: cells (whether primary, induced pluripotent, transformed or immortal) simply have to be grown in conditions that promote cell–cell adhesion while allowing gas, nutrient, signal, and metabolite exchange. Downstream analysis can become more complicated because many assays (like phase contrast microscopy) cannot be used, but their replacements have been in use for many years. Most importantly, there is a huge gain in value in obtaining data that is more representative of the organism in vivo. It is the goal of the protocols presented here to make the transition to a new dimension as painless as possible. Grown optimally, most biopsy derived organoids will retain patient phenotypes, while cell (both stem cell, induced or otherwise or immortalized) derived organoids or spheroids will recover tissue functionality.

Key words Protocols, Spheroids, Organoids, Clinostat, Bioreactor, Cell lines, Stem cells, Biopsy, Coculture, High yield, Long-term culture, High reproducibility

1 Introduction

Growing cells in "three dimensions" as organoids or spheroids brings a host of advantages. Both self-organize, recapitulating to some degree structures and functionality seen in vivo. This makes them a potent tool for dissecting healthy processes (e.g., cell–cell interactions and integration in organogenesis, organ function and decline in aging [1–5] and repair [6–9] as well as disease (both innate or induced [10–13] and the search for, and effects of, compounds [14–16] or other organisms [17–19] on these mimetic tissues. However, it is still early days and there are many caveats to be overcome. These drawbacks stem from variability in the cell types present (in numbers and functionality); their ability to achieve

Tiziana A.L. Brevini et al. (eds.), *Next Generation Culture Platforms for Reliable In Vitro Models: Methods and Protocols*, Methods in Molecular Biology, vol. 2273, https://doi.org/10.1007/978-1-0716-1246-0_2,

long term homeostasis and recover from perturbations (of any type); the fact that they only recapitulate part of an organism and will be lacking "input" from tissues not present; and very importantly reproducibility of tissue construction in and between labs [20–24].

All of these considerations point to the critical need for a comprehensive set of easy-to-use protocols that can be used both: to produce large numbers of highly reproducibly sized, shaped and aged 3D structures; and a set of standard assays that can be used on them. This is the aspiration of this paper. While different cell types will have different requirements, we hope that the protocols presented here will provide a coherent and practical starting point.

The transition from growing cells in 2D to 3D introduces a number of challenges [25]. The first of these relates to diffusion. In 2D cultures, every cell is roughly equally exposed to equal amounts of oxygen, nutrients and waste products. This is not true in a 3D culture. The 3D nature of a spheroid or an organoid means that some cells are exposed to the surrounding environment in a way similar to that seen in 2D cultures. However, other cells are "buried" to differing degrees by the cells above them or at the surface. These buried cells will receive less oxygen and nutrients and will be exposed to higher levels of secreted compounds (messengers, growth factors, ECM, etc. due to the greatly expanded level of communication between cells) and more CO_2 and other waste products.

Secondly, if the shape of the spheroids or organoids varies, then the proportion of cells in comparable environments will differ. Different environments will give rise to differences in gene expression which means that differently shaped 3D structures will respond differently, increasing variability.

Thirdly, previous studies have illustrated that cells undergo a far-reaching structural transformation [26] and metabolic reprogramming [27] as they adapt to life in 3D. As one example the doubling time falls from about 1 day to more than 60 days (after 40 days in 3D culture) [28]. For HepG2/C3A and many other cell types [29], this transformation typically takes about 18 days before they have recovered the ability to mimic in vivo functionality. Diffusion gradients play an important role in driving this transformation and thus size or shape differences will affect the rate at which this transformation occurs. Consequently, reproducibility takes on a whole new dimension as it is necessary to be able to replicate the age, size, shape, and metabolic functionality in each batch of 3D structures needed within an individual experiment and between different experiments. Since cultures can take 18 days or more before they are fully functional, this also raises the skill-set requirements from cell biologists to be able to maintain highly reproducible, infection-free cultures for extended periods. Using the protocols described below it is possible to reproducibly obtain spheroids which 21 days after initiation vary in size by only 21% (SD) [30]. Many 3D cell biologists need to keep cultures alive for weeks, months, or even over a year to allow for comprehensive tissue development. The protocols presented below have been developed for this purpose.

The final challenge facing 3D cell biologists is that spheroids and organoids are not so amenable to the tools normally used: after 18 days in culture spheroids can contain more than 80,000 cells and so are not transparent or translucent [28]. This impedes or precludes the use of standard phase contrast microscopic, luminescence or fluorescence assays (not only light but also reagents may not be able to penetrate the mimetic tissues constructed).

Mature 3D structures can also be considerably more enzyme resistant and more difficult to dissolve for subsequent analysis. Thus, cell biologists are resorting to approaches typically used by pathologists for investigating tissue biopsies.

2 Materials

2.1 Cell Culture Preparation

1. *Equipment*: Sterile bench; sterile 15 and 50 mL conical centrifuge tubes and racks; electrical pipette pump; pipettes (for 5, 10, 25, and 50 mL); Electronic or manual single channel pipette, 10–200 μL and 100–1000 μL and tips, standard cell culture incubator; water bath (37 °C); centrifuge; polystyrene box with dry ice; timer; hemocytometer; sterile 15 mL conical centrifuge tubes, 1.8 mL cryotubes, micro test tubes and racks; electrical pipette pump; sterile cell culture flask (T25 or T75). Optical inverted microscope with camera and objectives (4× or 10×); hemocytometer; automated cell counter and tips for 60 μL volume; freezer at −80 °C; freezing container with isopropanol; timer.

2. *Reagents*: Cell culture media DMEM (low glucose, with pyruvate, no glutamine, no phenol red); FBS: fetal bovine serum; MEM nonessential amino acids solution; penicillin–streptomycin (10,000 U/mL); GlutaMAX™ Supplement; Hanks' Balanced Salt Solution (HBSS) with Ca^{2+} and Mg^{2+}; HBSS (without Ca^{2+} and Mg^{2+}); Trypan Blue Solution (0.4%); trypsin–EDTA (0.5%); dimethyl sulfoxide (sterile-filtered, ≥99.7%); ethanol (70% v/v in water); isopropanol (≥99.0%).

2.2 ClinoReactor™ Bioreactor

1. *Equipment*: Sterile bench; electrical pipette pump; sterile disposable pipettes (for 50 mL); electronic or manual single channel pipette (10–200 μL) and tips; ClinoStar™; ClinoReactor™ (10 mL); syringes (sterile 10 mL) and needles (1.2 × 40 mm, or 2.1 × 80 mm recommended).

2. *Reagents*: Cell culture media, for example DMEM (low glucose, with pyruvate, no glutamine, no phenol red); HBSS (without Ca^{2+} and Mg^{2+}); ethanol (70% v/v in water).

2.3 Spheroid Culture

1. *Equipment*: Sterile bench; standard cell culture incubator and ClinoStar™; ClinoReactor™ (10 mL); water bath (37 °C); optical inverted microscope with camera and objectives (4× or 10×); centrifuge (rotor compatible with 15 mL conical centrifuge tubes, max speed set to 140 g; temperature set to 22 °C); vacuum centrifuge; timer; automated cell counter and tips for 60 μL volume; 15 mL tubes, sterile 15 mL conical centrifuge tubes, micro test tube (sterile, 1.5 mL) and racks; electrical pipette pump; sterile disposable pipettes (5, 10, and 25 mL); electronic or manual single channel pipettes, (0.1–2.5 μL or 0.1–10 μL, 10–200 and 100–1000 μL) and tips; gel loading tips (sterile 0.1–10 μL); tips (sterile, wide-bore or precut, 200 μL); gel loading tips (sterile, 0.1–10 μL); syringes (sterile, 10 mL) and needles (sterile, 1.2 × 40 mm, or 2.1 × 80 mm recommended). AggreWell™ 400 plate; PVC blocks (5 × 5 × 1 cm, rounded edges); Parafilm M; forceps (Sterile flat-ended); petri dishes (Sterile, 100 mm); Acrodisc® PF Supor® membrane syringe filter; cryomicrotome; microscope slides; coverslips.

2. *Reagents*: Cell culture media (e.g., DMEM (low glucose, with pyruvate, no glutamine, no phenol red)), HBSS (without Ca^{2+} and Mg^{2+}); trypsin–EDTA (0.5%); ethanol (70% v/v in water); AggreWell™ Anti-adherence (Rinsing) Solution; QGel™; QGel™ Buffer A; sodium alginate; cross-linker solution (50 mM $CaCl_2$ and 150 mM NaCl); sterile water; ethanol (70% v/v in water).

2.4 Spheroid Visualization

1. *Equipment*: ultramicrotome; diamond knife (45°); EM grids; transmission electron microscope cryomicrotome; microscope slides coverslips; micro test tubes; confocal microscope; washing baths; oven (60 °C); timer.

2. *Reagents*: HBSS (without Ca^{2+} and Mg^{2+}); formaldehyde (containing about 15% methanol); Tissue-Tek O.C.T; liquid nitrogen; 0.1 M tris-base buffered glycine pH 7.4; Triton X-100 (0.2% in PBS); BSA (5% in PBS); primary and secondary antibodies according to what you wish to investigate; DAPI; fluorescence-free mounting medium; glutaraldehyde (25% in H_2O); sodium cacodylate; osmium tetroxide; uranyl acetate (0.5% in 50 mM maleate pH 5.2); ethanol (96% in water and 100%); propylene oxide; Epon-substitute embedding medium.

2.5 Functional Assays

1. *Equipment*: Sterile bench; fume cupboard; fluorometer (with excitation/emission filters of 355/460); centrifuge; Scepter handheld automated cell counter, and 60 μM tips; plate shaker;

laboratory balance; sterile disposable pipettes (5 mL and 10 mL); electronic or manual 8-channel (5–100 (or 10–300) μL and 100–1000 μL) single channel pipettes, and tips; centrifuge tubes (50 mL) and racks, cuvettes or microtiter plate, ice bath, water bath (23 °C) and water bath (or block at 50 °C), freezer (−20 °C), aluminum foil; screw-capped bottle (500 mL); glass measuring cylinder (100 mL); volumetric flask (200 mL); magnetic stirrer and bar; thermometer (0–50 °C); filtration unit; vacuum suction point (preferably in the fume cupboard); 15 and 50 mL conical centrifuge tubes (unsterile and sterile), micronic tubes (1.4 mL), micro test tubes (unsterile and sterile) and racks; 15 and 50 mL tubes for freezer; shaking table; timer; electrical pipette pump; multichannel pipetting 50 mL reservoir; optical assay 96-well plates (black and white); sealing sheet, OneTouch Vita (Mediq Denmark).

2. *Reagents*: DNA Quantitation Kit ((Sigma-Aldrich, DNAQF contains: bisBenzimide H 33258 (Hoechst 33258); fluorescent assay buffer; calf thymus DNA Standard); pure water; Serdolyte MB-1 resin; urea (ultrapure); thiourea; chaps; Pharmalyte carrier ampholytes (pH ranges 3–10 and 6–11); dithiothreitol (DTT); pure H_2O. BSA standard (Sigma, P0834-10X1ML, 2 mg/mL); human serum albumin protein standard QC solution (Sigma P8369, concentration, 0.5 mg/mL); ProStain™ Protein Quantification Kit (Active Motif, 15,001, contains Dilution Buffer); methanol; ProStain™ Protein Quantification Kit, Dye Reagent (0.5 mg Lyophilized); lysis buffer B (*see* Subheading 3.18).

Adenosine 5' Triphosphate (ATP) standard (Sigma, FLAAS-1VL, approx. 1.2 mg/vial = 2 μmol ATP/vial); Cell-Titer-Glo® luminescent cell viability assay (Promega, G7571); HBSS (without Ca^{2+} and Mg^{2+}); pure water; cell culture media (*see* Subheading 2.3, **item 2**); heat treated medium; Toxi-Light™ BioAssay Kit (Lonza, LT07-117); AK detection reagent (Lonza, LT27-061); AK assay buffer (Lonza, LT27-066); CytoTox-Glo™ Cytotoxicity Assay Kit (Promega, G9290, contains digitonin and assay buffer); lysis buffer (*see* Subheading 3.18); Amplex Red Cholesterol Assay Kit (Invitrogen A12216); OneTouch Vita Quality Control solution (QC, Mediq Denmark, 64-07-081): D-glucose standard, (Sigma, G3285 1 mg/mL solution); Urea Fluorometric Assay Kit (BioNordika Denmark, 700,620, includes: urea standard, urea assay buffer 10×, urease assay reagent, urea ammonia detector, ethanol assay reagent); HPLC-grade water.

3 Methods

3.1 Preparing Growth Media

1. Thaw all the ingredients that are stored at −20 °C preferably in the fridge overnight. If necessary, they can be thawed at room temperature or in cold water.

2. Wipe and switch on the sterile hood and wait 15 min.

3. Clean the bottle with medium and tubes with medium supplements with a tissue soaked in 70% of ethanol before placing it in the sterile hood.

4. Mix all the medium components in a suitable sterile bottle starting with addition of the basal medium according to Table 1.

5. Mix well by pipetting up and down many times and then aliquot the supplemented medium into sterile falcon 15 mL and/or 50 mL conical centrifuge tubes.

6. Label the tubes with the right medium name, date and initials.

7. Place the medium in a fridge at 4 °C (*see* **Note 1**).

8. Register the medium stock in the cell culture media registration sheet.

3.2 Thawing Cells

1. Warm up the cell culture medium to 37 °C by placing its aliquots (approx. 15 mL/20 mL per one T25/T75 flask) for at least 15 min in a water bath (*see* **Note 2**).

2. Transfer the cryovial with cells from the cell tank filled with liquid nitrogen into the polystyrene box filled with the dry ice and close the box with a lid (*see* **Note 3**).

3. Loosen the lid on the cryovial to take off overpressure and make sure all the liquid nitrogen has evaporated before closing it again.

4. Place the closed tube in the water and thaw the frozen pellet AS QUICKLY AS POSSIBLE. Gently agitate the tube to speed up the process but avoid shaking not to disturb the cells.

Table 1
Mixing table for DMEM-based medium

DMEM medium mL	100	200	400	500
87.5% DMEM basal medium mL	87.5	175	350	437.5
1% nonessential amino acids mL	1	2	4	5
10% fetal bovine serum mL	10	20	40	50
0.5% penicillin–streptomycin mL	0.5	1	2	2.5
1% GlutaMAX mL	1	2	4	5

5. Wipe the tube carefully with a tissue soaked in disinfectant (e.g., 70% EtOH).

6. Prepare a sample for cell viability testing by mixing the cell suspension gently and transferring 20 μL of it into a 1.5 mL sterile micro test tube that is then placed in the incubator until the cell thawing procedure is finished.

7. Transfer the rest of cell suspension/freezing medium to the already prepared 15 mL conical centrifuge tube.

8. Add 9 mL of the cell culture medium to the tube.

9. Centrifuge the cells at $140 \times g$ for 5 min at room temperature.

10. Add the cell culture media to a cell culture flask (e.g.: 4 mL/ 9 mL per T25/T75 flask).

11. Remove the supernatant from 15 mL conical centrifuge tube.

12. Resuspend the cell pellet in 1 mL of the cell culture medium.

13. Transfer the cell suspension to a flask prefilled with the medium culture and place it in the cell culture incubator (*see* **Note 4**).

14. Add 30 μL of HBSS with Ca^{2+} and Mg^{2+} and 50 μL of trypan blue to the micro test tube containing 20 μL of the cell suspension.

15. Incubate for 3 min and then pipet 10 μL of the cell suspension to a hemocytometer cell counting chamber.

16. Count total number of cells and then count only the dead cells (stained in blue).

17. Calculate the viability: (alive cells/total cells) × 100% = % of viable cells.

3.3 Sub-culture of Cells in 2D

1. Thaw and/or warm up all the reagents (except the trypsin solution) in the water bath at 37 °C. Note: It is recommended to defrost the trypsin solution at room temperature in order to avoid its gradual autodigestion that occurs at 37 °C.

2. Switch on the sterile hood and wipe down surfaces with 70% ethanol.

3. Clean the bottle with medium and all tubes with the ingredients with a tissue soaked in 70% of ethanol before placing it in the sterile hood.

4. Perform a visual inspection of cells grown in cell culture flasks using the optical microscope and select a culture flask that is 70–90% confluent.

5. Tilt the flask and suck out all medium from the corner of the flask using a pipette pump equipped with 5 mL or 10 mL sterile pipette. Note: Avoid touching the cell monolayer growing on the bottom/floor of the flask with the pipette.

Table 2
Reagent volumes for trypsinization of different sizes of culture flask

Flask type	HBSS mL	0.05% trypsin (mL)	FBS (mL)	Medium wash (mL)	Final volume (mL)
T25	2 × 5	3	1	5	9
T75	2 × 15	5	3	5	13
T150	2 × 30	10	5	10	25
T175	2 × 35	15	10	10	35

6. Carefully add a suitable amount of HBSS (*see* Table 2) in the corner of the flask and carefully wash all sides of the flask (floor, walls and ceiling) by gently swirling and rotation of the flask.

7. Remove HBSS from the flask and repeat the washing step.

8. Remove HBSS and carefully add the suitable amount of trypsin to the flask by dripping directly on the cell monolayer (floor) (*see* **Note 5**).

9. Close the flask and incubate in the sterile hood for 3–4 min.

10. Detach cells from the bottom surface of the flask by gently occasionally tapping the flask against the hand.

11. When cells are fully detached add the suitable amount of FBS to the flask to stop the enzymatic activity of trypsin.

12. Transfer the cell suspension to a sterile 15 mL conical centrifuge tube.

13. Wash the flask with the suitable amount of medium and transfer it to the tube.

14. Centrifuge the cells for 5 min at $140 \times g$, 22 °C.

15. Remove the supernatant and resuspend the cell pellet in a suitable amount of medium (*see* **Note 6**).

16. Prepare new cell culture flasks by filling them with an appropriate amount of fresh medium.

17. Mix the cell suspension well and transfer the appropriate number of cells into the new flasks. Note: For subculturing C3A cells every fourth day, the recommended seeding density is equal to 0.3×10^5 cells/cm^2.

18. Gently swirl the flasks back and forward, and left-right a couple of times to ensure equal redistribution of cells within the cell growth area (*see* **Note 7**).

19. Place the new culture flasks in the cell culture incubator.

20. Register the subculturing in the cell culture maintenance registration sheet.

Table 3
Preparation of different volumes of freezing medium

Freezing medium	10	20	50	100
95% cell culture medium mL	9.5	19	47.5	95
5% dimethyl sulfoxide mL	0.5	1	2.5	5

3.4 Preparation of Freezing Media for Cryopreservation of Cells

1. Prepare the sterile hood.
2. Clean the bottle with medium and the DMSO ampoules with a tissue soaked in 70% of ethanol before placing it in the sterile hood.
3. Prepare an appropriate amount of a freezing medium by adding DMSO to the cell culture medium at 5% (v/v) final concentration according to Table 3 (*see* **Note 8**).
4. Aliquot the supplemented medium into sterile falcon 15 mL and/or 50 mL conical centrifuge tubes.
5. Label the tube(s) with the freezing medium name, date and initials.
6. Place the freezing medium in a fridge at 4 °C (*see* **Note 9**).
7. Register the freezing medium stock in the laboratory registration sheet.

3.5 Preparation of Cell Culture Stocks for Storage in Liquid Nitrogen

1. Thaw an appropriate amount of a freezing medium and bring it to 4 °C before use (and update the stock list).
2. Sub-cultivate the flat culture cells according to SOP for the relevant cell line.
3. Resuspend cell pellet in COLD freezing medium.
4. Count the cells using either a hemocytometer or an automated cell counter.
5. Add additional freezing medium to the cell suspension so that its final cell concentration is either 3×10^6 or 6×10^6 cells/mL.
6. Aliquot cell suspension into cryotubes by adding 1 mL of the suspension to each tube.
7. Write the ampoule number, cell-line and date on each tube.
8. Place the cryotubes in the container filled with isopropanol.
9. Put the container in the −80 °C freezer.
10. The next day move the cells to the tank with liquid nitrogen.
11. Register the cryovials with cells in the cell culture stocks registration sheet.

3.6 Preparation of the ClinoReactor™ Bioreactor [29]

1. ClinoReactor™ vessels should be washed and preequilibrated before use to prepare the membrane for use (*see* **Note 10**).

2. Clean the ClinoReactor™ package by spraying it with 70% ethanol solution and transfer it into the sterile hood.

3. Open the packages and place ClinoReactor™ vessels on your working space.

4. Wash the cell chamber of the ClinoReactor™ by filling it with 8–9 mL of sterile water, closing the chamber with the plug, and rotating the vessel in your hand (*see* **Note 11**).

5. Using 10 mL syringe with a blunt or flat-nosed needle (or plastic tip or pipette), remove water from the cell chamber.

6. Repeat the two preceding washing steps twice.

7. Fill the cell chamber with 8–9 mL of the medium that you will be using (the cell chamber should not be filled completely; leaving air bubbles helps with membrane pore equilibration). Close the chamber with the plug. Sterilize the area around the top port with 70% EtOH (*see* **Note 12**). Remove any remaining ethanol before putting the ClinoReactor™ back into the incubator.

8. Place the ClinoReactor™ on a drive axel in the ClinoStar™ CO_2 incubator (37 °C, 5% CO_2) and let them rotate overnight at 15 rpm.

3.7 Preparation of Self-Aggregated Spheroids from a Single-Cell Suspension

1. This procedure is valid for the C3A (HepG2/C3A) cell line and it can be used as guidelines for creation of procedures valid for other cell types (*see* **Note 13**).

2. Remove the medium from the prepared (equilibrated) ClinoReactor™ (*see* Subheading 3.6) via the top port.

3. Fill with approximately 5 mL of fresh warm medium.

4. Close and gently turn it upright.

5. Open the ClinoReactor™ bottom port.

6. Transfer an appropriate volume of the single cell suspension (prepared according to Subheading 3.3) into each ClinoReactor™ (via the bottom port) (*see* **Note 14**).

7. Close the bottom port and place the ClinoReactor™ in standing position with top access port up.

8. Use the top port to overfill the chamber (so that there is some medium in the port cup) with the remaining fresh warm medium and remove air bubbles (*see* **Note 15**). Close with the plug, remove excess medium from the port cup and clean this area with approximately 200 µL of ethanol. Remove leftover ethanol and place the ClinoReactor™ on its drive axel.

9. Adjust speed of rotation according to the size of the cells. Use an initial speed of 2.5 rpm for the HepG2/C3A cell line. Both faster and slower speeds lead to cell clumping.

10. Observe the ClinoReactor™ during the first 48 h for sizable cell clump formation. If present, remove the clump from the culture vessel (*see* **Note 16**).

11. Change medium after 72 h of culture (*see* Subheading 3.13).

12. At this stage it is expected that the spheroid population will be uneven in size (*see* Fig. 2 and **Note 17**).

13. Control and if necessary, adjust the rotation speed daily.

14. With time, the shape of spheroids became more spherical and after approximately 3 weeks the spheroid population should contain similar-sized, round spheroids (Fig. 1). If the sizes vary too much it is possible to select spheroids or organoids of similar sizes (or remove those of unwanted sizes).

3.8 Preparation of Spheroids Using AggreWell™ Plates [30]

1. **Day-1**: Equilibrate the number of 10 mL ClinoReactor™ vessels needed (*see* Subheading 3.6).

2. Prepare the AggreWell™ plate by washing wells once with 0.5 mL of the AggreWell™ Anti-adherence rinsing solution and then remove the liquid (*see* **Note 18**).

3. Add 1 mL of the cell culture medium to each well and centrifuge the plate using maximum rotor speed (approximately $3000 \times g$) for 5 min to remove any small air bubbles from the microwells.

4. Check for the presence of air bubbles in wells using an inverted microscope. If any bubbles are present, centrifuge the plate again for another 3 min at $3000 \times g$.

5. Remove the medium from wells, add 1 mL of fresh medium to each well and place the plate in the incubator.

6. Collect the cells from the cell culture flask and count the cells (*see* Subheading 3.3 and **Note 9**).

7. Transfer the plate from the incubator under the sterile hood and add desired amount of the prepared cell suspension to the prepared wells. For example: an AggreWell™ 400 plate has approximately 1200 microwells per well with so to obtain 800 cells in each microwell:

$$c1000 = \frac{1200 \text{ microwells} \times 800 \text{ cells}}{1,522,000 \text{ cells/mL}} = 0.631 \text{ mL/well}$$

8. Centrifuge the plate at $120 \times g$ (recommended speed for safe cyto-spin of C3A cells) for 3 min.

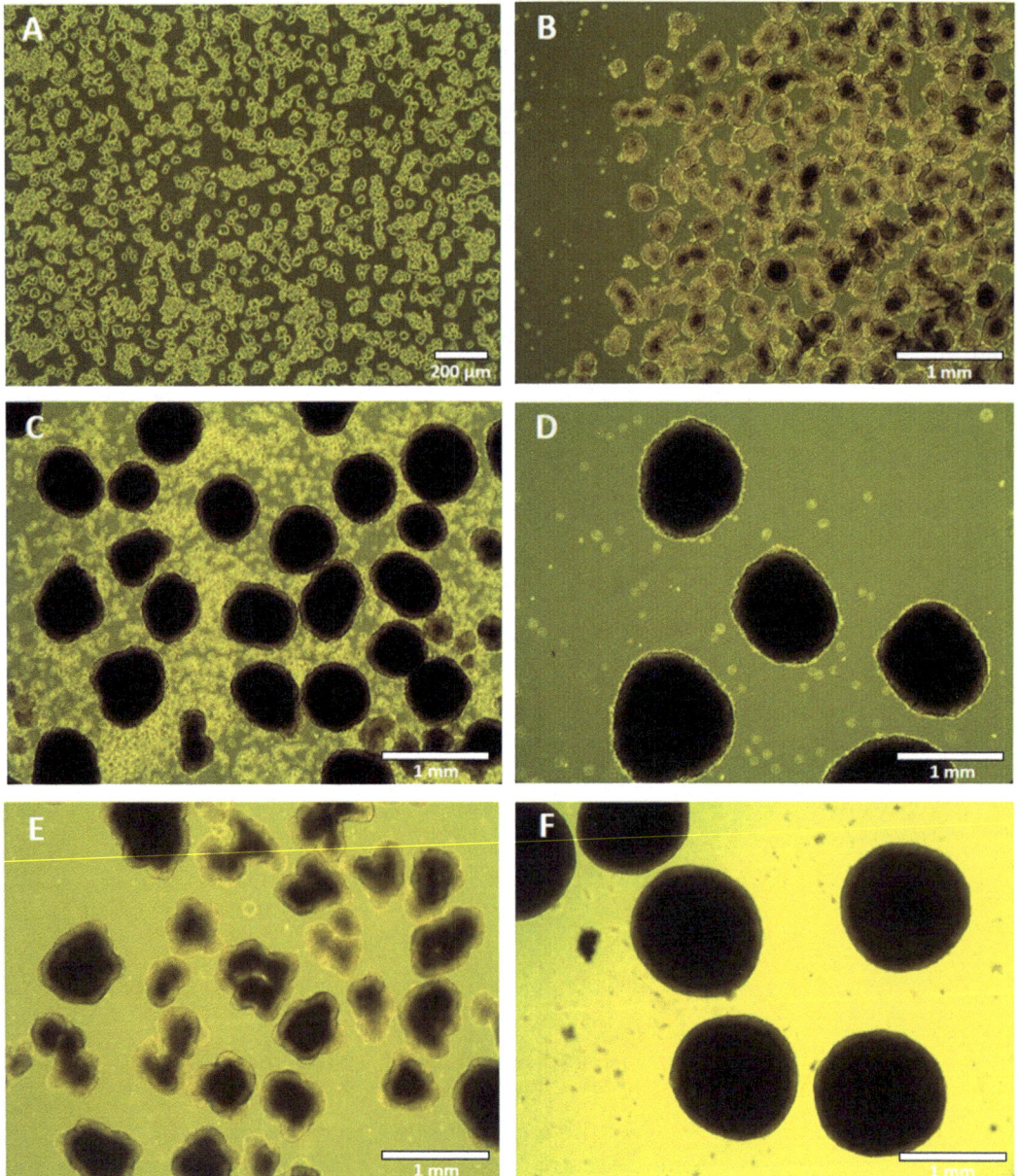

Fig. 1 ClinoReactor™ spheroid cultures, started as single cell suspension of of C3A cells (**a–d**) and grown for 3, 10, 22, and 29 days respectively; and INS-1E cells (**e** and **f**) grown for 7 and 21 days

9. Check the AggreWell® plate under the optical microscope (*see* **Note 19**) before placing it in the CO_2 incubator (37 °C, 5% CO_2) overnight (*see* **Note 20**).

10. **Day 0**: Removal of cells from AggreWell™ plate and transfer to ClinoReactor™s. Gently pipet the medium in the wells with preseeded cells up and down using a manual pipette equipped

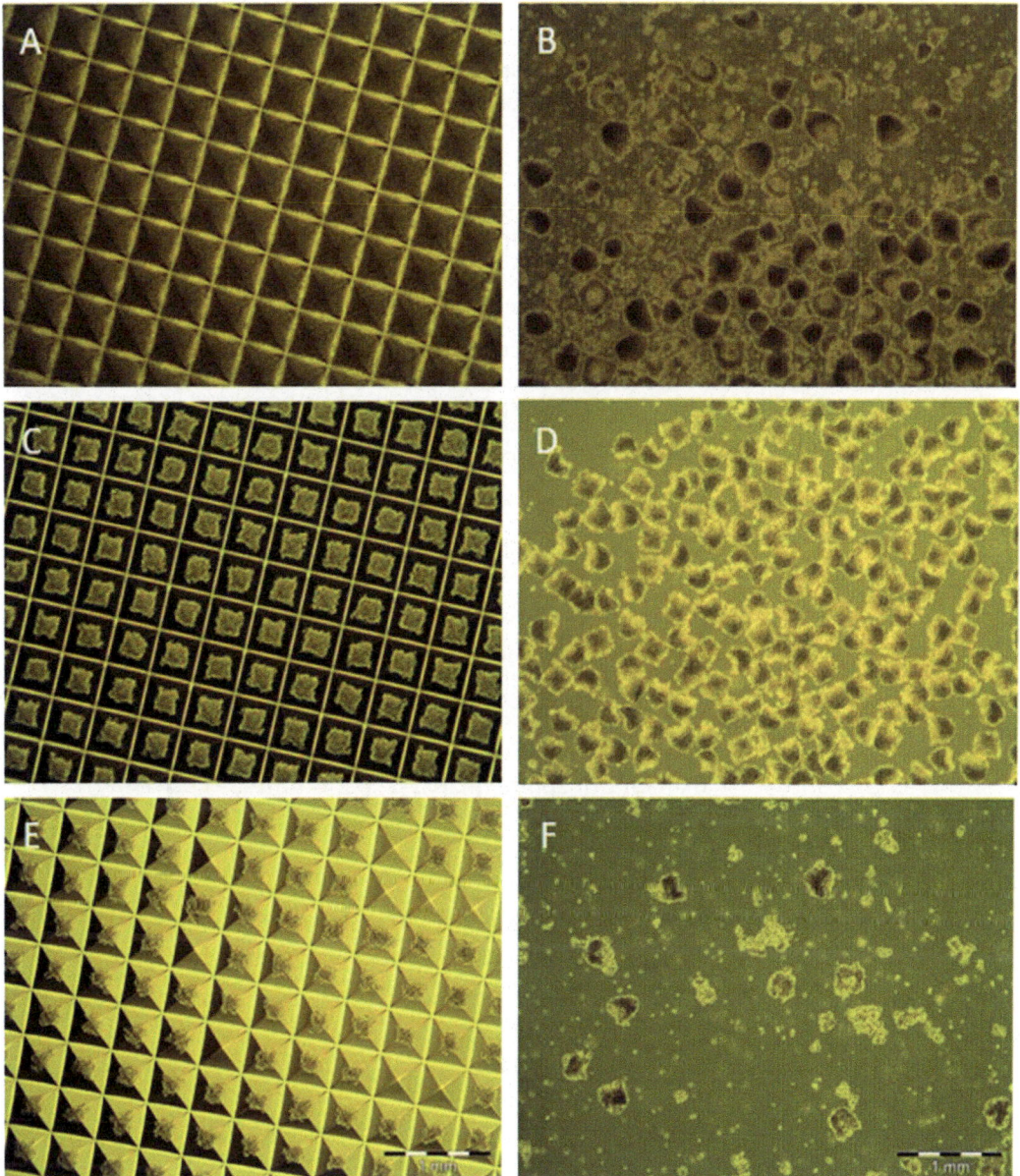

Fig. 2 Preparation of C3A spheroids based on AggreWell™ micropattern plates. (**a**) overfilled microwells result in (**b**) cellular aggregates which disintegrated during removal from the AggreWell™ plates. (**c**) optimally filled microwells result in (**d**) many good aggregates. (**e**) too few cells in microwells result in (**f**) a poor yield of aggregates. C3A aggregates need 18 days to mature to functional spheroids

with 10–200 µL volume wide-bore or cut tip to detach the spheroids from the bottom of the AggreWell™ plate. Figure 2 illustrates the consequences of having too many, the correct, or too few cells in an AggreWell™ plate and its consequences.

11. Collect the medium and spheroids in a small petri dish.

12. Wash the well with 0.5 mL fresh medium and transfer it to the petri dish. Add another 0.5 mL medium and observe the well under the microscope to see that at least 95% of the spheroids have been transferred (if not, repeat the washing procedure). Also, observe the quality of spheroids and take microphotographs.

13. Observe if any spheroids look ragged or are clumping. Try to remove these with a disposable pipette tip before they are transferred to the ClinoReactor™. This will reduce the risk of them clumping in the ClinoReactor™.

14. When all the spheroids have been collected in the petri dish, the ClinoReactor™ is now prepared for use.

15. Remove the medium from the equilibrated ClinoReactor™ using the top port.

16. Fill the ClinoReactor™ with approximately 5 mL of fresh warm medium.

17. Close the ClinoReactor™ and gently turn upright.

18. Open the ClinoReactor™ by removing the bottom plug.

19. Concentrate spheroids in the center of the petri dish by gently shaking it in circular motions and then transfer spheroids into the ClinoReactor™ using a manual pipette equipped with 10–200 µL volume wide-bore or cut tip (with an opening diameter of approximately 1.5–2.0 mm).

20. Overfill the ClinoReactor™ with the fresh warm medium, check for and remove air bubbles. Close with the plug, remove leftover medium from the area around the plug and clean this area with approximately 200 µL of ethanol. Remove leftover ethanol and place the ClinoReactor™ in the CO_2 incubator.

21. Adjust the rotation speed according to the size of the spheroids (*see* **Note 21**).

22. The ClinoReactor™s are now ready to be used in a project.

3.9 Aliquoting Q-Gel™

1. Prepare five micro test tubes properly labeled with the QGel™ lot number and expiry date.

2. Puncture the cap of the tubes with a small needle three times so the water can escape when lyophilizing (*see* **Note 22**).

3. Dissolve the QGel™ powder in 550 µL of sterile water (*see* **Note 23**).

4. Aliquot into portions of 100 µL into the five micro test tubes.

5. Close the tubes and snap-freeze them in liquid nitrogen.

6. Centrifuge the tubes in the vacuum centrifuge overnight (use maximum speed) (*see* **Note 24**).

7. After overnight lyophilization, close the tubes with their original lids and the aliquots can be used or stored at −20 °C.

3.10 Preparation of Spheroids Using Q-Gel™

1. 5 × 5 × 0.5 cm PVC blocks are washed with detergent, rinsed with deionized water and afterward sterilized in the autoclave. The blocks should be stored sterile packed until needed.

2. Cut 8 × 8 cm squares of Parafilm® M foil and sterilize (one square per block).

3. Use 14 cm petri dishes to prepare following sequence of wash solutions: bath (1) 70% ethanol bath; bath (2) 100% Hanks' solution; bath (3) 100% Hanks' solution; bath (4) drying space (empty petri dish(es)).

4. Remove the paper back from the Parafilm® M square, and using flat-ended forceps, wash the Parafilm® M squares in bath 1 (70% ethanol). Let the ethanol stop dripping from the paraffin before immersing it into the bath 2 (Hanks). Subsequently wash the Parafilm® M square in the bath 3 (Hanks), let the Hanks' solution stop dripping from the Parafilm® M surface and leave the square to dry completely in bath 4.

5. Repeat the procedure until you have all the Parafilm® M squares you need.

6. Unpack a sterile PVC block under sterile conditions and carefully wrap dry Parafilm® M around it so that the Parafilm® M foil creates flat even surface on the PVC casting block (*see* **Note 25**).

7. Place the casting blocks covered with Parafilm® M into a 10 mL petri dishes. Press it firmly so the block sticks to the petri dish bottom.

8. Fill the petri dish with Hanks' solution or sterile water so the level of the liquid is approximately half of the height of the casting block (*see* **Note 26**).

9. Bring the QGel™ aliquots and QGel™ buffer A to room temperature.

10. Cut the top off several gel loading tips so they fit on a 10 µL automated pipette, if needed, and set the pipette to load 1 µL ten times (*see* **Note 27**).

11. After counting the cells, prepare as many portions of the suspended cells, (each containing 3×10^5 cells) as needed (one cell portion per one QGel™ 100 µL aliquot, *see* **Note 28**).

12. Centrifuge the cell portion at $140 \times g$ for 5 min and remove supernatant. Do this immediately before preparing the spheroids.

13. Add 80 µL of buffer A and 20 µL of culture media to the cell pellet and mix it gently but well.

14. Transfer this mix to the tube with lyophilized QGel™ aliquot and mix it well. Make sure the QGel™ is completely dissolved. This should not take more than 2 min.

15. Quickly, using the 10 µL automated pipette and the cut gel loading tip, make 1 µL drops of the QGel™-cell mix on top of prepared casting blocks (covered with Parafilm® M). Make lines of 10 drops to make it easier to count (*see* **Note 29**).

16. Register the time on top on the petri dish and place the casting blocs with spheroids in the incubator (37 °C, 5% CO_2, 95% relative humidity) for 45 min.

17. After 45 min incubation check if the spheroids polymerized correctly. Use a syringe needle, to touch one of the drops gently to check if it is solid. If not, leave for 15 min more. The drops should not incubate more than 1 h, as this may harm the cells.

18. Cut the edges of the Parafilm® M layer on the casting blocks so the part with the QGel™-spheroids can be lifted out and placed, with the QGel™-spheroids facing down, in a new 96 mm petri dish filled with warm media. The spheroids should soak in the medium for about 2 min. This will wet the QGel™-spheroids and make it easier to shave them off the Parafilm® M.

19. Use the scalpel to shave off the QGel™-spheroids from the Parafilm® M. Make sure that the scalpel blade and the spheroids are wet all the time.

20. Take representative microphotographs of the created QGel™-spheroids.

21. Load the QGel™spheroids into a prepared ClinoReactor™ (*see* Subheading 3.6).

22. Adjust the speed of ClinoReactor™ vessel (*see* **Note 30**).

3.11 Preparation of Alginate-Encapsulated Spheroids [31]

1. Prepare the required number of ClinoReactor™s (*see* Subheading 3.6).

2. Prepare Parafilm® M covered PVC blocks as described in **steps 1–8**.

3. Prepare a 2.5% solution of sodium alginate in PBS and autoclave it (*see* **Note 31**).

4. Prepare the cross-linker solution and filter through an 0.2 µm membrane syringe filter.

5. Trypsinize a vigorously growing flask of cells (*see* **Note 32**), dilute to 2000 cells/µL and centrifuge at $140 \times g$ for 5 min at 37 °C.

6. Discard the supernatant (check that it contains only a few cells or cellular debris) and gently resuspend the pellet in prewarmed (37 °C) sodium alginate solution.

7. Working quickly with an automatic pipette, pipet 1 μL droplets of the cell/alginate suspension onto the Parafilm® M covered PVC blocks.

8. Add 0.5 μL cross-linker solution to each of these droplets, cover the petri dish and let stand (at room temperature) for 5 min to set (*see* **Note 33**).

9. Wash the alginate spheroids off the Parafilm® M with complete culture medium and transfer 300 spheroids to each ClinoReactor™.

10. Set initial rotation speed of the drive axel to 16 rpm.

3.12 Preparation of Spheroid or Organoid Cocultures

1. There are a wide variety of ways of coculturing mixtures of cell types. Here are a few ideas based on the protocols presented.

2. Prepare the required number of ClinoReactor™s (*see* Subheading 3.6).

3. Following the protocol described in Subheadings 3.8, 3.10, or 3.11 but prepare cell suspensions from two different cell types and mix them before pipetting into ClinoReactor™s for self-aggregation, AggreWell™ plates, Q-gel™, or alginate for spheroid or organoid formation (*see* **Note 34**).

4. Form spheroids or organoids with one cell type (*see* Subheadings 3.8, 3.10, or 3.11) and then add a single cell suspension (*see* Subheading 3.3) to the culture media in the ClinoReactor™. The individual cells will attach to the existing cells and coat the spheroids.

   OR

5. Form spheroids with one cell type, take them out onto Parafilm® covered PVC block and drip a cell-alginate mixture containing another cell type onto the spheroid and follow quickly with the cross-linker. Wash the alginate-layered spheroids off with complete culture medium and transfer to a fresh ClinoReactor™ (*see* **Note 35**).

3.13 Changing the Media [30]

1. Operations where you manipulate the cultures (like changing the media) are the most probable cause of infections. Therefore, be completely clear in what you intend to do, prepare your working area with what you will need and work smoothly through the steps without rushing.

2. Preheat fresh cell culture medium to 37 °C (using water bath or cell culture incubator) and check its pH (*see* **Note 36**).

3. Prepare the medium change station by placing tube holder with two 50 mL tubes in a rack close to your working area (*see* Fig. 3 and **Note 37**).

4. Transfer the medium containers into sterile bench.

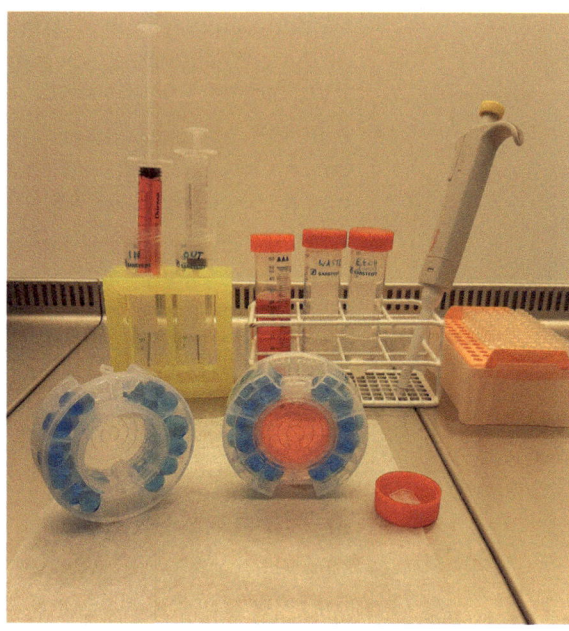

Fig. 3 Workstation prepared to change the media in the Clinoreactor™. Note that the top plug has been removed and has been placed in the lid of the centrifuge tube

5. Fill a syringe with approximately 12 mL of the medium and place it in the 50 mL tube (marked with "IN") standing in a rack.

6. Take a 10 mL ClinoReactor™ out of the incubator and place it on its feet on a sterile surface in the sterile bench and wait approximately 1 min. to allow spheroids to settle down to the bottom (small spheroids take longer).

7. Remove the top plug from the ClinoReactor™ and place it on a sterile surface (*see* **Note 38**).

8. Remove spent media from the cell chamber (usually 9.5–9.8 mL) using another 10 mL syringe (marked "OUT") equipped with a long syringe needle (*see* **Note 39**). Avoid removing any of the spheroids or damaging the sides of the chamber.

9. Place the "OUT"-syringe in the 50 mL tube standing in the rack (marked "OUT").

10. Overfill the ClinoReactor™ slowly with fresh medium (use the medium and the "IN" syringe prepared before) (*see* **Note 40**). Avoid disturbing the spheroids.

11. Check for eventual air bubbles to get out by gently tapping on the side of the ClinoReactor™ and waiting until they float to the top.

12. Close the chamber with the plug by placing it into the port.

13. Remove any remaining medium from the cup around the top plug of the ClinoReactor™ using a 200 μL micropipette.

14. Clean the cup of the ClinoReactor™ port with a 70% ethanol solution using a 200 μL micropipette. Remove any remaining ethanol before replacing the ClinoReactor™ into the ClinoStar™.

15. After 15–30 min check the ClinoReactor™ for air bubbles in the cell chamber. If any, remove them by following **steps 10–14** and adding some medium to help the air bubbles out.

16. Eject the spent medium from the "OUT" syringe into a petri dish and use an inverted microscope to check for free-floating cells, as well as the quality of accidentally aspirated spheroids.

3.14 Splitting a Spheroid or Organoid Culture [30]

1. The spheroids need to be split into 2 or 3 ClinoReactor™s when their reach around 60% density (after approximately 10 days of culturing, but this may vary depending on the cell type) (*see* **Note 41**).

2. The day before, preequilibrate a new ClinoReactor™(s) according to Subheading 3.6.

3. Warm cell culture media (approximately 12 mL/ClinoReactor™) to 37 °C.

4. Place all disposables required for medium change under the sterile hood.

5. Take the new equilibrated ClinoReactor™(s) out of the incubator, place it on your sterile working space and empty the cell chamber(s). Add around 5 mL of fresh medium so there is something to transfer the spheroids or organoids into.

6. Label the ClinoReactor™ vessels (*see* **Note 42**) and leave them in a sterile place.

7. Take the ClinoReactor™ with spheroids out of the ClinoStar™ incubator and place it in your sterile working space.

8. Wait for the spheroids to settle on the bottom, visualize how much medium you should remove (1/2 or 2/3) and then remove it.

9. Gently turn the ClinoReactor™ upright and remove the bottom plug. If you do this carefully, media will not come out of the top port.

10. Use a 1 mL wide-bore or cut tip to transfer an appropriate number of spheroids to a new ClinoReactor™, then close the bottom port.

11. Overfill the ClinoReactor™ slowly with fresh medium (use the medium and the syringe prepared before). Avoid disturbing the spheroids.

12. Check for eventual air bubbles: gently tap the side of the ClinoReactor™ and wait until they float to the top.

13. Close the chamber with the plug by placing it into the port.

14. Remove any remaining medium from the cup around the top plug of the ClinoReactor™ using a 200 μL micropipette.

15. Clean the cup of the ClinoReactor™ with a 70% ethanol solution using a 200 μL micropipette. Remove any remaining ethanol before putting the ClinoReactor™ back into the incubator.

16. Place the ClinoReactor™s in the incubator and check the rotor speed.

17. Register the split in the maintenance registration sheet.

18. After 15–30 min check the ClinoReactor™ for a presence of air bubbles in the cell chamber. If any big bubbles appear, remove them by following **steps 11–15**, adding some medium to help the air bubbles out.

3.15 Visualization of Spheroids or Organoids. Preparation for Immuno-microscopy [26]

1. Prepare a 4% formaldehyde solution in HBSS and place on ice (*see* **Note 43**).

2. Warm HBSS to 37 °C in the incubator.

3. Stop and open a ClinoReactor™, select a suitable number of spheroids and transfer them to a micro test tube.

4. Wash the spheroids four times with warm HBSS.

5. Fix the spheroids for 30 min at 37 °C in the 4% freshly made ice-cold formaldehyde.

6. Replace the formaldehyde solution with Tissue-Tek O.C.T and snap-freeze in liquid nitrogen (*see* **Note 44**).

7. Prior to sectioning, warm the samples to −18 °C in the cryomicrotome.

8. Sixteen micrometer sections are cut with a cryomicrotome, placed on glass slides, and stored at −80 °C until needed.

9. Sections are washed for 15 min at room temperature with 0.1 M tris-base buffered glycine pH 7.4.

10. Sections are then permeabilized with 0.2% Triton X-100 in PBS for 5 min at room temperature and unspecific binding sites blocked with 5% BSA in PBS.

11. The primary antibody (diluted appropriately in 1% BSA in PBS) is applied overnight at 4 °C.

12. Excess antibody is removed by washing three times in PBS.

13. The second antibody (diluted in PBS) is applied for 1 h at room temperature.

14. If required, the sections can be counterstained with DAPI for 5 min at room temperature.

15. Mount a coverslip with fluorescence-free mounting medium and collect images using a confocal microscope.

3.16 Visualization of Spheroids or Organoids. Preparation for Electron Microscopy [29]

1. Prepare a 2% glutaraldehyde solution in 0.1 M cacodylate buffer pH 7.2 (*see* **Note 45**).

2. Wash spheroids in HBSS by rotating in a ClinoReactor™ for 15 min at 37 °C.

3. Fix the spheroids in the 2% glutaraldehyde/cacodylate buffer pH 7.2 for at least 24 h while rotating in the ClinoReactor™ at 37 °C.

4. Wash in 0.1 M cacodylate buffer pH 7.2 for 5 min and then post-fix in 1% OsO_4 in 0.1 M cacodylate buffer pH 7.2 for 1 h while rotating in the ClinoReactor™ at 37 °C.

5. Wash the fixed spheroids successively while rotating in the ClinoReactor™ at 37 °C in:
   (a) 0.1 M cacodylate buffer pH 7.2 for 5 min.
   (b) 50 mM maleate pH 5.2 for 5 min.
   (c) 0.5% uranyl acetate in 50 mM maleate pH 5.2 for 1 h.

6. Dehydrate through a series of increasing strength ethanol/water solutions (e.g., 50%, 70%, 80%, 90%, 96%, and 100% (×2)) for 15 min (each step) at room temperature (*see* **Note 46**).

7. Wash with and transfer to propylene oxide for 15 min at room temperature.

8. Embed in an Epon-substitute embedding medium and allow to polymerize at 60 °C for at least 48 h.

9. Cut ultrathin sections (40–50 nm) with an ultramicrotome and a 45° diamond knife and capture them on EM grids prior to transmission electron microscopy.

3.17 Assay: DNA Content [26]

1. Thaw the fluorescent assay buffer stock (×10), mix until homogeneous and store at 2–8 °C.

2. Thaw the calf thymus DNA standard and incubate at 50 °C for 15–30 min. Take an aliquot (e.g., 120 µL) and place on ice at 2–8 °C. Freeze the rest of the standard and store at −20 °C (*see* **Note 47**). Using sterile techniques, dilute the DNA standard to give two concentrations (10 mg/mL and 100 mg/mL) as described in the Table 4 below:

3. Dilute an aliquot of the 10 mg/mL bisBenzimide H 33258 (Sigma cat. no. B1302) tenfold with water to give a concentration of 1 mg/mL and store in the dark at 2–8 °C. Prepare

Table 4
Preparation of DNA standards

	10 μg/mL DNA stock solution (μL)	100 μg/mL DNA stock solution (μL)
1 mg/mL DNA standard, DNA	10	100
10× fluorescent assay buffer	100	100
Molecular biology grade water	890	800
Total volume	1000	1000

Table 5
Preparation of bisBenzimide H 33258 reagent

	0.1 μg/mL bisBenzimide H 33258 (mL)	1 μg/mL bisBenzimide H 33258 (mL)
1 mg/mL bisBenzimide H 33258	0.003	0.030
10× fluorescent assay buffer	3	3
Molecular biology grade water	27	27
Total volume	30	30

sufficient amounts of this dye solutions for two standard curves (e.g., 30 mL) and for the sample determinations (e.g., no of samples × 2 + 2 mL) (*see* Table 5 and **Note 48**).

4. Turn on the fluorometer at least 30 min before use and set the excitation wavelength to 360 nm and the emission wavelength to 460 nm (*see* **Note 49**).

5. Pipet 2 mL of the appropriate dye solution (prepared in Subheading 3.17, **step 3**)directly into the cuvette and place the cuvette in the sample chamber ("Blank").

6. Read the blank emission at ambient temperature.

7. With the cuvette still in the fluorometer, add the appropriate DNA solution according to Tables 6 or 7 to the cuvette (*see* **Note 50**). Mix the solutions in the cuvette by pipetting up and down 3–4 times and then read the emission fluorescence (*see* **Note 51**).

8. Wash the cuvette and repeat **steps 5–7** with the remaining DNA standards (*see* **Note 52**) and the unknown samples using fresh dye solutions each time.

9. Prepare a calibration curve by plotting total DNA concentration versus relative fluorescence units.

Table 6
Reaction mixture using 0.1 μg/mL bisBenzimide H 33258 solution for unknown DNA concentrations in the range of 10–500 ng/mL

Sample[a]	10 μg/mL DNA stock solution (μL)	100 μg/mL DNA stock solution (μL)	0.1 μg/mL bisBenzimide H 33258 solution (mL)	Final amount of DNA in 2 mL reaction (ng)
1	0	0	2	Blank
2	2	–	2	20[a]
3	5	–	2	50
4	10	–	2	100
5	–	2	2	200
6	–	5	2	500
7	–	10	2	1000

[a]An additional standard of 10 ng/2 mL may be detected using some fluorometers (i.e., for a sample "1.5" use 1 μL DNA stock [10 μg/mL])

Table 7
Reaction mixture using 0.1 μg/mL bisBenzimide H 33258 Solution for unknown DNA concentrations in the range of 100 ng to 5 μg/mL

Sample[a]	100 μg/mL DNA stock solution (μL)	1 mg/mL DNA stock solution (μL)	0.1 μg/mL bisBenzimide H 33258 solution (mL)	Final amount of DNA in 2 mL reaction (μg)
1	0	0	2	Blank
2	2	–	2	0.2
3	5	–	2	0.5
4	10	–	2	1
5	–	2	2	2
6	–	5	2	5
7	–	10	2	10[a]

[a]An additional standard of 20 μg/2 mL may be detected using some fluorometers (i.e., for a sample 8 use 20 μL DNA stock 1 mg/mL). The ability to read DNA standard points at the upper and lower limits of the standard curve is dependent on the type of fluorometer used

10. Determine the least squares regression equation for the line generated by the standard samples. The linear equation is

$$y = mx + b$$

where y = emission expressed in relative fluorescence units (RFU); m = the slope; x = DNA concentration, and b = the intercept.

11. Interpolate the unknown samples.

12. Move the data-file to corresponding project folder.

3.18 Lysis Buffer B for Protein Determination [30]

1. Carry out as many of these procedures in the fume cupboard as possible because of the hazardous chemical used. Dry Urea and Thiourea under vacuum for about 30 min prior use (*see* **Note 53**).

2. Weigh urea and thiourea in a 500 mL screw topped bottle and add 100 mL pure water.

3. Dissolve by stirring in a water bath (max. 30 °C) (*see* **Note 54**).

4. When it is dissolved, add 2 g of Serdolyte MB-1 resin, stir gently for at least 20 m and filter the solution through a Millipore Stericup® using vacuum (to remove the Serdolyte resin) (*see* **Note 55**).

5. Keep it for a while on vacuum because it is very important to get all of the solution through.

6. Add the remaining chemicals to the Stericup® flask.

7. Transfer the solution into a 200 mL measuring cylinder and add pure H_2O up to the volume 200 mL (*see* **Note 56**).

8. Store Lysis buffer B frozen in aliquots of 1, 10, and 50 mL in tubes at $-20/-80$ °C.

9. Label the box with correct hazard symbol(s) and the yellow label for carcinogenic hazard.

3.19 Assay: Protein Content Using Prostain™ Kit [30]

1. Bring all components to room temperature for 45 min before use (*see* **Note 57**). Leave dilution buffer in the water bath at 23 °C and the dye reagent stock on the shaking table at speed 200 rpm (*see* **Note 58**). Switch on the fluorometer at least 30 min before use.

2. Leave samples, lysis buffer B (*see* Subheading 3.18), human serum albumin solution (acting as a quality control (QC)) and BSA standard on the shaking table at speed 1000 rpm (*see* **Note 59**).

3. Dilute BSA standard in low protein-binding micro test tubes according to Table 8 starting with the dilution buffer, then Lysis buffer B and finally the BSA standard (*see* **Note 60**).

4. Centrifuge samples at $14,000 \times g$ at room temperature for 15 min.

5. Dilute samples and human serum albumin QC as described in Table 9, starting with the dilution buffer, then the Lysis buffer B and finally the experimental or QC sample:

6. Mix standards, QC, and samples shortly with a vortex mixer.

Table 8
Preparation of standard dilution series of BSA

Std. No.	Dilution buffer μL	Lysis B buffer μL	BSA standard μL	BSA standard μg/mL	BSA standard μg/well
Std 1	465	5	30	120	12
Std 2	470	5	25	100	10
Std 3	475	5	20	80	8
Std 4	480	5	15	60	6
Std 5	485	5	10	40	4
Std 6	490	5	5	20	2
Std 7	495	5	0	0	0

Table 9
Mixing protocol for samples and BSA

	Dilution factor x	Dilution buffer (μL)	Lysis B buffer (μL)	Sample (μL)
S1–S24	100	495	0	5
QC	10	445	5	50

7. Add 100 μL of standards, QC, and samples to the appropriate wells (*see* **Note 61**). See plate setup in Table 10.

8. Prepare a dye reagent working solution in a 15 mL conical centrifuge tube covered with aluminum foil (to protect it from light) according to the Table 11 and add 100 μL into each well. If a full plate is used in the experiment, it is recommended to prepare the amount of a dye reagent working solution for 100 wells (*see* **Note 75**).

Table 10
Standard plate layout for protein determination

ID	1	2	3	4	5	6	7	8	9	10	11	12
A	Std1	Std1	Std1	S1	S1	S1	S9	S9	S9	S17	S17	S17
B	Std2	Std2	Std2	S2	S2	S2	S10	S10	S10	S18	S18	S18
C	Std3	Std3	Std3	S3	S3	S3	S11	S11	S11	S19	S19	S19
D	Std4	Std4	Std4	S4	S4	S4	S12	S12	S12	S20	S20	S20
E	Std5	Std5	Std5	S5	S5	S5	S13	S13	S13	S21	S21	S21
F	Std6	Std6	Std6	S6	S6	S6	S14	S14	S14	S22	S22	S22
G	Std7	Std7	Std7	S7	S7	S7	S15	S15	S15	S23	S23	S23
H	QC	QC	QC	S8	S8	S8	S16	S16	S16	S24	S24	S24

Table 11
Dilution series for dye reagent

No of wells	1	10	40	50	70	100	110
Dye reagent stock solution (µL)	11	110	440	550	770	1100	1210
ddH$_2$O (µL)	99	990	3960	4950	6930	9900	10,890

9. Mix the dye reagent and samples with an 8-channel manual or electronic pipet up and down 4–5 times. Start incubation time (30 min), when you have mixed the standard rows. The whole plate should be mixed within approximately 1 min.

10. Remove air bubbles in the assay plate by centrifugation at $3000 \times g$ for 2 min at room temperature (*see* **Note 62**).

    OR

11. Remove air bubbles by blowing ethanol vapor over the wells (*see* **Note 63**).

12. Leave the plate for the rest of the incubation time in a fluorescence plate reader.

13. Adjust gain to the highest concentration of the BSA standard. Measure fluorescence and save the readout.

14. Move the data-file to corresponding project folder.

3.20 Functional Assay: Intracellular ATP [30]

1. Take the HBSS and CellTiter-Glo luminescent viability reagent out of the freezer and let it equilibrate to room temperature (in a water bath at 23 °C for at least 15 min). Swirl the contents of the two bottles gently and leave them in the water bath until needed.

Table 12
Preparation of standard dilution series of ATP

ID	ATP μL	HBSS μL	Std. Conc. mMol/L	Final Std. Conc.[a] mM	Final Std. Conc.[a] mM
Std 1	50 (2 mM stock)	450	0.2	0.1	0.1
Std 2	75 of Std1	525	2.50×10^{-2}	1.25×10^{-2}	0.0125
Std 3	75 of Std2	525	3.12×10^{-3}	1.56×10^{-3}	0.0015625
Std 4	75 of Std3	525	3.91×10^{-4}	1.95×10^{-4}	0.0001953125
Std 5	75 of Std4	525	4.88×10^{-5}	2.44×10^{-5}	0.00002441406
Std 6	75 of Std5	525	6.10×10^{-6}	3.05×10^{-6}	0.00000305175
Std 7	75 of Std6	525	7.63×10^{-7}	3.81×10^{-7}	0.00000038146
Std 8	0	525	0	0	0.0

[a]Final Std. Conc. corresponds to a final standard concentration of ATP standards on a plate after addition of CellTiter-Glo™ luminescent viability assay solution

2. Take the ATP standard out of the freezer and let it thaw on ice until further use (*see* **Note 64**)

3. Prepare the standards in HBSS according to Table 12 (*see* **Note 65**):

4. Transfer spheroids (from 2 to 6 depending on their size) from the ClinoReactor™ into a white opaque 96-well plate. Transfer each sample in to one well onto the 96-well plate (*see* **Note 66**). See a suggested plate setup in Table 13.

5. Remove the treatment medium from all the wells carefully, while paying attention not to disrupt the cells in the wells. If cells are loose, a small volume of medium can be allowed to remain in the wells.

Table 13
Standard plate layout for ATP assay

ID	1	2	3	4	5	6	7	8	9	10	11	12
A	Std1	Std1	Std1									
B	Std2	Std2	Std2						S1	S1	S1	
C	Std3	Std3	Std3									
D	Std4	Std4	Std4						S2	S2	S2	
E	Std5	Std5	Std5									
F	Std6	Std6	Std6						S3	S3	S3	
G	Std7	Std7	Std7									
H	Std8	Std8	Std8									

6. Add 100 μL of room temperature HBSS without calcium and magnesium ions to wells containing samples (*see* **Note 67**).

7. Add 100 μL of ATP standards to the wells according to plate setup in Table 13.

8. Add 100 μL of CellTiter-Glo luminescent viability assay solution (prewarmed to room temperature) to wells containing samples or ATP standards.

9. Pipet up and down (about 4–6 times) to facilitate complete lysis of the cells in the sample wells. Check with a microscope if all spheroids have disintegrated and cells have lysed.

10. Remove air bubbles in the assay plate by vacuum centrifugation at $100 \times g$ for 2 min at room temperature (*see* **Note 76**).

    OR

11. Remove air bubbles by blowing EtOH vapor over the wells (*see* **Note 77**).

12. Cover the assay plate with a sticker seal and wrap in aluminum foil. Place it on a shaking platform for 40 min at 300 rpm.

13. Switch on the luminometer at least 30 min before use.

14. Cover the plate bottom with a white label. Place the assay plate in a luminescence plate reader and adjust gain to the highest concentration of the ATP standard. Measure luminescence and save the readout.

15. Move data-file to corresponding project folder.

3.21 Functional Assay: Adenylate Kinase in the Medium [29]

1. Equilibrate reagents and samples to room temperature.

2. Prepare a standards dilution series in heat treated medium according to Table 14 (*see* **Note 68**):

Table 14
Preparation of standard dilution series of heat-treated medium

Std No.	Std µL	Heat treated Medium µL	Std. Conc. cells/mL
Std1	17 of Std0*	183	3.65×10^5
Std2	100 of Std1	100	1.83×10^5
Std3	100 of Std2	100	$9,13 \times 10^4$
Std4	100 of Std3	100	4.56×10^4
Std5	100 of Std4	100	2.28×10^4
Std6	100 of Std5	100	1.14×10^4
Std7	100 of Std6	100	5.70×10^3
Std8	0	100	0

[a]Std0 corresponds to the concentration of cells grown in standard medium, then lysed in lysis buffer and equal to 4.29×10^6 cells / mL

Table 15
Standard plate layout for adenylate kinase assay

ID	1	2	3	4	5	6	7	8	9	10	11	12
A	Std1	Std1	Std1	S1	S1	S1	S9	S9	S9	S17	S17	S17
B	Std2	Std2	Std2	S2	S2	S2	S10	S10	S10	S18	S18	S18
C	Std3	Std3	Std3	S3	S3	S3	S11	S11	S11	S19	S19	S19
D	Std4	Std4	Std4	S4	S4	S4	S12	S12	S12	S20	S20	S20
E	Std5	Std5	Std5	S5	S5	S5	S13	S13	S13	S21	S21	S21
F	Std6	Std6	Std6	S6	S6	S6	S14	S14	S14	S22	S22	S22
G	Std7	Std7	Std7	S7	S7	S7	S15	S15	S15	S23	S23	S23
H	Std8	Std8	Std8	S8	S8	S8	S16	S16	S16	S24	S24	S24

3. Add 20 µL of standards and samples (S) to the appropriate wells according to plate setup presented in Table 15 (*see* **Note 69**).

4. Add 100 µL of room temperature adenylate kinase detection reagent to all wells, containing either samples or standards.

5. Mix well samples/standards with adenylate kinase detection reagent by gently pipetting the liquids up and down in wells (about 3–4 times) (*see* **Note 70**).

6. Start the incubation time when the last samples are mixed.

7. Switch on the luminometer at least 30 min before use.

8. Remove air bubbles in the assay plate by vacuum centrifugation at $100 \times g$ for 2 min at room temperature (*see* **Note 76**).
   OR

9. Remove air bubbles by blowing ethanol vapor over the wells (*see* **Note 77**).

10. Cover the plate bottom with a white label. Place the assay plate in a luminescence plate reader and incubate the samples for 20 min at room temperature (*see* **Note 71**).

11. Adjust the gain to the highest concentration of the standard. Measure luminescence and save the readout. Move data-file to corresponding project folder.

12. If the readouts for samples are higher than standard 1, it is recommended to dilute samples in sterile water and repeat the measurements.

3.22 Functional Assay: Cholesterol in the Medium [28, 29]

1. The test uses light sensitive reagents. All storage should be in dark and below $-20\ °C$ and where possible protect reagents from light.

2. Equilibrate all reagents from the kit and samples to room temperature (put onto shaking table for 1 h). Use a water bath at 22 °C for the 5× RB (reaction buffer) due to the volume of the reagent.

3. Prepare 20 mM Amplex Red reagent stock solution by mixing the following:
   (a) Component A (1 mg—all from the tube)
   (b) Component B (200 µL DMSO).

4. Store this stock in dark below $-20\ °C$ (*see* **Note 72**).

5. Prepare 1× Reaction Buffer working solution (1× RB) by mixing the following:
   (a) 2.5 mL of 5× reaction buffer (component E).
   (b) 10 mL of deionized H_2O (dH_2O).

6. Prepare 200 U/mL horseradish peroxidase (HRP) stock solution by mixing the following:
   (a) Component C.
   (b) 1 mL of 1× Reaction Buffer.

7. After component C is completely solubilized, aliquot into 100 µL portions in micro test tubes (50 µL is needed for 100 tests = one plate). Mark the tubes with blue dot and HRP symbol (·HRP).

8. Prepare 200 U/mL cholesterol oxidase stock solution by mixing the following:
   (a) Component F.
   (b) 265 µL of 1× Reaction Buffer (250 µL is written in manual but we found that it is necessary to add more on account of portioning/pipetting losses).

9. After component F is completely solubilized prepare 52 µL aliquots in micro test tubes (50 µL is needed for 100 tests = one plate) (*see* **Note 73**). Mark the tubes with red dot and CHOX symbol (·CHOX).

10. Prepare 200 U/mL cholesterol esterase stock solution by mixing the following:
    (a) Component G.
    (b) 250 µL of 1× Reaction Buffer.

11. After component G is completely solubilized prepare 20 µL portions in micro test tubes (5 µL is needed for 100 tests = one plate). Mark the tubes where marked with black dot and CHES symbol (·**CHES**).

12. Prepare immediately before use 20 mM H_2O_2 working solution by mixing the following:
    (a) 23 µL of 3% H_2O_2 stock solution (component D).
    (b) 977 µL of dH_2O.

13. Prepare a **positive control** by diluting the 20 mM H_2O_2 working solution to 10 µM in 1× Reaction Buffer.

14. Preparation of working solution dilution schema:
    (a) 10 µL of 20 mM H_2O_2 solution + 90 µL of 1× RB = 100 µL of 2 mM H_2O_2 solution.
    (b) 10 µL of 2 mM H_2O_2 solution + 90 µL of 1× RB = 100 µL of 0.2 mM H_2O_2 solution (200 µM).
    (c) 10 µL of 200 µM H_2O_2 solution + 190 µL of 1× RB = 200 µL of 10 µM H_2O_2 solution.

    Use 50 µL of these controls as the standards and samples per well.

15. Prepare a cholesterol reference standard curve: Dilute the appropriate amount of 2 mg/mL (5.17 mM) cholesterol reference standard (Component H) into 1× Reaction Buffer to produce cholesterol concentrations of 0 to 8 µg/mL (0 to ~20 µM) (*see* dilution scheme in Table 16). Use 1× Reaction

Table 16

Preparation of the cholesterol standard sample series

Std. No.	Std. (µL)	1× RB buffer (µL)	Std. conc. (µg/mL)	Finale conc. in assay plate (µg/mL)
A	4	996	8	4
B	200 std. A	200	4	2
C	200 std. B	200	2	1
D	200 std. C	200	1	0.5
E	200 std. D	200	0.5	0.25
F	200 std. E	200	0.25	0.125
G	200 std. F	200	0.125	0.625
H	0	200	0.0	0.0

Buffer without cholesterol as a negative control. A volume of 50 µL will be used for each reaction.

16. The cholesterol concentrations will be twofold lower in the final reaction volume. The cholesteryl esters are digested by cholesterol esterase to free cholesterol, which is then detected in the enzyme-coupled reaction with Amplex® Red reagent. The solution in the kit has been calibrated to yield the equivalent of 2 mg/mL cholesterol.

17. Dilute the samples ten times in water (20 µL samples + 180 µL water).

18. Add 50 µL of each standard, control, and sample to the assay plate (*see* Table 17).

19. Prepare the 5 mL (for one plate) of working solution (300 µM Amplex Red reagent) by mixing the following:
    (a) 4820 µL of 1× RB.
    (b) 5 µL of the cholesterol esterase stock solution (CHES).
    (c) 50 µL of the cholesterol oxidase stock solution (CHOX).
    (d) 50 µL of the HRP stock solution (HRP).
    (e) 75 µL of the Amplex Red reagent stock solution.

20. Add 50 µL of the working solution to each well on the assay plate.

21. Mix the plate shortly on the shaking table.

22. Incubate for 30 min (or longer) at 37 °C, protected from light. Cover plate with plate sealer to avoid evaporation (but remember to remove the shield before reading the plate!).

23. Switch on the fluorometer at least 30 min before use.

Table 17
Example of plate setup for determination of cholesterol

ID	1	2	3	4	5	6	7	8	9	10	11	12
A	4	4	4	1	1	1	81	81	81	161	161	161
B	2	2	2	11	11	11	91	91	91	171	171	171
C	1	1	1	21	21	21	101	101	101	030	030	030
D	0.5	0.5	0.5	31	31	31	111	111	111	Neg. cont	Neg. cont	Neg. cont
E	0.25	0.25	0.25	41	41	41	121	121	121	Pos. cont	Pos. cont	Pos. cont
F	0.125	0.125	0.125	51	51	51	131	131	131			
G	0.625	0.625	0.625	61	61	61	141	141	141			
H	0	0	0	71	71	71	151	151	151			

24. Read the assay plate on the FLUOstar Omega program: Amplex Red Cholesterol Assay. Use excitation wavelength of 530–560 nm and emission detection at 590 nm. Set the gain on the strongest standard sample (e.g., well A2) before reading.

25. Fit the standard curve to a liner fit and remember to correct for background fluorescence in samples by subtracting the media-control from the samples.

26. Move data-file to corresponding project folder.

3.23 Functional Assay: Glucose in the Media [27]

1. Bring frozen samples and QC to room temperature.

2. Place samples on the plate shaker for 1 h at 450 rpm.

3. For the fresh samples: collect supernatant samples into micro test tubes.

4. Insert a test strip into the test strip port with the contact bars facing the operator. This will turn on the meter (*see* **Note 74**).

5. When "APPLY SAMPLE" appears on the display, add 3 μL of the sample into the narrow channel on the strip.

6. Wait for the confirmation window to fill completely.

7. Read your result on the meter and write it down in a prepared spreadsheet.

8. Analyze the sample three times in order to establish reproducibility (*see* **Note 75**).

9. Remember to include into the result table (Excel sheet) the following:

    (a) OneTouch Vita instrument number (the one used for the measurements).

    (b) Test strip batch number.

    (c) OneTouch Vita control solution batch number and expected values.

    (d) Measurements results for QC, OneTouch Vita control solution and cell supernatants and/or medium samples.

    (e) Expected values for control medium samples.

10. Move data-file to corresponding project folder.

3.24 Functional Assay: Intracellular Glutathione [27]

1. For cells grown in 2D only: seed the cells overnight or as necessary (depending on cell type specifications). Pay attention to the following parameters:

    (a) The control samples are cells with water.

    (b) There should be a space between cell types, the control samples and if there are wells with a big difference in cell concentration to avoid signal cross talk.

    The cell concentration range for the assay should be from 1000 to 8000 cells per well (96-well plate and 200 μL of medium).

2. Equilibrate reagents and samples to room temperature.

3. Switch on the luminometer at least 30 min before use.

4. Prepare the reagents.

    For the total and oxidized glutathione lysis buffer and luciferin generation reagent follow Tables 18, 19, and 20. Calculate first how much you need of each reagent and add 3 reaction wells to the calculations just to be safe.

Table 18
Preparation of Total Glutathione Lysis Buffer (per reaction well)—for use in GSH samples and Standard

Component	Volume (μL)	Total volume (μL)
Luciferin-NT	1	50
Passive lysis buffer, 5x	9	
Water	40	

Table 19
**preparation of the Oxidized Glutathione Lysis Buffer (per reaction well)—
for use ONLY in GSSG**

Component	Volume (μL)	Total Volume (μL)
Luciferin-NT	1	50
NEM, 25 mM	0.5	
Passive lysis buffer, 5×	9	
Water	39.5	

Table 20
Preparation of the Luciferin Generation Reagent (per reaction well)

Component	Volume (μL)	Total volume (μL)
DTT, 100 mM	1.25	50
Glutathione-S-transferase	3.00	
Glutathione reaction buffer	45.75	

5. For the Luciferin Detection Reagent, transfer 50 mL of Reconstitution Buffer with Esterase to the amber bottle containing the lyophilized Luciferin Detection Reagent. Mix by inversion until the substrate is thoroughly dissolved and **DO NOT VORTEX**.

6. Prepare the standards by diluting 32 μL of 5 mM GSH with 468 μL of ddH$_2$O to make a 320 μM GSH solution (*see* Table 21). Perform 1:2 six serial dilutions by transferring 250 μL of the previous solution into a new micro test tube and adding 250 μL of ddH$_2$O (*see* **Note 76**).

7. Take photomicrographs of the cells before the reading (at least 1 image per cell concentration).

8. For unattached cells, take as much medium as possible out of the wells (*see* **Note** 77). Be sure that the cultures have as little medium left as possible before the adding the lysis reagent.

9. Attached cells should be washed twice with 100 μL of Hanks' solution (at 37 °C).

10. Add 50 μL of Total Glutathione lysis buffer into standard wells (*see* Table 22 for an example of the plate setup).

11. Add 50 μL of Total Glutathione lysis buffer into GSH sample.

12. Add 50 μL of Oxidized Glutathione lysis buffer into GSSG sample wells.

Table 21
Preparation of the standards STOCK 1:2 dilutions in water (per plate) 500 μL

ID	Stock (μM)	Stock (μL)	Water (μL)	Total volume (μL)	Volume left for assay (μL)	Stock conc. of Glutathione (μM)	Final assay conc.5 μL of stock +95 μL of reagents (μM)
1	320		0	500	250	320	16
2	320	250	250	500	250	160	8
3	160	250	250	500	250	80	4
4	80	250	250	500	250	40	2
5	40	250	250	500	250	20	1
6	20	250	250	500	250	10	0.5
7	10	250	250	500	500	5	0.25

Table 22
Example of plate setup for determination of total and oxidized glutathione

ID	1	2	3	4	5	6	7	8	9	10	11	12
A	Std1	Std1	Std1	S1	S1	S1	S9	S9	S9	S17	S17	S17
B	Std2	Std2	Std2	S2	S2	S2	S10	S10	S10	S18	S18	S18
C	Std3	Std3	Std3	S3	S3	S3	S11	S11	S11	S19	S19	S19
D	Std4	Std4	Std4	S4	S4	S4	S12	S12	S12	S20	S20	S20
E	Std5	Std5	Std5	S5	S5	S5	S13	S13	S13	S21	S21	S21
F	Std6	Std6	Std6	S6	S6	S6	S14	S14	S14	S22	S22	S22
G	Std7	Std7	Std7	S7	S7	S7	S15	S15	S15	S23	S23	S23
H	Std0	Std0	Std0	S8	S8	S8	S16	S16	S16	S24	S24	S24

13. Add 5 μL of the standard solutions to the wells with 45 μL Total Glutathione lysis buffer.

14. Shake the plates for 5 min.

15. Add 50 μL of Luciferin Generation Reagent to all sample and standard curve wells and incubate for 30 min.

16. Add 100 μL of Luciferin Detection Reagent in all sample well standard curve wells and equilibrate for 15 min.

17. If necessary, remove any bubbles from the wells by vacuum centrifuging the plate at $100 \times g$ for 1 min).

18. Read luminescence using the appropriate assay program (*see* **Note 78**).

19. Move data-file to corresponding project folder.

3.25 Functional Assay: Urea in the Media [27]

1. Equilibrate all reagents except urease, from kit and samples to room temperature (put into water bath at 22 °C for 30 min).

2. Predilute the standard from the kit in diluted assay buffer according to Table 23.

3. Add 150 μL of diluted assay buffer and 20 μL of standards (tubes StdA-H) per well on assay plate (*see* plate setup in Table 24).

4. Add 150 μL of diluted assay buffer and 20 μL of samples to assay plate (*see* plate setup S1–S12).

5. Add 170 μL of diluted assay buffer and 20 μL of samples to assay plate (*see* plate setup B1–B12).

6. Initiate the reactions by adding 20 μL Urease to all standard and sample wells. DO NOT add to sample background wells.

7. Cover plate with the plate cover and incubate at room temperature for 10 min.

8. Remove the plate cover and 10 μL Ammonia Detector to every well being used including the samples background wells.

9. Cover the plate again and incubate at room temperature for 15 min.

10. Remove plate cover and read the plate in the FLUOstar Omega (using the correct program) using excitation wavelength of 405–415 nm and emission wavelength of 470–480 nm. Remember to adjust the gain on highest standard.

Table 23
Preparation of the urea standard dilution series

Std. No.	Std. (μL)	Dilution buffer (μL)	Std. conc. (nM)	Final conc. in assay plate (nM)
StdA	0	1000	0	0
StdB	10	990	0.05	0.005
StdC	20	980	0.1	0.01
StdD	40	960	0.2	0.02
StdE	80	920	0.4	0.04
StdF	120	880	0.6	0.06
StdG	160	840	0.8	0.08
StdH	180	800	1.0	0.10

Table 24
Example of plate setup for determination of urea

A - H = Standards
S1 - S12 = Sample wells
B1 - B12 = Sample Background wells

ID	1	2	3	4	5	6	7	8	9	10	11	12
A	Std A	Std A	Std A	S1	S1	S1	S5	S5	S5	S9	S9	S9
B	Std B	Std B	Std B	B1	B1	B1	B5	B5	B5	B9	B9	B9
C	Std C	Std C	Std C	S2	S2	S2	S6	S6	S6	S10	S10	S10
D	Std D	Std D	Std D	B2	B2	B2	B6	B6	B6	B10	B10	B10
E	Std E	Std E	Std E	S3	S3	S3	S7	S7	S7	S11	S11	S11
F	StdF	StdF	StdF	B3	B3	B3	B7	B7	B7	B11	B11	B11
G	Std G	Std G	Std G	S4	S4	S4	S8	S8	S8	S12	S12	S12
H	Std H	Std H	Std H	B4	B4	B4	B8	B8	B8	B12	B12	B12

11. Subtract the fluorescence value of the standard A itself and the rest of the standards. This is the corrected fluorescence (CF). Plot the standard curve and use a linear regression.

12. Subtract the sample background fluorescence value from the sample value. This is the corrected sample fluorescence value (CSF). Calculate the results of the samples by using.

$$\text{Urea (mM)} = \left(\frac{\text{CSF} - (y - \text{intercept})}{\text{slope}} \right) \times \text{sample dilution}$$

13. The limit of detection for the assay is 0.05 mM (±0.025 mM) Urea.

14. Move data-file to corresponding project folder.

4 Notes

1. It is recommended to use the medium within a maximum of 2 weeks.

2. Part of the final medium can be added to the culture flask and placed in the incubator in advance for it to equilibrate with the CO_2 in the incubator). We recommend using a DMEM-based medium for C3A cells.

3. Follow the laboratory liquid nitrogen safety guidelines including wearing a face shield and a pair of goggles to protect the face and eyes from splatters of liquid nitrogen.

4. Change the medium the following day in order to remove any remaining cryoprotective additive from the freezing medium.

5. Portion trypsin into 1 mL portions and freeze at −70 °C. Before use dilute 10× by adding 9 mL of HBSS without Ca^{2+} and Mg^{2+} to the aliquot.

6. It is recommended to count the cells using either a haemocytometer or an automated cell counter. A T25 flask with C3A cells grown to 80–90% confluency will yield ~5×10^6 cells in total.

7. Note: do not use concentric movements as it will create uneven cell distribution.

8. Freezing medium can be prepared in advance and stored in the fridge until use.

9. It is recommended to use the thawed freezing medium within a maximum of 2 weeks.

10. We recommend performing this procedure 1 day before the start of your experiment.

11. Use a blunt 10 mL syringe with needle for filling the cell chamber—but a plastic pipette can also be used if one needs to avoid "sharps."

12. Use a 200 μL pipette or 2 mL syringe with a thin needle.

13. Self-aggregation is an excellent way to create spheroids or organoids from primary cells. Often, primary cells are obtained from the enzymatic digestion of tissue biopsies and as such contain more than one cell type. Self-aggregation is slower and gives cells extra time to organize themselves and or find the correct "neighbours."

14. For optimal results, we recommended to use a seeding density of 960,000 cells per ClinoReactor™ for the HepG2/C3A cell line. Other cell lines may require a different seeding density.

15. Overfill the chamber with media—this will make it easy to remove air bubbles. You need to remove bubbles because they will act like "sandpaper" and tear spheroids apart.

16. Small self-assembled C3A spheroids should be visible about the 48 h time point in the ClinoReactor™ vessel under the given conditions.

17. It is important to keep the ClinoReactor™ population not too dense to avoid clumping. The population can be divided into two ClinoReactor™ vessels or the number of the spheroids could be adjusted by preselecting spheroids of the desired size and discarding the others.

18. It is advisable to prepare 1 more well per condition than needed as a backup.

19. Check that the cells are in the centre of each well and that they are equally distributed in the wells.

20. The time needed for different cell lines varies from 4 to 36 h. "Sticky" cells need shorter times: find the best for your cell line which gives a good yield of roundish spheroids with least debris.

21. Spheroids containing 800 cells are usually started at 4 rpm and adjust after a few hours if needed.

22. The best way to do this is to make the holes in the lid of another tube and then cut it off and transfer the lid to the tube with the QGel™. This way, when the QGel™ is dry, the lid with holes can be removed and the lid attached to the tube can be used to close the tube.

23. Subheading 3.9, **steps 4–6** need to be done as quickly as possible to prevent Q-gel polymerization, so it is advisable to have liquid nitrogen prepared, in a small container, before starting Subheading 3.9, **step 4**.

24. Remember to fill the vacuum centrifuge freeze-dryer with ethanol before turning it on and precool before use. The Q-gel™ aliquots must be frozen when placed in the vacuum centrifuge.

25. Do not stretch the Parafilm™ M—it will lose some of its hydrophobicity and the dots will become flatter.

26. This solution will humidify the petri dish during the QGel™-- spheroids solidification and will prevent the star-shape like spheroid creation.

27. The tips can be prepared in advance and kept sterile until needed.

28. It is important that, from Subheading 3.10, **step 12**, the procedure is done one aliquot and cell suspension at a time. The gel polymerases quickly (in about 5–10 min) so it is

important to prepare the batches one at time. As soon as the cells + buffer A + medium are mixed with the QGel™, proceed as fast as possible.

29. Be aware of the pipette safety drops (the first and the last) and try not to load those on to the casting block because they will be different in size to the other drops. In case it happens, do not stop the loading the drops. Try to take these bigger drops when there is no more mix.

30. QGel™-spheroids are much lighter than the AggreWell spheroids, so the speed needs to be slower.

31. Alginate gels are especially useful when creating spheroids of cell lines which do not naturally aggregate or aggregate poorly, like the LS180 cell line. Alginates cross-linked with Ca^{2+} has been popularized for in vitro cell culture and tissue engineering applications due to their biocompatibility, low toxicity, relatively low cost and mild gelation properties. They have a relatively inert aqueous environment, high gel porosity (which allows for high diffusion rates) and are biodegradable.

32. A flask with cells at 80% confluency works well.

33. Working quickly with an automatic pipette, it is possible to cast 300 spheroids in 6 min. Concentrate on quality of the droplets rather than on numbers. Discard the first and last droplets in case their volumes are inaccurate.

34. Run a pilot experiment using different ratios of cells (e.g., 90:10, 75:25, 60:40, 50:50, 40:60, 25:75, 10:90, or whatever makes sense for your cell combination) in the initial mix. Cells may replicate at different speeds so in order to achieve the ratio seen in vivo may mean starting with a different ratio.

35. This approach will help to keep the different cells physically in different layers.

36. If necessary, adjust the pH of a bicarbonate-based media by reequilibrating the media inside the cell culture incubator.

37. For convenience you can mark the tubes one with IN and second with OUT.

38. We use the inside top of a 50 mL sterile tube (see Fig. 3). Avoid touching with gloves the bottom part of the plug that is positioned inside the cell chamber and will have direct contact with the cell culture medium and the spheroids.

39. Use a long plastic tip if you need to avoid "sharps."

40. Overfilling of the ClinoReactor™ with the medium should be clearly visible in the port cup. This greatly facilitates removing the air bubbles from the ClinoReactor™.

41. The split should be preferably done in connection with a medium change so that the cells are not disturbed unnecessarily.

42. The ClinoReactor™s run for several days and so we typically write the ID on the ClinoReactor™ (e.g., date for setting up/date for splitting (we use the same letter as mother culture)/type of the medium/name of the experiment (if relevant), e.g., 200,630/200711_A1/DMEM/C3A_ Seeding protocol optimization).

43. Use a formaldehyde that contains about 15% methanol: this prevents the formation of paraformaldehyde in stock solutions).

44. If your spheroids are small, place a small drop (ca. 20 μL) of Tissue-Tek in the bottom of the tube before you put the samples in. This way they will all be in the same plane and you will get a better section for staining.

45. Cacodylate buffer reduces acidification during fixation. pH 7.2 is close to the basic end of the buffering capacity of cacodylate so take care when titrating with NaOH. The use of pH 7.2 maximizes the buffering capacity available.

46. Avoid letting the sample dry out between dehydration steps.

47. This may be aliquoted because it is not recommended to freeze and thaw the DNA standard for more than 4 cycles. Incubate each thawed calf thymus DNA standard aliquot again at 50 °C for 15–30 min before use.

48. These solutions must be made fresh, just prior to use and stored in the dark.

49. Setting the fluorometer slits at 2.5 nm for the 0.1–5 μg/mL range and 5 nm for the 10–500 ng/mL range can increase the range of DNA detected. Average results from 5 cycles with 10 individual measurements per cycle should be used to calculate the DNA content of the sample. The sensitivity of plate readers varies between different fluorometers. It is recommended to use 200 μL of 2 μg/mL bisBenzimide H 33258 Solution. The range of DNA concentrations to be measured should be between 0.1 and 10 ng/μL.

50. The volume of the DNA solution added to the dye solution should not exceed 10 μL.

51. This is a bit complicated but will reduce photobleaching of the reaction mixture and increase the reproducibility of the data.

52. Work with increasing DNA amounts to minimize any cross-contamination.

53. This is a very good buffer for dissolving proteins—but should not be warmed up above 35 °C because the urea will start to

decompose and give ions that will attack proteins. Urea and thiourea can be stored in a desiccating flask or container (ideally under partial vacuum at 4 °C) to make sure that they are absolutely dry.

54. The water bath can conveniently be a glass beaker on a heated magnetic stirrer (use a thermometer-regulated heater if possible). Urea starts to break down at temperatures above 35 °C.

55. Do not leave it to stir for longer than 1 h as the Serdolyte resin will start to dissolve.

56. Check that the graduation on the cylinder is accurate—or make a mark. It can be difficult with these reagents to use a volumetric flask—find a short one.

57. Add 12.5 mL methanol to the Active Motif, 15001 brown container with the 0.5 mg dye and dissolve.

58. 45 min is enough if the sample contains less than 10 µg/mL protein. If the sample contains more than 10 µg/mL protein, then leave samples in lysis buffer overnight (at maximum 28 °C) on the shaker to ensure all proteins are lysed before testing. The Lysis buffer is effective up to 32 °C, so it is recommended to work below that temperature.

59. Just before use dilute stain with dilution buffer and aliquot 20 mL into 50 mL conical centrifuge tubes).

60. When adding the lysis B buffer, it is recommended to exchange tip in between pipetting of each standard dilution tube. The lysis B buffer tends to "stick to the tip," which makes the pipetting inaccurate when using the same tip for several tubes.

61. It is recommended to use an electronic multichannel pipette for this procedure if one is available.

62. It is recommended to start centrifuge, allow it to reach 200 rpm and then turn on the vacuum. When vacuum is complete (indicated by the suction noise stopping) wait approximately 60 s and stop the centrifuge. When the rotor has stopped, turn the vacuum valve slowly back into the open position (i.e., no vacuum in the centrifuge).

63. Use a squeeze bottle with freshly removed ethanol and a draw tube (or you can pull out the draw tube clearly out of the ethanol), so that by applying gentle pressure to the bottle, only the vapor is expelled but not any liquid ethanol.

64. Preparation of ATP aliquots: add 1 mL of sterile water to a vial with ATP standard, mix gently. Aliquot (approx. 70 µL/vial) and store at −20 °C.

65. Leave the standards on ice until further use. Mix thoroughly each standard before preparing its serial dilution.

66. It is recommended to use 1 mL wide-bore or cut tips for the transfer to preserve the spheroid or organoid structure. If ATP

measurements are performed for cells grown in 2D, no transfer of cells is required since cells can be grown directly in the white opaque 96-well plate.

67. It is recommended to use electronic multichannel pipette for the steps in this procedure, if available.

68. To heat treat the medium: incubate the medium for 2 days at 37 °C and then centrifuge it for 15 min at maximum speed in the bench centrifuge (aliquot and freeze at −20 °C). Use the same medium as the one used in the actual experiment. Mix thoroughly each standard before preparing its serial dilution.

69. Use an electronic pipette for the steps in this procedure, if available.

70. Set an electronic multichannel pipette to 100 μL.

71. It is important that the incubation time is 20 min every time, because the signal will change over time and results from plates with different incubation times are not comparable. It is also recommended to run the assay at room temperature every time since higher temperatures (e.g., 35 °C) affects the final readout.

72. One batch should be enough for 250 assays (75 μL per 100 assays).

73. Component F can be very hard to dissolve—take your time.

74. To improve comparability, use the same lot number for all measurements in a project and keep strips in a closed bag or box once you start to use a pack.

75. To avoid evaporation from your samples, remove the sheet seal from the plate in sections as you measure the samples.

76. This amount of standard solutions is enough for 16 plates (250 μL/15 μL [i.e., 5 μL in triplicate]). If more plates are being assayed, one should consider making a solution of 320 μM of GSH with a higher volume. Prepare the GSH standard curve dilutions no longer than 30 min before performing the assay.

77. Unattached or primary cells are not washed because of the risk to lose them.

78. This assay preserves proportionality and stability for up to 20 h, so you need not read the plate immediately. Typical program parameters would be: Positioning delay = 0.5 s; No. of kinetic windows = 1; No. of cycles = 5; Measurement start time = 0.3 s; Measurement interval time = 1.00 s; Cycle time = 210 s; reading direction = horizontal; Additional shaking = none.

Acknowledgments

The authors would like to acknowledge the valuable contribution that the late Vasco Botelho Carvalho made in developing some of the protocols described here. The authors are members of "CellFit" EU COST Action CA16119 platform. KW, BK, and SJF are employed by CelVivo ApS.

References

1. Barkauskas CE, Chung MI, Fioret B, Gao X, Katsura H, Hogan BL (2017) Lung organoids: current uses and future promise. Development 144(6):986–997. https://doi.org/10.1242/dev.140103

2. Morizane R, Bonventre JV (2017) Kidney organoids: a translational journey. Trends Mol Med 23(3):246–263. https://doi.org/10.1016/j.molmed.2017.01.001

3. Pasca SP (2018) The rise of three-dimensional human brain cultures. Nature 553 (7689):437–445. https://doi.org/10.1038/nature25032

4. Artegiani B, Clevers H (2018) Use and application of 3D-organoid technology. Hum Mol Genet 27(R2):R99–R107. https://doi.org/10.1093/hmg/ddy187

5. Xie J, Xu X, Yin P, Li Y, Guo H, Kujawa S, Chakravarti D, Bulun S, Kim JJ, Wei JJ (2018) Application of ex-vivo spheroid model system for the analysis of senescence and senolytic phenotypes in uterine leiomyoma. Lab Investig 98(12):1575–1587. https://doi.org/10.1038/s41374-018-0117-5

6. Laschke MW, Menger MD (2017) Life is 3D: boosting spheroid function for tissue engineering. Trends Biotechnol 35(2):133–144. https://doi.org/10.1016/j.tibtech.2016.08.004

7. Llonch S, Carido M, Ader M (2018) Organoid technology for retinal repair. Dev Biol 433 (2):132–143. https://doi.org/10.1016/j.ydbio.2017.09.028

8. Perkhofer L, Frappart PO, Muller M, Kleger A (2018) Importance of organoids for personalized medicine. Per Med 15(6):461–465. https://doi.org/10.2217/pme-2018-0071

9. Toda S, Blauch LR, Tang SKY, Morsut L, Lim WA (2018) Programming self-organizing multicellular structures with synthetic cell-cell signaling. Science 361(6398):156–162. https://doi.org/10.1126/science.aat0271

10. Nath S, Devi GR (2016) Three-dimensional culture systems in cancer research: focus on tumor spheroid model. Pharmacol Ther 163:94–108. https://doi.org/10.1016/j.pharmthera.2016.03.013

11. Dayem AA, Lee SB, Kim K, Lim KM, Jeon TI, Cho SG (2019) Recent advances in organoid culture for insulin production and diabetes therapy: methods and challenges. BMB Rep 52(5):295–303

12. Tiriac H, Plenker D, Baker LA, Tuveson DA (2019) Organoid models for translational pancreatic cancer research. Curr Opin Genet Dev 54:7–11. https://doi.org/10.1016/j.gde.2019.02.003

13. Clevers H (2016) Modeling development and disease with organoids. Cell 165 (7):1586–1597. https://doi.org/10.1016/j.cell.2016.05.082

14. Wang Z, Wang SN, Xu TY, Miao ZW, Su DF, Miao CY (2017) Organoid technology for brain and therapeutics research. CNS Neurosci Ther 23(10):771–778. https://doi.org/10.1111/cns.12754

15. Vorrink SU, Zhou Y, Ingelman-Sundberg M, Lauschke VM (2018) Prediction of drug-induced hepatotoxicity using long-term stable primary hepatic 3D spheroid cultures in chemically defined conditions. Toxicol Sci 163 (2):655–665. https://doi.org/10.1093/toxsci/kfy058

16. Lauschke VM, Hendriks DF, Bell CC, Andersson TB, Ingelman-Sundberg M (2016) Novel 3D culture Systems for Studies of human liver function and assessments of the hepatotoxicity of drugs and drug candidates. Chem Res Toxicol 29(12):1936–1955. https://doi.org/10.1021/acs.chemrestox.6b00150

17. Gripon P, Rumin S, Urban S, Le Seyec J, Glaise D, Cannie I, Guyomard C, Lucas J, Trepo C, Guguen-Guillouzo C (2002) Infection of a human hepatoma cell line by hepatitis B virus. Proc Natl Acad Sci U S A 99 (24):15655–15660. https://doi.org/10.1073/pnas.232137699

18. Yin Y, Zhou D (2018) Organoid and Enteroid modeling of salmonella infection. Front Cell

Infect Microbiol 8:102. https://doi.org/10.3389/fcimb.2018.00102

19. Ramani S, Crawford SE, Blutt SE, Estes MK (2018) Human organoid cultures: transformative new tools for human virus studies. Curr Opin Virol 29:79–86. https://doi.org/10.1016/j.coviro.2018.04.001

20. Dye BR, Hill DR, Ferguson MA, Tsai YH, Nagy MS, Dyal R, Wells JM, Mayhew CN, Nattiv R, Klein OD, White ES, Deutsch GH, Spence JR (2015) In vitro generation of human pluripotent stem cell derived lung organoids. Elife 4:e05098. https://doi.org/10.7554/eLife.05098

21. Ramaiahgari SC, Waidyanatha S, Dixon D, DeVito MJ, Paules RS, Ferguson SS (2017) Three-dimensional (3D) HepaRG spheroid model with physiologically relevant xenobiotic metabolism competence and hepatocyte functionality for liver toxicity screening. Toxicol Sci 160(1):189–190. https://doi.org/10.1093/toxsci/kfx194

22. Bergmann S, Lawler SE, Qu Y, Fadzen CM, Wolfe JM, Regan MS, Pentelute BL, Agar NYR, Cho CF (2018) Blood-brain-barrier organoids for investigating the permeability of CNS therapeutics. Nat Protoc 13(12):2827–2843. https://doi.org/10.1038/s41596-018-0066-x

23. Cederquist GY, Asciolla JJ, Tchieu J, Walsh RM, Cornacchia D, Resh MD, Studer L (2019) Specification of positional identity in forebrain organoids. Nat Biotechnol 37(4):436–444. https://doi.org/10.1038/s41587-019-0085-3

24. Kang HM, Lim JH, Noh KH, Park D, Cho HS, Susztak K, Jung CR (2019) Effective reconstruction of functional organotypic kidney spheroid for in vitro nephrotoxicity studies. Sci Rep 9(1):17610. https://doi.org/10.1038/s41598-019-53855-2

25. Wrzesinski K, Fey SJ (2015) From 2D to 3D - a new dimension for modelling the effect of natural products on human tissue. Curr Pharm Des 21:5605–5616

26. Wrzesinski K, Rogowska-Wrzesinska A, Kanlaya R, Borkowski K, Schwammle V, Dai J, Joensen KE, Wojdyla K, Carvalho VB, Fey SJ (2014) The cultural divide: exponential growth in classical 2D and metabolic equilibrium in 3D environments. PLoS One 9(9): e106973. https://doi.org/10.1371/journal.pone.0106973

27. Wrzesinski K, Fey SJ (2018) Metabolic reprogramming and the recovery of physiological functionality in 3D cultures in microbioreactors. Bioengineering (Basel) 5(1):22. https://doi.org/10.3390/bioengineering5010022

28. Wrzesinski K, Magnone MC, Visby Hansen L, Kruse ME, Bergauer T, Bobadilla M, Gubler M, Mizrahi J, Zhang K, Andreasen CM, Joensen KE, Andersen SM, Olesen JB, SdM OB, Fey SJ (2013) HepG2/C3A 3D spheroids exhibit stable physiological functionality for at least 24 days after recovering from trypsinisation. Toxicol Res 2(3):163–172. https://doi.org/10.1039/c3tx20086h

29. Wrzesinski K, Fey SJ (2013) After trypsinisation, 3D spheroids of C3A hepatocytes need 18 days to re-establish similar levels of key physiological functions to those seen in the liver. Toxicol Res 2(2):123–135. https://doi.org/10.1039/c2tx20060k

30. Fey SJ, Wrzesinski K (2012) Determination of drug toxicity using 3D spheroids constructed from an immortal human hepatocyte cell line. Toxicol Sci 127(2):403–411. https://doi.org/10.1093/toxsci/kfs122

31. Smit T, Calitz C, Willers C, Svitina H, Hamman J, Fey SJ, Gouws C, Wrzesinski K (2020) Characterization of an alginate encapsulated LS180 spheroid model for anticolorectal cancer compound screening. ACS Med Chem Lett 11(5):1014–1021. https://doi.org/10.1021/acsmedchemlett.0c00076

Chapter 3

Protocol to Study the Role of Extracellular Vesicles During Induced Stem Cell Differentiation

Kelly C. S. Roballo, Carlos E. Ambrosio, and Juliano C. da Silveira

Abstract

Extracellular vesicles (EVs) are vesicles released by cells, which due to their cargo and cell membrane proteins induce changes in the recipient cells. These vesicles can be a novel option to induce stem cell differentiation. Here we described a method to induce mesenchymal stem cell differentiation (MSC) into neuron-like cells using small EVs from neurons. First, we will describe a method based on neurons to induce adipocyte derived stem cells differentiation, a type of MSC, by coculturing both using inserts. Secondly, we will describe a follow-up method by using only isolated neuron-derived small EVs to directly induce ADSC differentiation in neuron-like cells. Importantly, in both methods it is possible to avoid the direct cell-to-cell contact, thus allowing for the study of soluble factors role during stem cell differentiation.

Key words Transwell, Extracellular vesicles, Exosomes, Cell differentiation, Indirect cell communication

1 Introduction

The process of stem cell differentiation to other cell types can be induced chemically, and/or biologically [1]. After cell differentiation induction, mesenchymal stem cells (MSC) can differentiated in several cell types such as neuron-like cells, osteoblast-like cells, chondrocyte-like cells, adipocyte-like cells [2–5]. For MSC, several protocols were tested for inducting cell differentiation in vitro, and they only differ in cell differentiation rate after induction [2–5]. Most of these protocols start with stem cells primary cell culture, differentiation, and characterization of the differentiated cells [2–4].

Recently studies have been showing that small EVs are able to modify recipient cells [3, 6–8]. Small EVs are vesicles secreted by cells, with a range of 50–200 nm in size, and it is becoming evident that they strongly influence cellular behavior [6–8], for example inducing cell differentiation [3].

Tiziana A.L. Brevini et al. (eds.), *Next Generation Culture Platforms for Reliable In Vitro Models: Methods and Protocols*, Methods in Molecular Biology, vol. 2273, https://doi.org/10.1007/978-1-0716-1246-0_3,

Here, we describe a method for stem cell differentiation, using derived EVs from a specialized cell type (neurons) to induce cell differentiation of a multipotent cell type culture (adipose derived stem cells). First, we will describe a coculture methodology system using inserts (transwell) to prevent direct cell–to–cell contact, as an alternative for studying the in vivo situation with an in vitro model [9]. Neurons are coculture with adipocyte derived stem cells (ADSCs), a type of MSC, and neuronal differentiation of ADSCs are analyzed.

Secondly, we will describe a method using small EVs from neurons to induce neuronal differentiation of ADSC [3]. For this last methodology, small EVs can be isolated, characterized and tracked in vitro. We believe that this protocol can be used for other types of stem cell differentiation experiments such as osteogenic, chondrogenic and adipogenic linages. Where neurons and neuron-derived small EVs could be replace for osteoblasts, chondrocytes, or adipocytes and respective small EVs.

2   Materials

2.1   Primary Cell Culture, Coculture with Inserts, and Cell Differentiation Validation

1. Sterile supplies: 15 and 50 mL polypropylene conical tubes, 15 mL syringe, 0.20 μm filter units, 1 and 2 mL microtubes, 100 μm cell strainer, 15 mL polystyrene conical tubes, hematocytometer, ophthalmic scissors and forceps, number 4 or 5 blade, blade cable number 4 or 5, 24-well tissue culture plate, 100 mm petri dish, T-25 cell culture flasks, inserts (transwells; 400 nm, ThinCert™ inserts), histology slides and glass covers for 24-well plate, mounting media.

2. Nonsterile supplies: freezing apparatus (e.g., Mr. Frost).

3. Equipment: Inverted light microscope, fluorescence microscope, shaking incubator at 37 °C, water bath at 37 °C, centrifuge with 4 °C and room temperatures and at least $16,500 \times g$ of speed, biosafety hood, 5% CO_2 humidified incubator.

4. Reagents: 70% ethanol, 0.5% trypsin–EDTA solution, sterile bovine serum albumin (BSA), sterile type 1 collagenase, sterile Dulbecco's Modified Eagle Media (DMEM) high glucose, sterile fetal bovine serum (FBS), paraformaldehyde 4%, 1× penicillin–streptomycin antibiotic solution, trypan blue, sterile KnockOut™ serum replacement (or FBS depleted of EVs), sterile lipofectamine 2000, lentivirus vector encoding GFP (Addegene # 14883), lentivirus vector encoding PalmTdTomato (tracking EVs), sterile phosphate buffered saline (PBS) (commercial solution of PBS is available and can be use in this protocol), 1% Triton X-100 in PBS, primary antibodies (1:500 anti-beta tubulin III; 1:400 anti-SNAP25), secondary antibodies (1:1000 goat anti-rabbit FITC; 1:500 goat anti mouse

Alexa Fluor 647 or PE), Hoechst (Trihydrochloride, Trihydrate,) or DAPI (DAPI 4′,6 Diamidine-2′-phenylindole dihydrochloride), 4% paraformaldehyde, pH 7.2–7.4.

5. Buffers: PBS: 37 mM NaCl, 2.7 mM KCl, 10 mM Na_2HPO_4, 2 mM KH_2PO_4, pH 7.4. Sterilize using a 0.20 μm filter. Keep at 4 °C when not in use.

6. Collagenase solution: Take 10 mg of type 1 collagenase and dissolve in 10 mL PBS. After sterile filtration with 0.20 μm filter, warm the solution to 37 °C in a water bath. This solution should be prepared within 1 h before to be used. Mesenchymal stem cell and neurons medium for primary culture: DMEM high glucose with 10% FBS, and 1× antibiotic solution (penicillin–streptomycin).

7. Mesenchymal stem cell and neurons medium for coculture and extracellular vesicles isolation (experimental media): DMEM high glucose with 15% KnockOut™ serum replacement (or FBS depleted of EVs), and 1× antibiotic solution (penicillin–streptomycin).

2.2 Extracellular Vesicles Isolation, Characterization, and Cell Differentiation Method

1. Sterile supplies: 1.5 mL and 15 mL centrifuge tubes, 6 mL ultracentrifuge tube, 15 mL syringe, and 0.20 μm filter.

2. Non-sterile supplies: copper grid coated with Pioloform, filter paper.

3. Equipment: centrifuge with 4 °C and room temperatures and at least 16,500 × g of speed, biosafety hood, NanoSight NS300 (Malvern Instruments, Malvern, UK) or other similar, transmitted electron microscope (TEM), ultracentrifuge.

4. Reagents and Buffers: sterile ExoQuick, 100 nm of beads (Malvern Instruments, Malvern, UK), sterile phosphate buffered saline (PBS) (commercial solution of PBS is available and can be use in this protocol), ultrapure water.

5. Fixative solution for TEM: 2.5% glutaraldehyde, 0.1 M cacodylate and 4% paraformaldehyde, pH 7.2–7.4.

3 Methods

Carry out all procedures at sterile conditions under a laminar flow hood unless otherwise specified. For this protocol, a specified cell type (e.g., neurons, chondrocytes, myocytes, cardiomyocytes) is used as an induction method for cell differentiation, and a multipotent stem cell type (e.g., mesenchymal stem cells, embryonic stem cells) are used as the cell to be differentiated. Neurons and mesenchymal stem cells will be used as an example.

3.1 Cell Isolation and Primary Cell Culture

3.1.1 Neurons Isolation and Culture

1. Proceed with animal euthanasia using anesthesia or CO_2 chamber as approved by the AVMA Guidelines on Euthanasia and local ethical committee (*see* **Note 1**).

2. To isolate neurons from mice brain, shave the top of the animal's head, make an incision using number 4 or 5 blade to access the skull, and further occipital bone.

3. Cut the occipital bone (*see* **Note 2**) and isolate hippocampus region using sterile scissors.

4. Place hippocampus at a 100 mm petri dish with 5 mL of primary culture medium. Mince hippocampus into small pieces (1–2 mm^2) using blade.

5. With 1 mL pipette move hippocampus pieces and medium from the petri dish to a 15 mL centrifuge tube (*see* **Note 3**). Centrifuge at $300 \times g$ at room temperature for 5 min.

6. Remove supernatant and add 5 mL of 0.5% trypsin–EDTA solution, mix with 1 mL pipette (*see* **Note 4**). Centrifuge at $300 \times g$ at room temperature for 5 min.

7. Remove supernatant and add 5 mL of fresh media primary culture media, mix pellet with medium.

8. Count isolated cells using hemocytometer chamber and trypan blue to determinate cell concentration and % of live cells.

9. Plate 0.7×10^6 neurons in T-25 cell culture flask.

10. Check cells under light microscope every day. Change media every 2–3 days after cell attachment.

3.1.2 Mesenchymal Stem Cell Isolation and Culture

1. Before mesenchymal stem cells (MSC)isolation from adipose tissue (ADSCs), warm collagenase solution, primary cell culture media, and PBS using the 37 °C water bath.

2. Proceed with animal euthanasia using anesthesia or CO_2 chamber as approved by the AVMA Guidelines on Euthanasia and local ethical committee (*see* **Note 1**), shave abdominal region, and sterilize with 70% ethanol.

3. With sterile scissors, open abdominal cavity, and isolate retroperitoneal fat tissue (*see* **Note 5**). Weight fat tissue.

4. Use sterile forceps to transfer the tissue to the bottom of a 100 mm petri dish containing enough PBS (5–6 mL) and 1% of antibiotics to rinse the tissue.

5. Transfer rinsed tissue to other sterile 100 mm petri dish and mince the adipose tissue into small pieces using sterile scissors or blade (1–2 mm^2).

6. With 1 mL pipette transfer pieces to a sterile 50 mL centrifuge tube. Use for each 1 g of adipose tissue a total of 10 mL of warmed collagenase solution.

7. Sterilize outside of 50 mL with 70% ethanol. Place 50 mL centrifuge tube in a shaker at 37 °C incubator. Shake at 100 rpm for 60–90 min, if shaker is not available place bottles in 37 °C incubator, and swirl bottles every 3–5 min. After 60–90 min the solution should be almost liquid but containing some cell debris.

8. Filter solution through a sterile 100 μm filter into a sterile 50 mL conical tube. Centrifuge tube at 450 × g at room temperature for 5 min.

9. Cell pellet will have a dark brown color at the top part with light brown color in the bottom. Using 1 mL pipette, without disturb the pellet aspirate most of the supernatant. Leave a small solution so that the pellet (containing cells) is not disturbed.

10. Add 5 mL primary cell culture media and mix well. Count cells.

11. Plate 0.7×10^6 isolated ADSCs in T-25 cell culture flask.

12. Twenty-four hours after ADSCs isolation, aspirate the media from flask, add fresh primary cell culture media. Evaluate cells using an inverted light microscope.

13. Change media every 2–3 days until cells are 80–90% confluent. When cells reach confluence, transfer to a new flask or store as frozen stocks as passage one (P1) or differentiated.

3.2 Coculture with Inserts (Transwell)

Indirect cell contact can be achieved using an insert system to separate fully differentiated cells from the stem cells to induce stem cell differentiation. This process is partially mediated by EVs released from specialize cells in the culture media as well as other soluble factors mimicking in vivo conditions without the direct cell-to-cell contact.

3.2.1 EVs and Cell Label for Tracking

1. To coculture neurons with ADSCs, and tracking EVs, both cells are transfected with different plasmids, using lentivirus system.

2. Neurons are transduced with PalmtdTomato to label neuron derived EVs. The ideal plasmid vector should contain tandem dimer Tomato (tdTomato) fused at NH_2-termini with a palmitoylation signal (PalmtdTomato) [10] (*see* **Note 6**).

3. Neurons are culture with experimental media and transduced with lentivirus vector encoding PalmtdTomato for 24 h. Media is changed and experiments performed (*see* **Note 6**).

4. ADSCs are transduced with a lentivirus vector encoding GFP (Addgene # 14883) to label cells.

5. ADSCs are culture with experimental media and transduced with lentivirus vector encoding GFP for 24 h. Media is changed and experiments performed (*see* **Note 6**).

3.2.2 Transwell
Coculture

1. After transfections, neurons and ADSCs are quantified and concentrated in the following concentrations: neurons (10^5 cells/½ mL) and ADSCs (10^4 cells/½ mL) (*see* **Note 7**).

2. To test the differentiation rate of ADSCs. Coculture can be carried for 3, 7, and 14 days (time of cell induction) (*see* **Note 7**). One well per replicate is used, and three wells per time point. A total of 15 wells will be used (including controls).

3. Seed ADSCs GFP+, and neurons[PalmtdTomato] only to be used as control groups.

4. For immunocytochemistry analysis, ADSCs should be culture in glass coverslips. Add one single glass coverslip in each well that will be used of a 24-well tissue culture plate.

5. Seed ADSCs in the bottom of 24-wells tissue culture plate. Use one well per experimental replicate, and time point.

6. Place an insert in each well containing ADSCs. Seed neurons[PalmtdTomato] in the insert membrane (*see* **Note 7**).

7. Add 1 mL experimental media per well (*see* **Note 8**).

8. To validate and quantify cell differentiation rate immunocytochemistry is perform (described below). The fluorescence levels [3, 11, 12], and percentage of positive cells for each specific marker are quantified to determinate cell differentiation.

3.3 Validation of Stem Cell Differentiation

3.3.1 Immunocytochemistry

1. To determine cell phenotype changes after coculture and validate ADSCs differentiation, immunocytochemistry is performed.

2. Each protein of interest is evaluated in replicates. First, fix cultured cells in 4% paraformaldehyde, pH 7.2–7.4 for 12 min.

3. Wash cells carefully in PBS for 10 min, three times. Then, incubate cell with 1% Triton X-100 in PBS for 30 min.

4. Remove 1% Triton X-100 in PBS, and block samples at room temperature with 1% BSA diluted in PBS for 1 h.

5. After blocking, incubate cells with primary antibody (1:500 rabbit polyclonal anti-beta tubulin III; 1:400 goat polyclonal anti-SNAP25) overnight at 4 °C (*see* **Note 9**).

6. Wash cells three times with PBS and incubated with secondary antibody (1:1000 goat anti-rabbit FITC; 1:500 goat anti mouse Alexa Fluor 647 or PE) for 1 h at room temperature.

7. Wash cells another three times in PBS for 10 min, and perform nuclei staining with Hoechst (Trihydrochloride, Trihydrate,) or DAPI (DAPI 4′,6-Diamidine-2′-phenylindole dihydrochloride) in the concentration of 1:1000 (diluted in PBS).

8. Negative controls are performed by omitting the primary antibodies.

9. Acquire images using a fluorescence microscope at 10× and 20× magnification. Quantify fluorescence levels, and percentage of positive cells for each specific marker to determinate cell differentiation rate.

3.4 Extracellular Vesicle Isolation, Characterization, and Cell Differentiation Method

3.4.1 EVs Isolation from Cell Culture Media

1. To isolate and characterize EVs. Media from neurons (from 10^6 cells), ADSCs (from 10^6 cells), and coculture media of the three periods (days 3, 7, 14 of culture or coculture) and replicates are isolated, and each sample is placed in a 5 mL centrifuge tube.

2. To isolate EVs, each media from each cell type is centrifuge three times (*see* **Note 10**).

3. First media is centrifuge at $300 \times g$ for 10 min, to remove life cells. Supernatant is transferred for another 5 mL centrifuge tube.

4. Second supernatant is centrifuge at $2000 \times g$ for 10 min, to remove dead cells and cell debris. Supernatant is removed and transfer for another 5 mL centrifuge tube.

5. Last, supernatant is centrifuged at $16,500 \times g$ for 30 min.

6. Remaining pellet will contain large extracellular vesicles (*see* **Note 10**).

7. Supernatant is transfer for 15 mL syringe with 0.20 μm filter attached and filtered in a 15 mL tube (*see* **Note 11**).

8. ExoQuick TC tissue culture is added to the filtered medium, to isolate small vesicles (50–200 nm of size).

9. For each mL of media 100 μL of ExoQuick TC is added. Media with ExoQuick TC is storage overnight (>16 h) at 4 °C.

10. Upon storage media with ExoQuick TC is centrifuged at $16,500 \times g$ for 30 min. Remaining pellet contain small vesicles (*see* **Note 12**).

3.4.2 EVs Characterization

1. After isolating small EVs, surface proteins called ALIX, CD63, CD9, HSP70, cytochrome, calnexin, and COX IV should be analyzed by western blot (not described), to validate the isolation process. Alix, CD63, CD9, and HSP70 should be present in the EVs and cell pellets, while cytochrome, calnexin, COX IV should not be present in EVs and only present in cell pellets.

2. For nanoparticle quantification of isolated small EVs, NanoSight NS300 with a laser of 405 nm and NanoSight NTA software v2.3 is used (*see* **Note 13**).

3. Diluted small EVs in PBS (1:100 μL). Record at least five videos of 30 s with 14-camera level at 37 °C for each sample after calibration with 100 nm of beads to ensure equipment accuracy as indicated by the manufacturer.

4. The final concentration and size are defined from the average of the recorded movies.

3.4.3 Transmission Electron Microscopy for EVs Analysis

1. For morphology analysis, small EVs are analyzed using transmission electron microscopy.

2. Isolate EVs from neurons, ADSCs, and cocultured media.

3. Fix EVs with 2.5% glutaraldehyde, 0.1 M cacodylate, and 4% paraformaldehyde pH 7.2–7.4 for 2 h at room temperature. For 10 µL of EVs add 100–200 µL of the previous solution (*see* **Note 14**).

4. Weight the samples before next step for balancing (*see* **Note 15**).

5. Ultracentrifuge fixative solution with EVs at 100,000 × *g* for 80 min in a 6 mL ultracentrifuge tube (*see* **Note 15**).

6. Wash ultracentrifuge tube with 4 mL of ultrapure water.

7. Ultracentrifuge at 100,000 × *g* for 80 min. Remove supernatant.

8. Suspend pellet in 100 µL of ultrapure water.

9. With a 10 µL pipette add, in small drops (5–10 µL), the fixed EVs solution to a copper grid coated with Pioloform. Let it dry for 5 min at room temperature.

10. Slowly, remove the excess liquid with wet filter paper.

11. Add 2% of uranyl acetate drops (5–10 µL) to the copper grid. Let it dry for 3 min at room temperature.

12. Look the grid in the transmission electron microscopy FEI at 200 kV, Tecnai20, LAB6.

13. EVs should have circular shapes.

3.4.4 Cell Derived Small EVs as Cell Differentiation Inductor

After testing cells capacity to induce stem cell differentiation using the transwell system. Next step is to test if isolated neuron-derived EVs can induce stem differentiation without coculture system. For that the following culture condition will be used: ADSCs cultured with isolated small EVs from neurons[PalmtdTomato].

1. To test the capacity of small EVs induction of differentiated ADSCs, stem cells are analyzed at 3, 7, and 14 days (time of cell induction) (*see* **Note 16**).

2. First, for immunocytochemistry and confocal analysis, ADSCs should be culture in glass coverslips. Add one single glass coverslip in each well of a 24-well plate.

3. A total of 10^4 cells/½ mL ADSCs GFP+ is seed in each well of a 24-well plate, and 0.5 mL of experimental medium is added per well. One well per replicate is used, and at three wells per time point. A total of 15 wells will be used. Each time point should be performed in one simple plate. Seed ADSCs GFP+, and neurons only to be used as control groups.

4. Isolated small EVs from neurons (10^5 cells) culture media can be added to ADSCs GFP+.

5. Immunocytochemistry and confocal analysis are performed in ADSCs at 3, 7, and 14 days (time of cell induction), to determine differentiation rate.

3.4.5 Confocal Microscopy Analysis for EVs In Vitro Tracking in Recipient Cells

1. Pre-seeded ADSCs (10^4 cells/½ mL) on coverslips placed inside the wells of a 24-well plate and culture for (3, 7, 14 days) with media contain EVs from neurons^{PalmtdTomato}.

2. Wash coverslips with PBS three times and fix with 4% PFA solution for 12 min.

3. Incubated coverslips with 1 μg/mL DAPI for 5–10 min in dark room temperature.

4. Wash coverslips with PBS three times and mount on slides for confocal imaging.

5. Acquire images at $63\times$ magnification with immersion lens.

4   Notes

1. Recommended to utilize humanize euthanasia using CO_2 chamber or use other procedures recommended by the ethical committee for experiments with animals from your local institution.

2. To avoid brain damage, when occipital bone is exposed, there is a thin line that is the easiest region to break the skull.

3. For a high quality of neurons culture, neurons should be isolate within 30 min from brain isolation to final step of primary culture.

4. Pipetting should be performed carefully and mixing up and down should be performed for 30 s.

5. Retroperitoneal fat tissue is along the dorsal abdominal wall, near the kidney. Other options of fat tissue in mouse are gonadal, inguinal, mesenteric (containing multiple lymph nodes), and suprascapular (brown fat) adipose tissue.

6. If plasmid of palmTdTomato is not available, other labels can be used for EVs tracking such as: PKH26, PKH67, and CellVue®.

7. Neurons and ADSCs should be plated at the same time. Seed specialize cells (neurons) in the top of the well, and stem cells (ADSCs) in the bottom.

8. Add cell culture media slowly in the plate, and 250–500 μL in the transwell. Add the lid to the plate and avoid leave the plate out of the incubator. To change cell culture media with the

transwell, first with a 100 μL pipette remove the media outside of the transwell avoiding touching the bottom of the plate, then change the pipette tip and remove media present within the transwell.

9. Other antibodies can be used for neuronal differentiation of MSC such as: nestin, map2, synaptophysin. For other cell differentiation types use more than one antibody to characterize differentiated cells.

10. There are other methods to isolate EVs, such as ultracentrifugation, density gradient, the isolation method should be adequate for each experiment. For example, if ultracentrifuge is not available ExoQuick TC could be used.

11. Before starting the filtration, place the 15 mL tube in a rack for centrifuge tubes, open the tube, remove syringe's plunger, attach the 0.20 μm filter in syringe, slowly add media to the syringe, place back the plunger, and very slowly pull the plunger flange down. Do not apply to much force in the plunger otherwise the filter can be damaged.

12. EVs can be storage at 4 °C or −20 °C for a short time, and long time at −80 °C. If it will be the case, EVs can be diluted in a small quantity of PBS, 0.9% normal saline solution.

13. If Nanosight is not available, other equipment can be used such as: Particle Metrix' ZetaView.

14. Small EVs can be storage with fixative at 4 °C for 24 h.

15. Before place samples in the ultracentrifuge weight them to guarantee the perfect balance between opposite tubes. Add more fixative if needed for balance weights.

16. The time of differentiation should be adequate for each stem cell line. If cell' doubling time is less than 24 h may the induction be faster than described in this protocol.

Acknowledgments

We are grateful for the funding supporting the research involving extracellular vesicles from the University of Sao Paulo. Additional supported by São Paulo Research Foundation (FAPESP) grants: #2015/17897-9, #2014/22887-0, #2015/21829-9, #2017/21266-0 and #2013/08135-2, the National Council for Scientific and Technological Development—CNPq grant number #420152/2018-0 and CAPES—Finance Code 001. The Authors are observer members of the COST Action CA16119 In vitro 3-D total cell guidance and fitness (CellFit).

References

1. Hwang NS, Varghese S, Elisseeff J (2008) Controlled differentiation of stem cells. Adv Drug Deliv Rev 60(2):199–214. https://doi.org/10.1016/j.addr.2007.08.036

2. Ciuffreda MC, Malpasso G, Musarò P et al (2016) Protocols for in vitro differentiation of human mesenchymal stem cells into osteogenic, chondrogenic and adipogenic lineages. Methods Mol Biol 1416:149–158. https://doi.org/10.1007/978-1-4939-3584-0_8

3. Roballo KCS, da Silveira JC, Bressan FF et al (2019) Neurons–derived extracellular vesicles promote neural differentiation of ADSCs: a model to prevent peripheral nerve degeneration. Sci Rep 9(1):11213. https://doi.org/10.1038/s41598-019-47229-x

4. Hanna H, Mir LM, Andre FM (2018) In vitro osteoblastic differentiation of mesenchymal stem cells generates cell layers with distinct properties. Stem Cell Res Ther 9(1):203. https://doi.org/10.1186/s13287-018-0942-x

5. Pistollato F, Canovas-Jorda D, Zagoura D, Price A (2017) Protocol for the differentiation of human induced pluripotent stem cells into mixed cultures of neurons and glia for neurotoxicity testing. J Vis Exp 9(124):55702. https://doi.org/10.3791/55702

6. Stahl DP, Raposo G (2019) Extracellular vesicles: exosomes and microvesicles, integrators of homeostasis. Physiology 34(3):169–177. https://doi.org/10.1152/physiol.00045.2018

7. Lee S, Won J, Lim GJ et al (2019) A novel population of extracellular vesicles smaller than exosomes promotes cell proliferation. Cell Commun Signal 17(1):95. https://doi.org/10.1186/s12964-019-0401-z

8. Doyle LM, Wang MZ (2019) Overview of extracellular vesicles, their origin, composition, purpose, and methods for exosome isolation and analysis. Cell 8(7):727. https://doi.org/10.3390/cells8070727

9. Bahmani L, Taha MF, Javeri A (2014) Coculture with embryonic stem cells improves neural differentiation of adipose tissue–derived stem cells. Neuroscience 272:229–239. https://doi.org/10.1016/j.neuroscience.2014.04.063

10. Lai C, Kim E, Badr C et al (2015) Visualization and tracking of tumour extracellular vesicle delivery and RNA translation using multiplexed reporters. Nat Commun 6:7029. https://doi.org/10.1038/ncomms8029

11. Roballo KCS, Bushman J (2019) Evaluation of the host immune response and functional recovery in peripheral nerve autografts and allografts. Transpl Immunol 53:61–71. https://doi.org/10.1016/j.trim.2019.01.003

12. de Oliveira VC, Gomes Mariano Junior C, Belizário JE, Krieger JE, Fernandes Bressan F, Roballo K et al (2020) Characterization of post-edited cells modified in the TFAM gene by CRISPR/Cas9 technology in the bovine model. PLoS One 15(7):e0235856 https://doi.org/10.1371/journal.pone.0235856

Chapter 4

A Simplified Method for Three-Dimensional (3D) Porcine Preantral Follicles Culture Utilizing Hydrophobic Microbioreactors

Malgorzata Duda, Lucia Gizler, and Gabriela Gorczyca

Abstract

The technological revolution in reproductive biology that started with artificial insemination procedures and embryo transfer led to the development of assisted reproduction techniques such as in vitro fertilization or even cloning of domestic animals by nuclear transfer from somatic cells. Currently, procedures of isolated immature ovarian follicles in vitro culture are becoming the prominent technology aimed to preserve or restore fertility especially of young oncological patients or those at risk of premature ovarian failure.

Here, we describe a protocol that can be applied for in vitro growth of porcine, preantral ovarian follicles in three-dimensional (3D) culture conditions. After enzymatic isolation from the ovarian cortex, preantral follicles are suspended in a drop of medium and enclosed with fluorinated ethylene propylene (FEP) powder particles (microbioreactors). Such microbioreactors maintain the 3D structure of the follicles during the whole process of in vitro growth what is crucial to ensure proper folliculogenesis progression and their ability to survive.

Key words Pig, Ovary, Follicle, In vitro follicle growth, 3D, Microbioreactors, FEP

1 Introduction

The process of follicular growth and development—folliculogenesis—begins already during the fetal life of the female and continues until the end of the ovarian hormonal activity. The proper progress of folliculogenesis from the stage of primordial follicles to the preovulatory ones requires the presence of multiple, interconnecting regulatory pathways to provide good-quality oocytes suitable for fertilization for the whole reproductive lifetime of a female [1]. The implement that allows for studying physiology of ovarian follicles irrespective of their complicated structure is the model of whole organ in vitro culture, which reconstructs conditions and complex interactions occurring in vivo. Cultures of such a kind are very sensitive objects to test the biological activity of various factors

Tiziana A.L. Brevini et al. (eds.), *Next Generation Culture Platforms for Reliable In Vitro Models. Methods and Protocols*, Methods in Molecular Biology, vol. 2273, https://doi.org/10.1007/978-1-0716-1246-0_4,
© The Author(s), under exclusive license to Springer Science+Business Media, LLC, part of Springer Nature 2021

and thanks to them one can follow in cultured structures reactions manifesting by an increased or decreased steroid hormones secretion, induction or inhibition of cell proliferation, and also induction or inhibition of apoptosis. Although the clinical application of in vitro growth of human follicles is still investigational, in a laboratory setting, it is a robust method that permits to study the basic biology of the ovary and the follicle in a controlled yet modifiable environment. Multiple culture systems have been invented that support the development of isolated preantral follicles [6–9]. Each of them has its own advantages and strategy to provide useful ways to understand follicles function. The majority of pioneering works on in vitro growth of follicles were using conventional two-dimensional (2D) culture systems. The work of Epigg and Schroeder [10], who managed to obtain live offspring after in vitro culture of isolated mouse preantral follicles, should be mentioned here. Next, they further reported that primordial follicles can be cultured to complete maturation using a combination of ovarian tissue culture and 2D follicle culture that generated developmentally competent oocytes [11]. Despite this, it has been recognized that 2D follicle culture such as in microdroplets of culture media covered with mineral oil on a culture plate [12] is not sufficient to sustain their appropriate, spherical architecture [13–15]. The follicles are flattened which causes breakdown of the surrounding basal lamina, then granulosa cells detach from each other and from the oocyte and disperse. Such structural remodeling interferes with the essential bidirectional communication between granulosa cells and the oocyte resulting in desynchronized growth and maturation defects of the oocyte [16].

In recent years substantial technical progress has been obtained in the form of 3D follicle culture, which better sustains spatial, three-dimensional follicle structure in vitro [17]. In 3D culture systems, isolated follicles are individually encapsulated in various types of substrates such as collagen, alginate, hyaluronic hydrogel or Matrigel [14, 18, 19]. It is worth to mention, that the most widely applied system for 3D follicle culture is encapsulation with alginate hydrogel [19, 20]. For example, encapsulation of mouse secondary preantral follicles within alginate beads resulted in their growth and development under these 3D culture conditions and live offspring have been obtained from oocytes matured in such a way [21–24]. Despite this, 3D follicle culture using alginate beads is usually considered a complicated and time-consuming procedure [25–26].

Here we describe a convenient and simple protocol that can be applied for in vitro growth and development of porcine, preantral ovarian follicles in 3D culture conditions. Preantral (primary and early secondary) follicles, after enzymatic isolation from the ovarian cortex, are suspended in a drop of medium and encapsulated with fluorinated ethylene propylene (FEP; a copolymer of hexafluoropropylene and tetrafluoroethylene) powder particles, that form microbioreactors defined as Liquid Marbles (LM). LM are a form of 3D bioreactors that have been previously shown to support, among others, growth of tumor spheroids [27], living microorganisms, and embryonic stem cells. LM have been also successfully used for sheep and porcine oocyte culture. In most experiments mentioned earlier, performed using LM, bioreactors were prepared using polytetrafluoroethylene (PTFE) powder with particle size of 1 μm. In the presented protocol we use FEP powder, which is very similar in composition and properties to the fluoropolymers PTFE. However, it is distinguished by greater durability of formed bioreactors and what is especially important, it is highly transparent. It is worth to mention, that this method is much simpler than either the 2D or 3D follicle culture systems described above. This method perfectly mimics the microenvironment inside the ovary in vivo. Thanks to this, it is possible to maintain the three-dimensional organization of the follicle and preserve the functional relationship between the oocyte and follicular cells surrounding it (*see* Fig. 1). Although it is still very challenging to generate high quality human oocytes in vitro, presented here 3D culture system

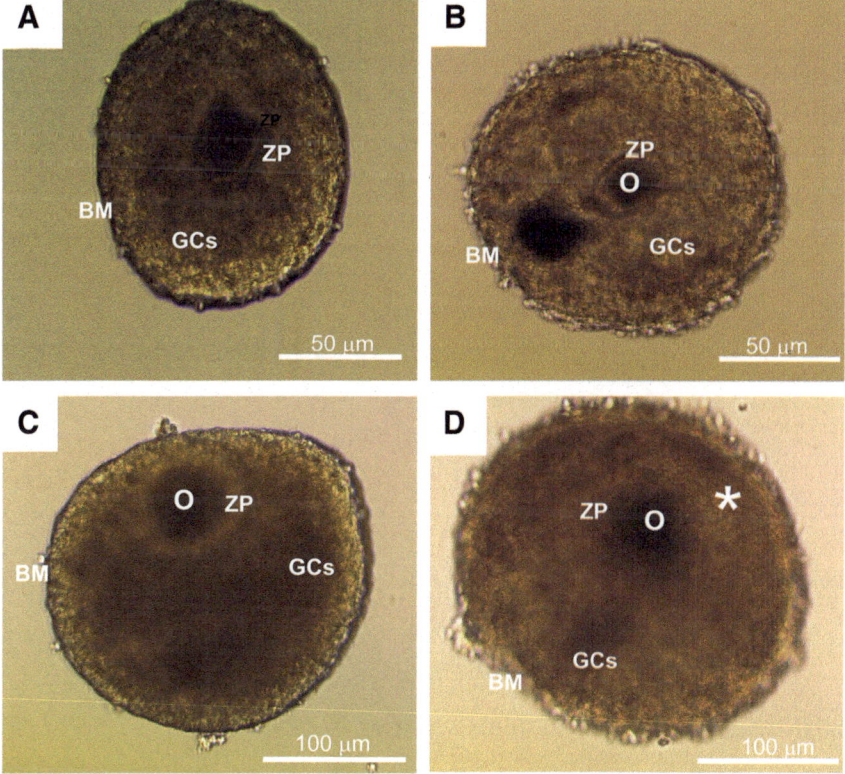

Fig. 1 Images of follicles growth in three-dimensional (3D) culture conditions. Representative images of a secondary follicle growing using the FEP microbioreactors on day: 2 (**a**), 6 (**b**), 8 (**c**), and 10 (**d**) of culture. 10 days of 3D culture resulted in antrum formation (shown with white asterisk). *BM* basement membrane, *ZP* zona pellucida, *O* oocyte, *GCs* granulosa cells

originally intended for use with porcine follicles can pave the way to improvement of similar technologies enabling fertility preservation in humans.

2   Materials

2.1   Porcine Ovary Handling and Preantral Follicle Isolation

1. Dulbecco's phosphate buffered saline (DPBS): 137 mM NaCl, 2.7 mM KCl, 0.9 mM $CaCl_2$ (dihydrate), 0.49 mM $MgCl_2$ (hexahydrate), 1.5 mM KH_2PO_4, 8.9 mM Na_2HPO_4.

2. Follicle handling medium (HM): 88% (v/v) DMEM/F12 Complete Medium (1:1 mixture of Dulbecco's Modified Eagle Medium; DMEM and Ham's F12 medium; F12), 10% (v/v) fetal bovine serum (FBS), 1% (v/v) L-glutamine solution, 1% (v/v) antibiotic–antimycotic solution (AAS).

3. Liberase™ Thermolysin High (TH) Research Grade stock solution: reconstitute enzymatic powder (1 vial with 5 mg) in 2 mL sterile DPBS. Aliquots of 100 μL could be stored at −20 °C up to 6 months. If stored at 4 °C, use within 2 weeks. For 10 mL of DPBS add 100 μL Liberase stock solution.

4. Deoxyribonuclease (DNase) I stock solution: 50 mg DNase I powder (2000 Kunitz/mg) dissolve in 1 mL of sterile water. For 10 mL of DPBS add 10 μL DNase stock solution. Aliquots of 100 μL could be stored at −20 °C up to 4 weeks (*see* **Note 1**).

5. 50 mL glass beaker.

6. 10 cm petri dish.

7. Sterile surgical blade 15C.

8. Insulin syringe (u-40) with needle.

9. Sterile 5 mL syringe.

10. Sterile syringe filter with 0.2 μm membrane pore size.

11. Tissue slicer coronal.

12. Sterile 125 mL Erlenmeyer flask with ventiled cap.

13. Pipettor with micromanipulation pipettes 170 μm inner diameter.

2.2   Microbioreactor Preparation

1. Fluorinated ethylene propylene (FEP) powder.

2. 30 mm petri dish.

3. 1000 μL pipette tip.

2.3   Preantral Follicle Culture

1. Follicle culture medium (CM): 77% (v/v) DMEM/F12 Complete Medium (1:1 mixture of Dulbecco's Modified Eagle Medium; DMEM and Ham's F12 medium; F12), 10% (v/v) fetal bovine serum (FBS), 1% (v/v) L-glutamine solution, 1%

(v/v) insulin, transferrin and selenium (ITS) solution, 1% (v/v) antibiotic–antimycotic antibiotic–antimycotic solution (AAS), 10% (v/v) porcine follicular fluid (pFF) and 50 μg/mL L-ascorbic acid, 100 mIU/mL follicle-stimulating hormone (FSH).

2. The porcine follicular fluid (pFF): aspirate from porcine antral follicles larger than 4–6 mm in diameter using the insulin syringe (u-40). Next, centrifuge obtained pFF ($100 \times g$ for 10 min at room temperature) and subsequently filter the supernatant using sterile syringe attached to 0.2 μm membrane pore filters. After filtration, snap freeze the pFF at −80 °C for further procedures of follicle culture.

3. ITS: 10 μg/mL insulin, 5.5 μg/mL transferrin, and 6.7 ng/mL sodium selenite.

4. L-Ascorbic acid stock solution at 50 μg/mL in sterile water: dissolve 500 mg L-ascorbic acid into 10 mL H_2O and sterile on filter. Aliquots of 100 μL store in sterile eppendorf tubes at −20 °C (*see* **Note 2**). For 10 mL of CM add 10 μL L-ascorbic acid stock solution.

5. FSH stock solution: dissolve 10 IU FSH in 10 mL of sterile DPBS and aliquots of 100 μL store in sterile eppendorf tubes at −20 °C. For 10 mL of CM add 10 μL FSH stock solution.

6. Sterile-filtered water.

7. Sterile cell nylon strainers: size 300 μm and 200 μm.

8. 60 mm IVF petri dish.

9. Sterile Pasteur pipette.

10. 30 mm sterile petri dish.

11. 150 mm sterile petri dish.

3 Methods

For optimal results during follicular handling, carry out all procedures in the handling medium (HM). Control the pH at the CO_2 level in the environment, and perform all procedures on a heating table with temperature control (37 °C) and under the laminar flow hood to minimize bacterial contamination.

The following procedures were approved by the Animal Welfare Committee at the Institute of Zoology and Biomedical Research at Jagiellonian University.

3.1 Isolation of Porcine Preantral Follicles (Fig. 2)

1. To collect material for the isolation of ovarian, preantral follicles, excise porcine ovaries from prepubertal gilts (approximately 6–7 months of age, weighing 70–80 kg) at a local

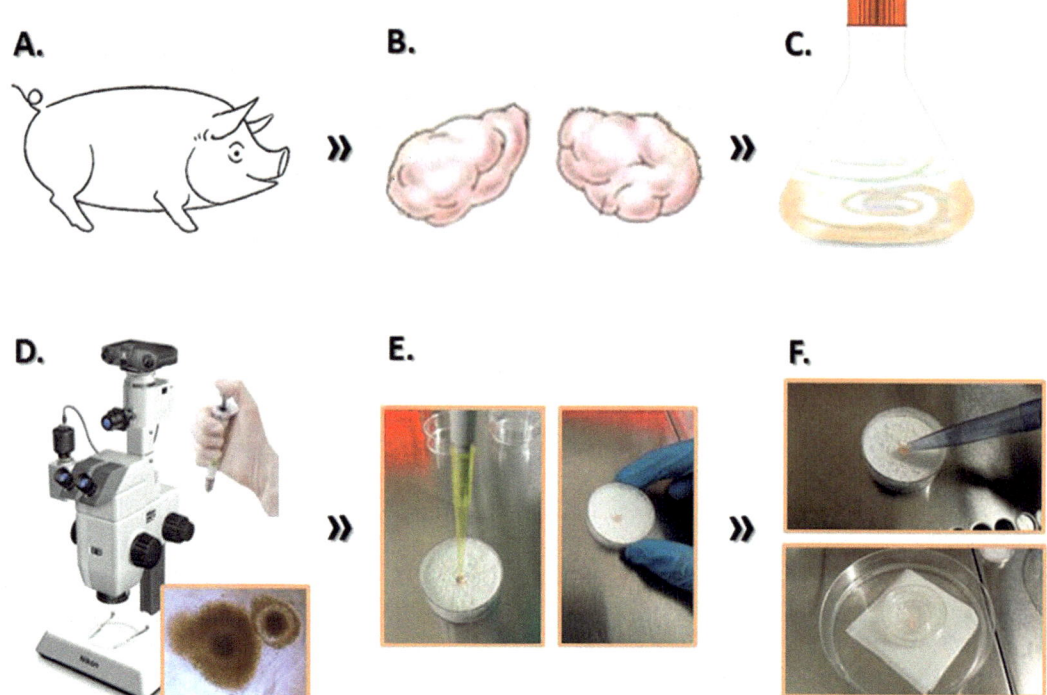

Fig. 2 Flowchart for preantral follicle isolation and encapsulation in FEP microbioreactors. Preantral follicles are isolated from a prepubertal pig (**a**). The ovaries are dissected (**b**) and after enzymatic digestion of pieces of the ovarian cortex (at temperature 37 °C for 2 h) (**c**), isolated preantral follicles (**d**) are transferred to the petri dish containing FEP powder. After gentle rotation of the plate follicles are encapsulated within FEP microbioreactors (**e**). Next, microbioreactors are transferred using a pipette with 1000 μL tip into the central well of IVF petri dish. Finally, such IVF petri dish is placed inside a humidity chamber (petri dish with a gauze soaked in sterile water) (**f**)

slaughterhouse. Choose approximately eight pig ovaries from four animals for follicles isolation in each experiment (*see* **Note 3**).

2. Place ovaries in a thermos with sterile Dulbecco's phosphate-buffered saline (DPBS; pH 7.4; 37 °C) containing 1% AAS. Ensure that the experimental material is transported to the laboratory within 1 h where it is rinsed twice with sterile DPBS containing antibiotics.

3. After rinsing the ovaries, transfer them to the beaker filled with HM medium and store in an incubator at 37 °C for the time of all manipulations.

4. To isolate preantral follicles, transfer one ovary to a sterile petri dish (10 cm in diameter) filled with 5 mL HM. After that, gently cut the ovary with a sterile surgical blade into two halves along the longitudinal axis. Next, using the same surgical blade, separate ovarian cortex from the ovarian medulla.

5. Cut obtained ovarian cortex into uniform-size pieces of ~1 mm^3 with a tissue slicer. Next, transfer the fragments to sterile 125 mL Erlenmeyer flask with ventiled cap, containing 10 mL of DPBS supplemented with 0.05 mg/mL Liberase™ TH Research Grade (*see* **Note 4**).

6. Perform enzymatic digestion of ovarian cortex pieces in an incubator at temperature 37 °C, with 150 rpm for 2 h. To remove damaged DNA fragments from the suspension, add DNase I solution (50 μg/mL) to the Liberase™ solution, for the last 15 min of incubation (*see* **Note 5**).

3.2 Recovery of Isolated Preantral Follicles

1. Terminate enzymatic digestion by adding an equal volume of cold DPBS (4 °C) containing 10% FBS. After that, put onto ice the resulting suspension and aspirate gently up and down with a 5 mL sterile Pasteur pipette. Next, filter it successively, through 300 and 200 microns nylon strainers.

2. In the further step, wash follicular suspension several times (three times at least) in sterile DPBS and recover it by centrifugation (10 min × 90 × g, at room temperature).

3. After the last centrifugation, discard the supernatant and transfer the pellet in a small volume of CM (4–5 mL) into a sterile petri dish for further procedures under a stereomicroscope.

4. Prepare 3–5 60 mm IVF petri dishes.

5. Add 1 mL of prepared earlier culture medium (CM) to the central wells and place 3–4 drops of HM (50 μL per drop) in their outer rings (*see* **Note 6**).

6. Then, pick up the follicles using a transfer pipettor taking care to avoid aspiration of the stromal cells. Place them in a drop of HM in the outer rings to rinse them briefly for 3–4 times (*see* **Note 7**).

7. At the end, individually transfer them into the central well. Store this IVF plates in an incubator for further procedures.

3.3 Follicles Encapsulation in Superhydrophobic Fluorinated Ethylene Propylene (FEB) Microbioreactors

1. Prepare a 30 mm petri dish containing 5 g of FEP powder bed-average particle size of 1 μm (*see* **Note 8**).

2. Distribute a single droplet of CM (~30 μL in volume) containing 3–4 preantral follicles onto the bed of the FEP powder. Rotate the plate gently in a circular motion to be sure that they completely covered the surface of the liquid drop and formed liquid marbles (LM).

3. Pick up formed LM using a pipette with 1000 μL tip (*see* **Note 9**).

4. Prepare several 60 mm IVF petri dishes. Add 3–4 mL of sterile water to the outer rings to prepare the humidity chamber. Next, place one LM into the central well (*see* **Note 10**).

5. Incubate marbles for 10 days at 37 °C in 5% CO_2 incubator (*see* **Note 11**).

6. Change the medium every 2 days following the procedure: apply 30 μL of CM on each LM, which will cause their spreading because direct liquid contact disrupts the hydrophobicity of the coated FEP powders. When the marble content dissolves, transfer follicles released from the bioreactor to a drop of fresh CM in the petri dish (*see* **Note 12**).

7. After several washes in CM (3–4 times) to remove FEP particles, transfer preantral follicles with 30 μL of fresh CM onto the FEP powder bed. Gently rotate the plate in a circular motion to ensure that the powder particles completely covered the surface of the liquid drop and formed a new LM.

8. Then follow the procedure from **steps 3** to **5** (*see* **Note 13**).

4 Notes

1. DNase I activity is strictly dependent on Ca^{2+} presence and is activated by Mg^{2+}, so it is critical to perform preantral follicles enzymatic isolation in DPBS supplemented with both of them.

2. Aliquots of L-ascorbic acid stock solution should be thawed only once.

3. For isolation procedures select ovaries preferably with antral follicles <2 mm.

4. For enzymatic treatment of ovarian cortex we selected the Liberase™ Thermolysin High (TH) Research Grade which belongs to the group of Liberase Research Grade Purified Enzyme Blends. They are mixtures of highly purified Collagenase I and Collagenase II, with a high concentration of Thermolysin in the case of Liberase™ TH. The working concentration was selected experimentally in order to obtain a digestion power equivalent to 1 mg/mL (0.15 Wunsch Units) collagenase type IA; it is 0.05 mg/mL. A higher Liberase concentration (0.08 mg/mL) was also tested but failed to yield more isolated follicles. The best results were obtained after 120 min of Liberase treatment.

5. The day before preantral follicle isolation thaw DNase I stock solution overnight at 2–8 °C.

6. Make sure that all procedures are carried out on thermostatically controlled table and follicles are maintained at 37 °C throughout their handling.

7. Only primary or early secondary follicles (diameters between 45 μm and 160 μm) with the following characteristics are

considered suitable for further procedures: the absence of the antral cavity, an intact basement membrane and a visible rounded and centrally located oocyte within the follicle.

8. It is important to use fresh FEP powder. The reused FEP tends to aggregate and forms clumps.

9. Cut the pipette tip at the edge to accommodate it to the diameter of the LM (4–5 mm). The approximate diameter of such prepared tip should be slightly bigger than that of the LM.

10. This procedure requires good manual skills and precision—if the marble is placed carelessly or falls even from a low height it will be destroyed.

11. To ensure optimal humidity and prevent evaporation of the LM, you can prepare an additional humidity chamber: put a sterile gauze soaked in sterile water into a petri dish (150 mm). Next, transfer three IVF petri dishes containing LM prepared in advance, close the chamber and place at 37 °C in 5% CO_2 incubator.

12. To precisely transfer released follicles, use a transfer pipettor.

13. The transparent coating of FEP powder allows to monitor LM content using a light microscope.

Acknowledgments

The authors are very grateful to Dr. Zbigniew Tabarowski (Department of Experimental Hematology, Institute of Zoology and Biomedical Research, Jagiellonian University) for critical revision of the manuscript. This work was supported by grant 2018/29/N/NZ9/00983 from National Science Centre Poland. GG and MD are members of the COST Actions CA16119.

References

1. Fortune JE, Ribera GM, Yang MY (2004) Follicular development: the role of follicular microenvironment in selection of dominant follicle. Anim Reprod Sci 82-83:109–126. https://doi.org/10.1016/j.anireprosci.2004.04.031

2. Telfer EE, Binnie JP, McCaffery FH et al (2000) In vitro development of oocytes from porcine and bovine primary follicles. Mol Cell Endocrinol 163:117–123. https://doi.org/10.1016/s0303-7207(00)00216-1

3. Wu J, Emery BR, Carrell DT (2001) In vitro growth, maturation, fertilization, and embryonic development of oocytes from porcine preantral follicles. Biol Reprod 64:375–381. https://doi.org/10.1095/biolreprod64.1.375

4. Abir R, Nitke S, Ben-Haroush A et al (2006) In vitro maturation of human primordial ovarian follicles: clinical significance, progress in mammals, and methods for growth evaluation. Histol Histopathol 21:887–898. https://doi.org/10.14670/HH-21.887

5. Kreeger PK, Deck JW, Woodruff TK et al (2006) The in vitro regulation of ovarian follicle development using alginate-extracellular matrix gels. Biomaterials 27:714–723. https://doi.org/10.1016/j.biomaterials.2005.06.016

6. Eppig JJ, Schroeder AC (1989) Capacity of mouse oocytes from preantral follicles to undergo embryogenesis and development to live young after growth, maturation, and fertilization in vitro. Biol Reprod 41:268–276. https://doi.org/10.1095/biolreprod41.2.268

7. Eppig JJ, O'Brien MJ (1996) Development in vitro of mouse oocytes from primordial follicles. Biol Reprod 54:197–207. https://doi.org/10.1095/biolreprod54.1.197

8. Oktem O, Oktay K (2007) Sphingosine-1-phosphate enhances human primordial follicle survival and blocks ovarian apoptosis in vitro. Paper presented at the ASRM 63rd Annual Meeting, Washington, DC, USA 15–17 Oct 2007

9. Belli M, Vigone G, Merico V et al (2012) (2012) towards a 3D culture of mouse ovarian follicles. Int J Dev Biol 56:931–937. https://doi.org/10.1387/ijdb.120175mz

10. Desai N, Alex A, Abdel Hafez F et al (2010) Three-dimensional in vitro follicle growth: overview of culture models, biomaterials, design parameters and future directions. Reprod Biol Endocrinol 8:119. https://doi.org/10.1186/1477-7827-8-119

11. Filatov MA, Khramova YV, Semenova ML (2015) In vitro mouse ovarian follicle growth and maturation in alginate hydrogel: current state of the art. Acta Nat 7:48–56

12. Abir R, Fisch B, Nitke S et al (2001) Morphological study of fully and partially isolated early human follicles. Fertil Steril 75:141–146. https://doi.org/10.1016/s0015-0282(00)01668-x

13. Cukierman E, Pankov R, Yamada KM (2002) Cell interactions with three-dimensional matrices. Curr Opin Cell Biol 14:633–639. https://doi.org/10.1016/s0955-0674(02)00364-2

14. Combelles CMH, Fissore RA, Albertini DF et al (2005) In vitro maturation of human oocytes and cumulus cells using a co-culture three-dimensional collagen gel system. Hum Reprod 20:1349–1358. https://doi.org/10.1093/humrep/deh750

15. Dorati R, Genta I, Ferrari M et al (2016) Formulation and stability evaluation of 3D alginate beads potentially useful for cumulus-oocyte complexes culture. J Microencapsul 33:137–145. https://doi.org/10.3109/02652048.2015.1134691

16. Sadr SZ, Fatehi R, Maroufizadeh S et al (2018) Utilizing fibrin-alginate and matrigel-alginate for mouse follicle development in three-dimensional culture systems. Biopreserv Biobank 16(2):120–127. https://doi.org/10.1089/bio.2017.0087

17. Pangas SA, Saudye H, Shea LD et al (2003) Novel approach for the three-dimensional culture of granulosa cell-oocyte complexes. Tissue Eng 9:1013–1021. https://doi.org/10.1089/107632703322495655

18. Xu M, West E, Shea LD et al (2006) Identification of a stage-specific permissive in vitro culture environment for follicle growth and oocyte development. Biol Reprod 75:916–923. https://doi.org/10.1095/biolreprod.106.054833

19. West ER, Xu M, Woodruff TK et al (2007) Physical properties of alginate hydrogels and their effects on in vitro follicle development. Biomaterials 28:4439–4448. https://doi.org/10.1016/j.biomaterials.2007.07.001

20. Xu M, Kreeger PK, Shea LD et al (2006) Tissue-engineered follicles produce live, fertile offspring. Tissue Eng 12:2739–2746. https://doi.org/10.1089/ten.2006.12.2739

21. Jin SY, Lei L, Shikanov A et al (2010) A novel two-step strategy for in vitro culture of early-stage ovarian follicles in the mouse. Fertil Steril 93:2633–2639. https://doi.org/10.1016/j.fertnstert.2009.10.027

22. Parrish EM, Siletz A, Xu M et al (2011) Gene expression in mouse ovarian follicle development in vivo versus an ex vivo alginate culture system. Reproduction 142:309–318. https://doi.org/10.1530/REP-10-0481

23. Arbatan T, Al-Abboodi A, Sarvi F et al (2012) Tumor inside a pearl drop. Adv Healthc Mater 1:467–469. https://doi.org/10.1002/adhm.201200050

24. Tian J, Fu N, Chen XD et al (2013) Respirable liquid marble for the cultivation of microorganisms. Colloids Surf B Biointerfaces 106:187–190. https://doi.org/10.1016/j.colsurfb.2013.01.016

25. Sarvi F, Jain K, Arbatan T et al (2015) Cardiogenesis of embryonic stem cells with liquid marble micro-bioreactor. Adv Healthc Mater 4(1):77–86. https://doi.org/10.1002/adhm.201400138

26. Ledda S, Idda A, Kelly J et al (2016) A novel technique for in vitro maturation of sheep oocytes in a liquid marble microbioreactor. J Assist Reprod Genet 33:513–518. https://doi.org/10.1007/s10815-016-0666-8

27. Gorczyca G, Wartalski K, Tabarowski Z et al (2020) Proteolytically degraded alginate hydrogels and hydrophobic microbioreactors for porcine oocyte encapsulation. J Vis Exp. https://doi.org/10.3791/61325

Chapter 5

Use of Transparent Liquid Marble: Microbioreactor to Culture Cardiospheres

Jeffrey Aalders, Laurens Léger, Davide Piras, Jolanda van Hengel, and Sergio Ledda

Abstract

Cells have a remarkable ability to self-organize and rearrange in functional organoids, this process was greatly boosted by the recent advances in 3D culture technologies and materials. Presently, this approach can be applied to model human organ development and function "in a dish" and can be used to predict drug response in a patient specific fashion.

Here we describe a protocol that allows for the derivation of functional cardiac mini organoids consisting of cocultured cardiomyocytes and cardiac fibroblast. Cells are suspended in a drop of medium and encapsulated with hydrophobic fumed silica powder nanoparticles. These nanoparticles are treated with hydrophobic chemicals, hexamethyldisilazane (nHMDS), and result in the formation of microbioreactors. These microenvironments are defined as "liquid marbles," stimulating cell coalescence and 3D aggregation. Then nHMDS shell ensures optimal gas exchange between the interior liquid and the surrounding environment. This microbioreactor makes working in smaller volumes possible and is therefore amenable for higher throughput applications. Moreover, the properties of liquid marble microbioreactors makes it an excellent culture technique for cocultures. Here we demonstrate how cocultures of cardiac fibroblast and cardiomyocytes in a cardiosphere can be a valuable tool to model cardiac diseases in vitro and to assess cell interactions to decipher disease mechanisms.

Key words Microbioreactor, nHMDS, Organoid, Cardiosphere, Drug screening, Multicellular spheroids, Three-dimensional culture, Coculture, Tissue engineering, Cardiac models

1 Introduction

Numerous attempts have been made in the last two decades to develop 3D cell culture models and led to the formation of structures referred to as organoids [1]. Bridging the gap between cell-based assays and animal studies was the main reason. With the aim of reducing experimental uncertainties arising from monolayer

---

Jolanda van Hengel and Sergio Ledda contributed equally to this work.

Tiziana A.L. Brevini et al. (eds.), *Next Generation Culture Platforms for Reliable In Vitro Models: Methods and Protocols*, Methods in Molecular Biology, vol. 2273, https://doi.org/10.1007/978-1-0716-1246-0_5,
© The Author(s), under exclusive license to Springer Science+Business Media, LLC, part of Springer Nature 2021

cultures and hence cutting the cost of subsequent drug screening processes. Today's methods are the product of what started in 1906 with the hanging drop tissue culture approach [2], but at present biologists have several types of 3D cell culture techniques at their disposal. 3D cell cultures can be subdivided in scaffold-free, such as suspension culture in low-adherence plates and spinner flasks or scaffold based, such as hydrogels and bioprinting. These 3D culture techniques often require expensive equipment, extensive training or lack reproducibility. Many types of mammalian cells can aggregate and differentiate into 3D multicellular spheroids when cultured in suspension or a nonadhesive environment. Compared to conventional monolayer cultures, multicellular spheroids resemble real tissues better in terms of structural and functional properties [3, 4].

Here we describe a protocol that allows for the derivation of functional cardiac organoids from a coculture of cardiomyocytes and cardiac fibroblasts that are encapsulated in superhydrophobic microbioreactors, which have been previously shown to support the growth of living microorganisms [5], tumor spheroids [6], fibroblasts [7], red blood cells [8], embryonic stem cells [9], pancreatic Mini-organoids [10], epigenetically erased fibroblasts [11], and oocytes [12]. Structures defined as liquid marbles (LM) are formed by enveloping cells suspended in a drop of medium, with hydrophobic fumed silica powder nanoparticles. The fumed silica powder was made more hydrophobic by treatment with hexamethyldisilazane (nHMDS), which results in powder with a particle size of 0.2–0.3 microns, that can be used to form an elastic shell with fine pores [13]. nHMDS particles adhere to the surface of the medium drop, isolating the liquid core from the supporting surface, while allowing an optimal gas exchange between the interior liquid and the surrounding environment. The coating material acts as a confined space, which is nonadhesive and allows cells to freely interact with each other [14]. This method makes it possible to scale down experiments and is therefore amenable for higher-throughput applications. Furthermore, working in smaller volumes allows to study the effect of paracrine/autocrine signaling of the rich environment established within the microbioreactor.

In the protocol described, cells suspensions of cardiomyocytes and cardiac fibroblasts are encapsulated, in order to stimulate them to coalesce and form cardiac spheroids. At the end of incubation of cocultured cardiomyocytes and cardiac fibroblast within the LM 3D structures are formed. These cardiospheres provide valuable study material for higher throughput applications. The 3D coculture approach using the LM technology provides an excellent opportunity for disease modeling to study cardiac diseases such as arrhythmogenic ventricular cardiomyopathy and Marfan syndrome and toxicological screenings. 3D cultures have been described in literature to more accurately represent the in vivo physiological status and thus offer more reliable results [15, 16]. Moreover, coculturing of multiple cell types in 3D further advances the reliability and more closely mimics in vivo tissues, that consist of a heterogeneous cell population. Furthermore, for disease modeling, this approach allows to combine various cell types from patients and healthy individuals. For instance, healthy cardiomyocytes can be combined with diseased cardiac fibroblasts, this can help to investigate the disease mechanisms that are involved and decipher how specific cell types contribute to the disease. Furthermore, the described protocol of LM preparation also shows how fresh medium and drug addictions, as well as coculture of different types of cells, could be achieved and how waste medium removal can be obtained by separating the formed LMs with a splitting technique. The described LM technology can be of great value in the field of 3D in vitro modeling since its ease of use, low-cost and good reproducibility.

2 Materials

2.1 Human Stem Cell Culture

1. H9 (WiCell, WA09).
2. CELLSTAR® 6-well culture plates culture dish (Greiner Bio One, cat no. 657160).
3. DMEM/F12 (Life Technologies, cat no. 31330038).
4. Geltrex® (Life Technologies, cat no. A1413302).
5. Essential 8 medium (Life Technologies, cat no. A1517001).
6. Penicillium/streptomycin (pen/strep) (Life Technologies, cat no. 15140-122).
7. Phosphate buffered saline (PBS) (Life Technologies, cat no. 10010-015).
8. Versene(EDTA) (Lonza, cat no. BE17-711E).
9. Cell scraper (Greiner Bio One, cat no.541070).
10. EVOS™ XL Core Cell Imaging System (Thermo Fisher Scientific).

2.2 Cardiomyocyte Differentiation and Culture

1. CELLSTAR® 12-well culture plates culture dish (Greiner Bio One, cat no. 665180).
2. DMEM/F12.
3. Geltrex.
4. Essential 8 medium.
5. Pen/Strep.
6. PBS.
7. Versene(EDTA).

8. Cell scraper.

9. NucleoCounter®NC-100™ (Chemometec) or Bürker counting chamber.

10. PSC Cardiomyocyte Differentiation Kit (Life Technologies, cat no. A2921201).

11. PSC Cardiomyocyte Maintenance Medium (Life Technologies, cat no. A2920801).

12. Multi Tissue Dissociation Kit 3 (Miltenyi Biotec, cat no. 130-110-204).

13. Fetal Bovine Serum (FBS) (Life Technologies, cat no. 10500-064).

14. 70 µm strainer (VWR, cat no. 732-2758).

15. RevitaCell (Life Technologies, cat no. A2644501).

2.3 Cardiomyocyte Purification

1. NucleoCounter or Bürker counting chamber.

2. Wash buffer (consisting of 2% FBS in PBS).

3. Human PSC-Derived Cardiomyocyte Isolation Kit (Miltenyi-Biotec, cat no. 130-110-188).

4. LS column (MiltenyiBiotec, cat no. 130-042-401).

5. MidiMACS separator system (MiltenyiBiotec, cat no. 130-042-302).

6. Cardiomyocyte maintenance medium.

7. FBS.

8. RevitaCell.

2.4 Cardiac Fibroblast Differentiation and Culture

1. CELLSTAR® 24-well culture plates culture dish (Greiner Bio One, cat no. 662160).

2. DMEM/F12.

3. Geltrex.

4. Essential 8 medium.

5. Pen/Strep.

6. PBS.

7. Versene(EDTA).

8. Cell scraper.

9. Nucleocounter or Bürker counting chamber.

10. PSC Cardiomyocyte Differentiation Kit.

11. DMEM (Life Technologies, cat no. 41965039).

12. FBS.

13. L-glutamine (Life Technologies, cat no. 25030-081).

14. Sodium Pyruvate (Life Technologies, cat no. 11360039).

Table 1
Antibodies used for immunofluorescent staining

Staining products	Dilution	Company (cat no.)	Remarks
Rabbit-anti-NKX2.5	1/100	Life Technologies (701622)	Primary Ab
Mouse-anti-TNNT2	1/250	Life Technologies (MA5-12960)	Primary Ab
Goat-anti-mouse IgG DyLight 488	1/500	Life Technologies (35503)	Secondary Ab
Goat-anti-rabbit IgG DyLight 594	1/500	Life Technologies (35561)	Secondary Ab

15. bFGF (PeproTech, cat no. 100-18B-10UG).

16. TrypLE (Life Technologies, cat no. 12563011).

17. 70 μm strainer.

18. RevitaCell.

2.5 Immuno-
fluorescent Staining
for Cardiac Markers

1. PBS.

2. 4% paraformaldehyde (PFA) in PBS.

3. 10% Triton X-100.

4. Goat serum (Life Technologies, cat no. 16210-064).

5. Bovine serum albumin (BSA).

6. 10% Tween20.

7. Primary and secondary antibodies (*see* Table 1).

8. Hoechst (Life Technologies, cat no. H3570).

9. Mounting medium with DABCO (Sigma-Aldrich, cat no. D27802).

10. Glass slide.

11. Coverslip.

12. Nail polish.

13. ZEISS LSM900 confocal microscope.

2.6 Microbioreactor
and Liquid Marble
Preparation
and Manipulation

1. Fumed silica powder nanoparticles treated with hexamethyldi-silazane (nHMDS) (Cabot, CAB-O-SIL® TS530) (*see* **Note 1**).

2. 35 mm petri dishes.

3. 1000 μl pipette tips cut at the edge.

4. Culture media.

5. Thin sterile pipette tips (1–20 μl).

6. Sterile surgical blades.

7. Bürker counting chamber.

3   Methods

All the procedures described below must be performed under a laminar flow hood in sterile conditions. Make sure that all culture procedures are carried out on thermostatically controlled stages and cells are maintained at 37 °C, 5% CO_2, and 5% O_2 throughout the experiments.

3.1   Ethics Statement

Experiments conducted with human stem cells were approved by the local ethical committee of Ghent University Hospital (EC UZG 2017/0855).

3.2   Stem Cell Culture

1. To prepare coating for stem cells, add 1.5 ml of 1:100 Geltrex in DMEM/F12 in a 6-well of a CELLSTAR® culture plate and incubate for 1 h at 37 °C.

2. Remove the 1.5 ml Geltrex suspension from the 6-well and add 1 ml Essential 8 medium supplemented with 1:100 pen/strep.

3. Add approximately 300,000 cells in a volume of at least 1 ml Essential 8 medium per 6-well (to have at least 2 ml medium per 6-well).

4. Ensure that the stem cells are evenly distributed over the well by moving the 6-well plate in a figure eight shape.

5. Incubate the cell culture at 37 °C and refresh the medium daily with Essential 8 medium.

6. When the confluency of the stem cell culture reaches 80%, the cells are ready to be passaged.

7. Wash the cell culture with 2 ml PBS per 6-well.

8. To dissociate the stem cells in clumps, add 1 ml Versene (EDTA) per 6-well and incubate for 3 min at 37 °C.

9. Remove Versene(EDTA) and add 2 ml Essential 8 medium supplemented with 1:100 pen/strep per 6-well.

10. Dislodge cells in clumps by using a cell scraper and moving three times horizontally over the surface and three times in the vertical direction, covering the complete surface of the 6-well.

11. Transfer the correct amount of cell suspension in accordance with the passaging ratio (normally between 1:4 and 1:6) to a freshly Geltrex coated 6-well.

12. Add up with Essential 8 medium supplemented with 1:100 pen/strep so to have 2 ml per 6-well and incubate at 37 °C.

3.3   Cardiomyocyte Differentiation

1. When the stem cell culture has a confluence of approximately 80% the cells are ready to be passaged.

2. Prepare a coating by adding 0.75 ml 1:100 Geltrex diluted in DMEM/F12 in a 12-well and incubate for 1 h at 37 °C.

3. Dissociate stem cells in clumps using Versene(EDTA). First, wash once with 2 ml PBS per 6-well, next, add 1 ml Versene/EDTA and incubate for 3 min at 37 °C.

4. Remove Versene(EDTA), add 2 ml Essential 8 medium per 6-well and dislodge the stem cells in clumps with the cell scraper.

5. Count cells using a nucleocounter (*see* **Note 2**).

6. Transfer 200,000 stem cells to a freshly Geltrex coated 12-well and add up to 1 ml Essential 8 medium.

7. Refresh the medium daily with Essential 8 medium.

8. When a confluency of approximately 60% is reached, refresh the medium with Essential 8 medium supplemented with 1:100 Geltrex to create a matrix overlay.

9. Change medium to Essential 8 medium the next day.

10. The following day, when a confluency of approximately 80% is reached, wash the cell with 1 ml PBS per 12-well.

11. Start the cardiac differentiation by adding 1 ml of cardiac differentiation medium A to the 12-well (take note of the time, *see* **Note 3**).

12. After exactly 48 h, change the medium to 1 ml cardiac differentiation medium B.

13. After 48 h and every other day, change the medium with 1 ml cardiac maintenance medium (the precise timing is no longer essential).

14. First beating can be observed starting from day 7 after cardiac induction with cardiac differentiation medium A.

3.4 Cardiac Fibroblast Differentiation

The cardiac fibroblast differentiation is based on a previous study [17].

1. When the stem cell culture has reached a confluence of approximately 80%, the cells are ready to be passaged.

2. Prepare a coating by adding 0.38 ml 1:100 Geltrex diluted in DMEM/F12 in a 24-well and incubate for 1 h at 37 °C.

3. Dissociate stem cells in clumps using Versene/EDTA. First wash once with 2 ml PBS per 6-well, add 1 ml Versene (EDTA) and incubate for 3 min at 37 °C.

4. Remove Versene(EDTA), add 2 ml Essential 8 medium per 6-well and dislodge the stem cells in clumps with the cell scraper.

5. Count cells using a nucleocounter.

6. Transfer 100,000 stem cells to a freshly Geltrex coated 24-well and add up to 0.5 ml Essential 8 medium.

7. Refresh the medium daily with Essential 8 medium.

8. When a confluency of approximately 80% is reached, wash the cell with 1 ml PBS per 12-well.

9. Start the differentiation toward cardiac fibroblasts by adding 0.5 ml differentiation medium A (precise timing is essential, take note of the start time).

10. After exactly 48 h, change the medium to 0.5 ml cardiac fibroblast basal medium (CFBM) supplemented with 10 ng/μl bFGF. CFBM consists of DMEM supplemented with 2% FBS, 1:100 L-glutamine, 1:100 sodium pyruvate, and 1:100 Pen/Strep.

11. Refresh the medium after 24 h with CFBM supplemented with 10 ng/μl bFGF.

12. Replace the medium after 24 h with fresh CFBM supplemented with 10 ng/μl bFGF (day 4) and refresh every other day up to day 20.

13. Replace medium from day 20 and onward with CFBM every other day. Cardiac fibroblasts were at least 20 days old for use in coculture experiments.

3.5 Dissociating Cardiomyocytes in Single Cell Suspension

1. Wash the cardiomyocyte cultures three times with 1 ml PBS per 12-well.

2. Add 400 μl dissociation mix per 12-well (this consists of 40 μl enzyme T and 360 μl buffer X) and incubate for 10 min at 37 °C.

3. Gently pipet up and down (approximately five times) using a 1000 μl micropipette to dislodge the cells and to create a single cell suspension.

4. Directly after, add 2 ml cardio maintenance medium supplemented with 20% FBS to inactivate the enzymes.

5. Transfer the cell suspension to a 70 μm strainer and collect the single cell suspension. Take a small aliquot to determine the number of cells.

6. Centrifuge the cell suspension for 5 min at 200 × g and remove the supernatant.

7. Resuspend the cell pellet in maintenance medium supplemented with 20% FBS and 1:100 RevitaCell so to have a concentration of 1×10^6 cells/ml.

3.6 Cardiomyocyte Purification

After cardiac differentiation, the purity of cardiomyocytes in the culture could be increased using magnetic assisted cell sorting (MACS). Perform procedures on ice to increase cell viability.

1. Dissociate cells into a single cell suspension by following Subheading 3.5 until **step 5**.

2. Count the number of cells in the suspension using a nucleo-counter or Bürker counting chamber (up to 5×10^6 cells can be purified in a single run).

3. Centrifuge cell suspension for 5 min at $200 \times g$ and remove supernatant.

4. Resuspend the cell pellet in 80 µl wash buffer.

5. Add 20 µl Non-Cardiomyocyte Depletion Cocktail, mix well and incubate for 5 min at 4 °C.

6. Add 1 ml wash buffer and centrifuge for 5 min at $200 \times g$ and remove supernatant.

7. Resuspend the pellet in 80 µl wash buffer.

8. Add 20 µl Anti-Biotin MicroBeads, mix well and incubate for 10 min at 4 °C.

9. Add 400 µl wash buffer and mix the cell suspension.

10. Place an LS column in the magnetic field of the MACS separator system.

11. Wash the column once with 3 ml wash buffer.

12. Add the cell suspension (500 µl) to the column and collect flow-through.

13. Add three times 3 ml wash buffer to the column and collect flow-through in the same collection tube.

14. Centrifuge the purified cardiomyocyte cell suspension at $200 \times g$ for 5 min and remove supernatant.

15. Resuspend the cell pellet in maintenance medium supplemented with 20% FBS and 1:100 RevitaCell.

3.7 Dissociating Cardiac Fibroblast in Single-Cell Suspension

1. Wash the cardiac fibroblast culture with 1 ml PBS per 24-well.

2. Add 250 µl TrypLE to each 24-well and incubate for 5 min at 37 °C.

3. Gently pipet up and down (approximately three times) to dislodge the cells and to create a single cell suspension.

4. Directly after, add 1 ml CFBM.

5. Transfer the cell suspension to a 70 µm strainer and collect the single cell suspension. Take a small aliquot to determine the number of cells.

6. Centrifuge the cell suspension for 5 min at $200 \times g$ and remove the supernatant.

7. Resuspend the cell pellet in maintenance medium supplemented with 20% FBS and 1:100 RevitaCell so to have a concentration of 1×10^6 cells/ml.

**3.8 Preparation
of Liquid Marbles
Microbioreactors
(Fig. 1)**

1. The cell suspensions used for the LM preparations are 50 μl droplets with 50,000 cells, this means a concentration of 1000 cells/μl or 1×10^6 cells/ml.

2. After dissociating cardiomyocytes and cardiac fibroblasts the cell suspension can be mixed by different ratios, for instance 1:1, meaning 25 μl + 25 μl, 50,000 cells in total (*see* **Note 4**).

3. Prepare a petri dish containing fumed silica nanoparticles treated with nHMDS powder bed with an average particle size of 0.2–0.3 μm (*see* **Note 5**).

4. Dispense 50 μl single droplet containing 50,000 cells onto the powder bed (*see* **Note 6**).

5. Gently rotate the plate in a circular motion to ensure that the powder particles completely cover the surface of the liquid drop and form a LM.

6. Pick up the marbles using a 1000 μl pipette tip, cut at the edge, to accommodate the diameter of the marble (*see* **Note 7**).

7. Place the marble in a new 35 mm petri dish (one marble/dish) (*see* **Note 8**).

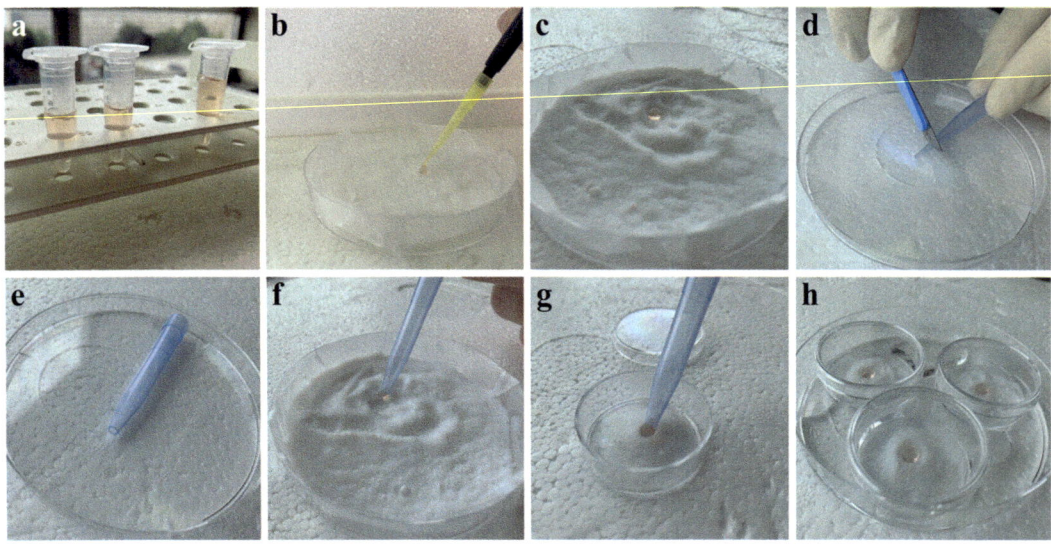

Fig. 1 Preparation of liquid marble. (**a**) Single cell suspensions of cardiomyocytes and cardiac fibroblasts are prepared and mixed to obtain a co-culture cell suspension. (**b**) A 50 μl drop of cell suspension is placed on top of a powder bed of fumed silica. (**c**) The marble is coated by moving the liquid drop in circular motions. (**d**) P1000 tip is cut at the end of the tip with a blade. (**e**) The tip is adjusted so that the diameter of the tip is a little smaller than the LM. (**f**) The LM is collected in the adjusted pipette tip, (**g**) and transferred to a 35 mm petri dish containing some fumed silica. (**h**) The 35 mm petri dishes are placed in a larger petri dish which is filled with distilled water to provide high humidity during incubation

8. Place three petri dishes containing the LMs in a larger petri dish and fill the bottom of the larger petri dish with distilled water to create a humid environment.

9. Incubate marbles for 24 h at 37 °C in 5% CO_2 and 5% O_2 incubator.

3.9 Harvesting Liquid Marbles

Cardiospheres were harvested from the LM 24 h after incubation.

1. To prepare the recipient (12-well plate), incubate a droplet of coating of 50 μl 1:10 Geltrex in the middle of the well for 1 h at 37 °C. A coverslip can be placed in the well to allow subsequent immunostaining. To minimize evaporation add sterilized water in the surrounding well spaces.

2. Remove half of the droplet of Geltrex coating (25 μl) before plating the LM (50 μl).

3. Transfer the LM using a 1000 μl micropipette, of which the tip was cut at the edge so that its diameter is slightly smaller than the diameter of the marble.

4. Gently place the LM on top of the Geltrex droplet so that the marble breaks, causing dispersion of the content of the LM (*see* **Note 9**).

5. Incubate the cardiospheres in the new recipient for 1 h at 37 °C to attach.

6. After 1 h, gently add cardio maintenance medium supplemented with 20% FBS and 1% RevitaCell to the recipient (1 ml/12-well).

7. Incubate the cardiospheres for 48 h at 37 °C so that the 3D structures can further attach and spread out.

8. After 48 h, change the medium with fresh cardio maintenance medium every other day.

3.10 Waste or Spent Medium Removal or Medium and Drugs Addition Through Injections with a Needle or Thin Tips (Fig. 2 Medium Removal)

1. Use sterile thin tips (used for 1–20 μl dispensing) or needle.

2. Insert gently the thin tip or needle inside the LM drops and aspirate the amount of spent medium (*see* **Note 10**).

3. Remove gently the tip/needle and discharge the collected sample in the testing tube.

4. Similar procedures can be done to inject fresh medium or specific drugs or cells (*see* **Note 11**).

3.11 Liquid Marble Splitting Method (See Fig. 2 Liquid Marble Splitting)

1. Put the LM drop (containing the spheroids) in a clean 35 mm petri dish.

2. Take a sterile blade which should be moved over the drop considering that the morphology of the LM is analogous to a spherical cap (*see* **Note 12**).

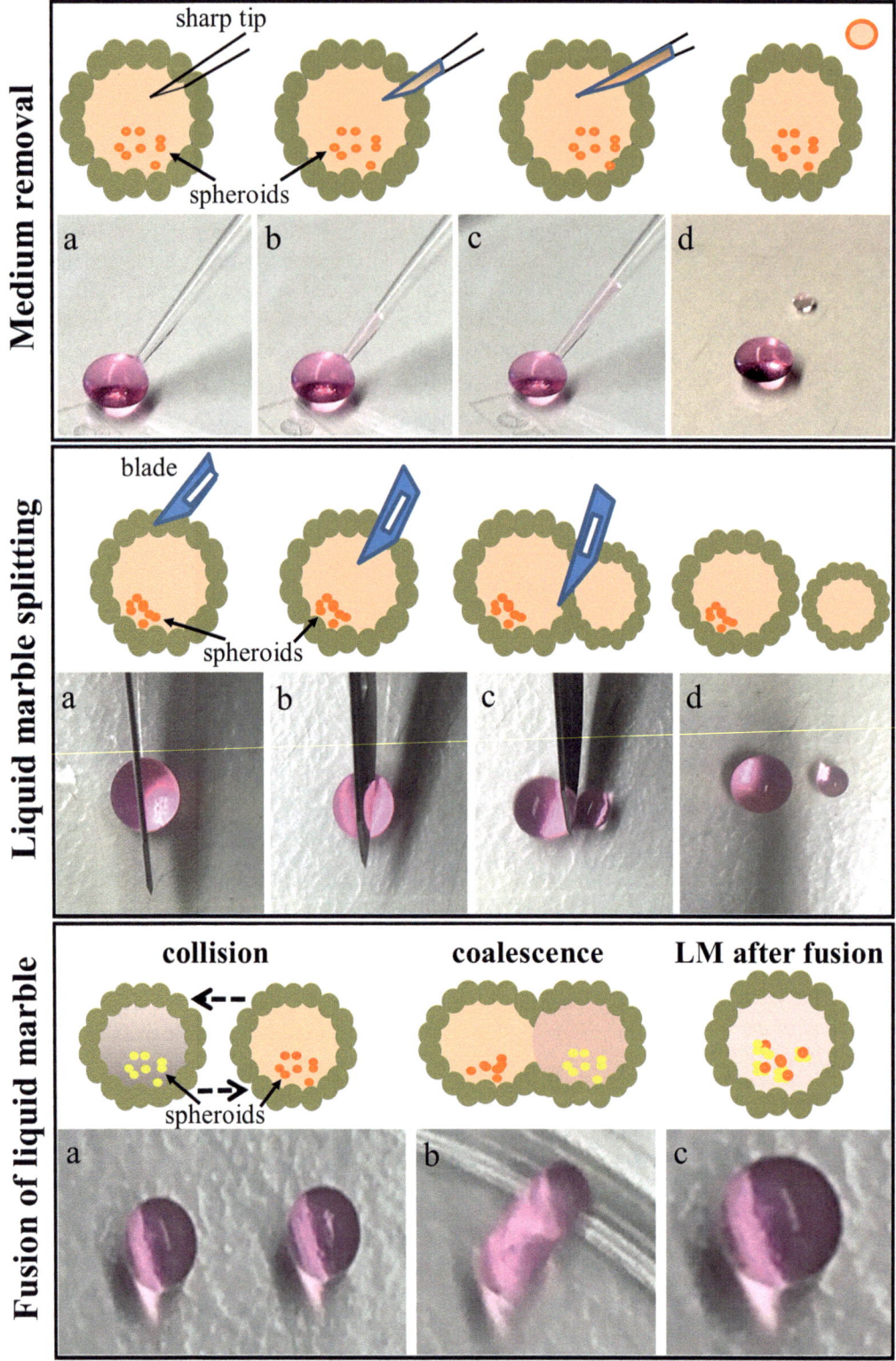

Fig. 2 Handling procedures of the liquid marble. Medium removal: medium was removed from the LM by opening the fumed silica shell with a thin pipette tip. The same procedure can be used to introduce medium,

3. Exert an external force over the surface of the LM, moving the surgical blade through the LM drop until the two sides are divided (the LM has sufficient elasticity and the coating nanoparticles will be coating the new drops maintaining the original spherical shape) (*see* **Note 13**).

4. Take out the drops using a 1000 μl tip of which the aspiration tip will be cut according to the LM drops diameter (*see* **Note 14**).

5. The LM drop containing the spheroids can be incubated to continue culture, while the drop with spent medium can be stored for subsequent analysis.

6. The amount of spent medium removal depend of the site in which the blade will be moved through the LM drop.

7. Bright-field (BF) and fluorescence microscopy can be used to screen the LMs by inspecting which LM contained the spheroid (*see* **Note 15**).

3.12 Coalescence of Liquid Marbles to Introduce Drugs, Fresh Medium or Cells (See Fig. 2 Fusion of Liquid Marble)

1. LM drops (with different volumes) are prepared as indicated in previous section (Subheading 3.8, **steps 3–7**). LM drops could contain fresh culture medium or culture medium supplemented with drugs or culture medium with cell suspension (*see* **Note 16**).

2. During LM drop preparation, put the single LM drops in separate 35 mm petri dish.

3. Using a 1000 μl cut tip, transfer the two drops in a new 35 mm petri dish.

4. Shake the petri dish to allow the collision between the two drops. This would help to overcome the energy barrier between the LMs hydrophobic layers of silica nanoparticles, preventing the formation of liquid bridges.

5. After liquid-to-liquid contact occurred, the coalescence of the LMs would be a spontaneous process and drops will be fused in a short timeframe (ms).

6. The new LM drop can be cultured in the incubator according to time required by the experiments; while maintaining high humidity.

Fig. 2 (continued) drugs or cells into the microbioreactor; **a, b, c, d** indicate the sequential phases. *Liquid marble splitting:* waste or spent media can be collected after splitting the LM using a blade to obtain drops of different size. LM drops maintain the spherical shape due to LM flexibility and coating property of superhydrophobic powder; **a, b, c, d** indicate the sequential phases. *Fusion of liquid marble:* addition of fresh media, drugs or cells by coalescence of two LM drops; after (**a**) the LM collision, (**b**) liquid bridges will be formed, (**c**) creating a new spherical LM

3.13 Immuno-fluorescent Staining for Cardiac Markers

1. Remove the medium from the cell culture, wash once with PBS.

2. Add 4% PFA and incubate at RT for 20 min to fix the cells (0.5 ml/12-well).

3. Wash the cells three times for 5 min with PBS.

4. Add permeabilization solution (consists of 0.1% Triton X-100 in PBS) and incubate for 30 min at RT (0.5 ml/12-well).

5. Remove permeabilization solution and add blocking solution (blocking solution consists of 5% goat serum in PBS) to the cells and incubate for 30 min at RT (0.5 ml/12-well).

6. Prepare primary antibodies (*see* Table 1) in dilution solution (consisting of 1% BSA + 0.05% Tween 20 in PBS) and incubate the cells overnight at 4 °C.

7. The next day, wash the cells three times for 5 min with permeabilization solution.

8. Prepare secondary antibodies (*see* Table 1) in dilution solution and incubate for 30 min at RT.

9. Wash the cells three times for 5 min with permeabilization solution.

10. Incubate the cells with 1:1000 Hoechst for 10 min at RT.

11. Wash the cells two times for 5 min with PBS.

12. Mount the cells using DABCO on a glass slide, place a cover glass on top and seal with nail polish. Fluorescent images were made using the ZEISS LSM900 confocal microscope.

3.14 Downstream Applications of Cardiospheres in the Liquid Marble

3D cocultured cardiomyocytes can be used in various downstream application for higher throughput analysis. In this protocol we show the attachment to a glass coverslip to follow the spreading of the 3D structure.

1. Purified cardiomyocyte and cardiac fibroblasts are cocultured in a LM to form cardiospheres within 24 h (*see* Fig. 3).

2. The content of the liquid marble is harvested according to previous Subheading 3.9 and attached to a glass coverslip in a 12-well.

3. Phase contrast images of the spreading 3D structures are recorded (after 24 h and 48 h), and confocal microscopy is applied to visualize the presence of cardiomyocytes (Fig. 4).

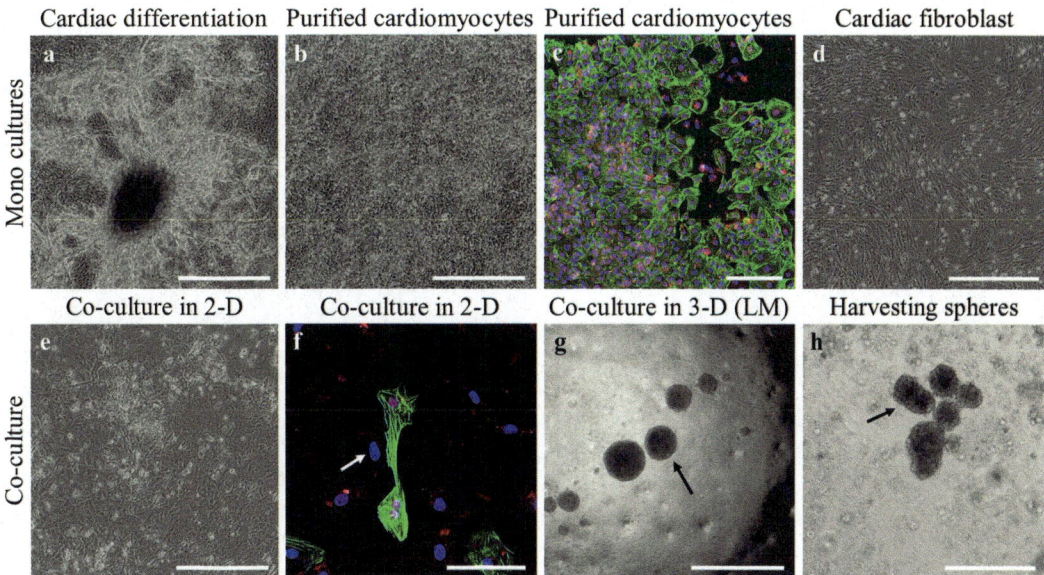

Fig. 3 Cell cultures of cardiomyocyts and cardiac fibroblasts. Cell cultures in mono- and coculture in 2D and 3D. (**a**) Following the cardiac differentiation protocol beating cell cultures are observed. (**b**) Using MACS cardiomyocyte population was purified. (**c**) Immunofluorescent staining of purified cardiomyocyte culture shows cardiac markers NKX2.5 (red), cTnT (green), and nuclei (blue). (**d**) Cardiac fibroblast obtained with the cardiac fibroblast differentiation from human stem cells. (**e**) Coculture between cardiomyocytes and cardiac fibroblast in 2D. (**f**) Immunofluorescent staining for cardiac markers NKX2.5 (red), cTnT (green) and nuclei (blue) of coculture shows that cardiomyocytes elongates along a cardiac fibroblast (indicated by the arrow). (**g**) 3D coculture of 25.000 cardiomyocytes and 25.000 cardiac fibroblasts, arrow indicates cardiosphere inside the LM. (**h**) Cardiospheres are harvested by breaking the LM, arrow indicates cardiosphere outside the LM. Bar in phase contrast images (**a**, **b**, **d**, **e**, **g**, **h**) represents 500 μm. Bar in confocal image c and f represent respectively 200 μm and 100 μm

4   Notes

1. It is crucial to use clean fumed silica, because previously used powder tends to aggregate and increase the size of nanoparticles.

2. CellCounter can be used for total cell count when using lysis buffer and stabilizing buffer. Dead cells can be counted by only adding stabilizing buffer and no lysis buffer. Total live cells can be derived by deducting the number of dead cells from total cell count.

3. Precise timing during early steps of the cardiac differentiation greatly impact differentiation efficiency, do not deviate more than 1 h from the stated times during the first 4 days of differentiation.

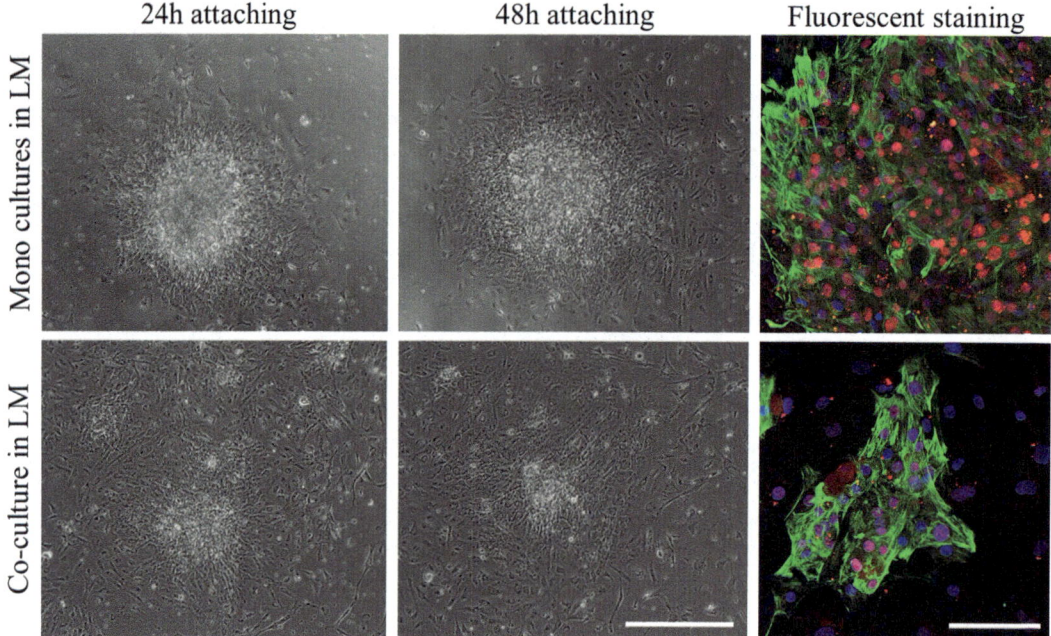

Fig. 4 Liquid marble outgrowth on glass coverslip. Outgrowth from attached cardiospheres in mono- and coculture. Phase-contrast images show the outgrowths of attached cardiospheres in 2D after 24 h and 48 h of attachment. Immunofluorescent staining of 48 h outgrowths for cardiac markers NKX2.5 (red), cTnT (green), and nuclei (blue) show cardiomyocytes present in mono- and coculture after 3D culture in LM. Coculture cardiospheres were observed to spread faster compared to monoculture cardiospheres. This could be explained by the matrix production of cardiac fibroblasts that aid in the attachment and spreading of the cardiomyocytes. Interestingly, the cardiomyocytes from the cardiospheres formed in the LM clustered together after attachment in 2D culture rather than dispersed. Bar in phase contrast images represents 500 μm and bar in confocal images indicates 100 μm

4. It is important to optimize the cell seeding density inside the LM and determine the ideal volume of LM for organoid production, based on the characteristics of the cells used during the experiment.

5. When preparing the LM, it is better to work in the lowest ambient humidity because the air humidity can affect the quality of the LM preparation.

6. To make the surface of the petri dish containing the fumed silica more hydrophobic it can help to place parafilm on the bottom, this avoids that the liquid drop attaches to the plastic of the petri dish.

7. The approximate diameter of the opening pipette tip is slightly smaller than the marble diameter, creating a friction fit to grip the marble inside the tip.

8. To avoid breaking of the LM when transferring to the new petri dish (35 mm) it helps to place some fumed silica powder in the petri dish.

9. Direct liquid contact disrupts the hydrophobicity of the fumed silica powders and breaks the marbles.

10. The amount of medium aspirated from the LM should be limited, otherwise the drop could collapse.

11. The amount of medium injected with pipette tips during spent medium removal or medium addition should be therefore extremely limited; the ultimate capacity of the LM is limited by the outer hydrophobic particles; the shell only tolerates slight expansion from the addition of an external solution; otherwise, surface defects are generated and the shell easily collapses.

12. During the splitting of LM, to remove spent or waste medium it is suggested to cover the blade with fumed silica nanoparticles, this could help performing the LM drop division.

13. Perform the procedures under stereomicroscope can help in the precise division and spheroids allocation in one LM drop.

14. The amount of spent medium removal is fully scalable and only depends by the portion of spherical cap that will be cut.

15. The fine coating of hydrophobic powders allows monitoring of the LM content using optical–fluorescent microscopy. At the same time, it ensures optimal gas exchange between the interior liquid and the surrounding environment thanks to the powder particle size.

16. Drops with fresh medium or cell suspension cannot contain large amounts of liquid (drops between 10 and 20 µl are suggested) otherwise the spherical shape of the LM cannot be maintained.

Acknowledgments

The authors are members of the COST Actions CA16119. JvH and this research are supported by a Ghent University grant (BOF 01 J13919). SL and DP are supported by MISE (MinistrodelloSviluppoEconomico) project n.F/200110/01-03/X45, "Agrifood," PON, FESR 2014-2020.

References

1. Simian M, Bissell MJ (2017) Organoids: a historical perspective of thinking in three dimensions. J Cell Biol 216(1):31–40. https://doi.org/10.1083/jcb.201610056

2. Harrison RG (1959) The outgrowth of the nerve fiber as a mode of protoplasmic movement. J Exp Zool 142:5–73. https://doi.org/10.1002/jez.1401420103

3. Lin RZ, Chang HY (2008) Recent advances in three-dimensional multicellular spheroid culture for biomedical research. Biotechnol J 3 (9–10):1172–1184. https://doi.org/10.1002/biot.200700228

4. Clevers H (2016) Modeling development and disease with organoids. Cell 165 (7):1586–1597. https://doi.org/10.1016/j.cell.2016.05.082

5. Tian J, Fu N, Chen XD, Shen W (2013) Respirable liquid marble for the cultivation of microorganisms. Colloids Surf B Biointerfaces 106:187–190. https://doi.org/10.1016/j.colsurfb.2013.01.016

6. Arbatan T, Al-Abboodi A, Sarvi F, Chan PP, Shen W (2012) Tumor inside a pearl drop. Adv Healthc Mater 1(4):467–469. https://doi.org/10.1002/adhm.201200050

7. Serrano MC, Nardecchia S, Gutiérrez MC, Ferrer ML, del Monte F (2015) Mammalian cell cryopreservation by using liquid marbles. ACS Appl Mater Interfaces 7(6):3854–3860. https://doi.org/10.1021/acsami.5b00072

8. Arbatan T, Li L, Tian J, Shen W (2012) Liquid marbles as micro-bioreactors for rapid blood typing. Adv Healthc Mater 1(1):80–83. https://doi.org/10.1002/adhm.201100016

9. Sarvi F, Jain K, Arbatan T, Verma PJ, Hourigan K, Thompson MC, Shen W, Chan PP (2015) Cardiogenesis of embryonic stem cells with liquid marble micro-bioreactor. Adv Healthc Mater 4(1):77–86. https://doi.org/10.1002/adhm.201400138

10. Brevini TAL, Manzoni EFM, Ledda S, Gandolfi F (2019) Use of a super-hydrophobic microbioreactor to generate and boost pancreatic mini-organoids. Methods Mol Biol 1576:291–299. https://doi.org/10.1007/7651_2017_47

11. Pennarossa G, Manzoni EFM, Ledda S, deEguileor M, Gandolfi F, Brevini TAL (2019) Use of a PTFE micro-bioreactor to promote 3D cell rearrangement and maintain high plasticity in epigenetically erased fibroblasts. Stem Cell Rev Rep 15(1):82–92. https://doi.org/10.1007/s12015-018-9862-5

12. Ledda S, Idda A, Kelly J, Ariu F, Bogliolo L, Bebbere D (2016) A novel technique for in vitro maturation of sheep oocytes in a liquid marble microbioreactor. J Assist Reprod Genet 33(4):513–518. https://doi.org/10.1007/s10815-016-0666-8

13. Nguyen N-K, Ooi CH, Singha P, Jin J, Sreejith KR, Phan H-P, Nguyen N-T (2020) Liquid marbles as miniature reactors for chemical and biological applications. Processes 8(7):793

14. Chen M, Shah MP, Shelper TB, Nazareth L, Barker M, Tello Velasquez J, Ekberg JAK, Vial ML, St John JA (2019) Naked liquid marbles: a robust three-dimensional low-volume cell-culturing system. ACS Appl Mater Interfaces 11(10):9814–9823. https://doi.org/10.1021/acsami.8b22036

15. Zuppinger C (2019) 3D cardiac cell culture: a critical review of current technologies and applications. Front Cardiovasc Med 6:87. https://doi.org/10.3389/fcvm.2019.00087

16. Giacomelli E, Meraviglia V, Campostrini G, Cochrane A, Cao X, van Helden RWJ, Krotenberg Garcia A, Mircea M, Kostidis S, Davis RP, van Meer BJ, Jost CR, Koster AJ, Mei H, Míguez DG, Mulder AA, Ledesma-Terrón M, Pompilio G, Sala L, Salvatori DCF, Slieker RC, Sommariva E, de Vries AAF, Giera M, Semrau S, Tertoolen LGJ, Orlova VV, Bellin M, Mummery CL (2020) Human-iPSC-derived cardiac stromal cells enhance maturation in 3D cardiac microtissues and reveal non-cardiomyocyte contributions to heart disease. Cell stem cell 26(6):862–879. e811. https://doi.org/10.1016/j.stem.2020.05.004

17. Zhang J, Tao R, Campbell KF, Carvalho JL, Ruiz EC, Kim GC, Schmuck EG, Raval AN, da Rocha AM, Herron TJ, Jalife J, Thomson JA, Kamp TJ (2019) Functional cardiac fibroblasts derived from human pluripotent stem cells via second heart field progenitors. Nat Commun 10(1):2238. https://doi.org/10.1038/s41467-019-09831-5

Chapter 6

Isolation, Culture, and Characterization of Primary Bovine Endometrial, Epithelial, and Stromal Cells for 3D In Vitro Tissue Models

Antonio Murillo and Marta Muñoz

Abstract

Efficient isolation, characterization, and culture of endometrial epithelial cells and stromal fibroblasts from calf uteri collected at the slaughterhouse is key to develop useful 3D culture tissue models to investigate uterine physiology and pathology without the need of performing invasive procedures to recover tissue samples.

Here we provide a detail methodology that gives consistently pure and viable populations of distinct primary bovine endometrial cells.

Key words 3D in vitro tissue models, Cell culture, Endometrium, Bovine, Immunohistochemistry

1 Introduction

Significant efforts have been dedicated in the last few years to propose advanced in vitro tissue culture models to bridge the gap between the in vivo situation and conventional over-simplified in vitro 2D monolayer models. A common feature in this new generation *of* in vitro models is their three-dimensional (3D) character, which allows incorporating cell–cell interactions and cell–extracellular matrix interactions, essential to preserve the cell phenotype, behaviour, and differentiation status. Going 3D also supports the preparation of multicellular coculture models, which is key to emulate the structure and the biochemical and biomechanical microenvironments of native tissues in vivo. Thus, 3D cell cultures are more physiologically relevant and predictive than 2D cultures [1].

The endometrium, the mucosal lining of the mammalian uterus, consists of a single-layered prismatic epithelium with or without cilia (depending on how far along the estrous cycle is)

Tiziana A.L. Brevini et al. (eds.), *Next Generation Culture Platforms for Reliable In Vitro Models: Methods and Protocols*, Methods in Molecular Biology, vol. 2273, https://doi.org/10.1007/978-1-0716-1246-0_6,
© The Author(s), under exclusive license to Springer Science+Business Media, LLC, part of Springer Nature 2021

and its basal lamina, uterine glands, and a specialized, cell-rich connective tissue (stroma) containing a rich supply of blood vessels. The isolation and in vitro culture of endometrial cells is an invaluable model for human and animal reproductive research. Applications range from the study of the normal cyclic physiological process to endometrial pathologies such as reproductive disorders or infectious diseases.

To date, there is no proper comprehensive and faithful in vitro model of the mammalian endometrium that emulates its complex microarchitecture in a 3D configuration. Developing such a model faces challenges that include amongst others choosing the most appropriate (1) source of cells, that is, primary cells vs. cell lines, (2) type of scaffold-based technology, that is, nanofiber vs. hydrogel scaffolds and (3) type of culture, that is, dynamic vs. static.

Here we provide a comprehensive protocol that leads consistently to the isolation of bovine primary epithelial and stromal cells in high quantity and purity appropriate for developing 3D scaffold based in vitro culture experiments.

2 Materials

2.1 Endometrial Cell Isolation

Prepare all solutions using ultrapure water and cell culture tested reagents. Prepare and store all reagents at room temperature (unless indicated otherwise). Ensure sterility. Diligently follow all waste disposal regulations when disposing waste materials.

1. Dulbecco's Phosphate Buffered Saline with antibiotics (DPBS-Ab): DPBS containing 100 μg/mL streptomycin, 100 IU/mL penicillin, 0.25 μg/mL amphotericin B (*see* **Note 1**).

2. Neutralizing solution: 10% heat-inactivated fetal bovine serum (FBS) in Hanks Buffered Saline Solution.

3. Tissue digestion solution: Hanks Buffered Saline Solution containing 0.1 mg/mL bovine serum albumin (BSA), 2.5 BAEE units/mL trypsin–EDTA, 0.5 mg/mL collagenase from *Clostridium histolyticum*, Type I-A, 0.1 mg/mL DNase I (*see* **Note 2**).

4. Cell culture medium: 10% FBS in Dulbecco's Modified Eagle's Medium/Nutrient Mixture F-12 Ham (DMEM/F12, 1:1) containing 100 μg/mL streptomycin, 100 U/mL penicillin, and 0.25 μg/ mL amphotericin B (*see* **Note 1**).

5. Accutase® solution.

6. Dissection kit.

7. Dissection tray.

8. Sterilized cloth surgical drape.

9. Tissue culture grade petri dishes: 100-mm.

10. Tissue culture grade flasks: 25- and 75-cm^2.

11. 4-well tissue culture grade plates.

12. Serological pipettes: 5-, 10-, and 25-mL.

13. Membrane filter: 0.2-, 0.45-μm.

14. Nylon mesh cell strainer: 40 μm.

15. Haemocytometer.

16. Portable laboratory refrigerator.

17. Shaking water bath.

18. Incubator (*see* **Note 3**).

2.2 Endometrial Cell Characterization

1. Phosphate buffered saline (PBS): Weigh 8 g NaCl, 0.2 g KCl, 1.44 g Na_2HPO_4, 0.24 g KH_2PO_4. Add salts one after the other to 800 mL of ddH_2O. Stir until dissolve. Adjust pH to 7.4 using HCl and NaOH. Bring volume to 1 L. Autoclave or filter sterilize.

2. Fixative: Add 4 g of paraformaldehyde to 50 mL of H_2O. Add 1 mL of 1 M NaOH and stir gently on a heating block at ~60 °C until the paraformaldehyde is dissolved. Add 10 mL of $10\times$ PBS and allow the mixture to cool to room temperature. Adjust the pH to 7.4 with 1 M HCl (~1 mL), then adjust the final volume to 100 mL with H_2O. Filter the solution through a 0.45-μm membrane filter to remove any particulate matter (*see* **Note 4**).

3. Permeabilization solution: 0.1% Triton X-100 in PBS.

4. Blocking solution: 5% normal goat serum in PBS.

5. DAPI stock solution (5 mg/mL): dissolve 10 mg of DAPI dihydrochloride in 2 mL of deionized water in dim light. For long-term storage the stock solution can be aliquoted and stored at −20 °C. When handled properly, DAPI solutions are stable for at least 6 months (*see* **Note 5**).

6. Sterilized coverslips.

7. Humid chamber.

3 Methods

3.1 Isolation and Primary Culture of Epithelial and Stromal Cells from Bovine Endometrial Tissue

1. Collect female reproductive tracts within 15 min of slaughter.

2. Identify tracts as being free from disease and visible infection.

3. Estimate the physiological status of the tissue was estimated based on the ovarian morphology described by Ireland et al. [2] (*see* **Note 6**).

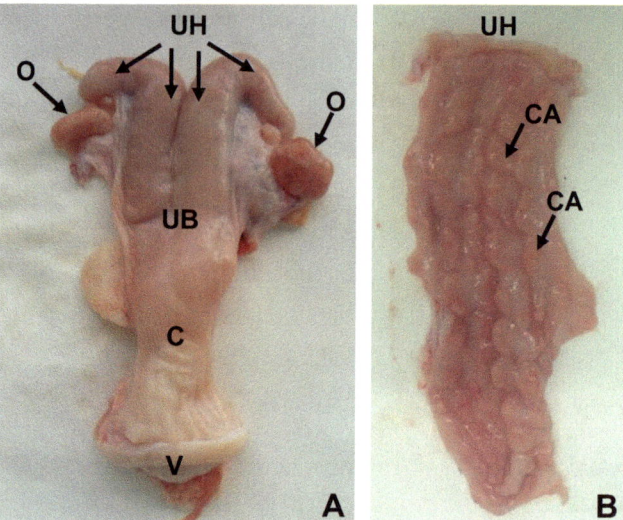

Fig. 1 Representative image of bovine reproductive tract used for cell isolation. (**a**) Anatomy of bovine reproductive tract. *UH* uterine horn, *O* ovary, *UB* uterine body, *C* cervix, *V* vagina. (**b**) Ipsilateral uterine horn longitudinally opened for endometrial tissue collection. *UH* uterine horn, *CA* Caruncle

4. Wash the external surface of the tract in 70% ethanol, wrap in a sterilized cloth surgical drape and place within a plastic bag.

5. Transport reproductive tracts to the laboratory within 90 min of collection in a portable laboratory refrigerator between 4 and 8 °C.

6. Upon arrival to the laboratory, wash the external surface of the tract again with 70% ethanol (Fig. 1a).

7. Excise the ipsilateral uterine horn (corresponding to corpus luteum presence in the ovary) and place in a safety cabinet, on a sterilized cloth surgical drape placed on a cooling block at 4 °C.

8. Open the horn longitudinally with sterile scissor and gently wash the exposed endometrium with DPBS supplemented with antibiotics (Fig. 1b).

9. Dissect the endometrial surface using sterile scissors or sterile scalpel and forceps. Be extremely careful to strip endometrial tissue from the underlying myometrium.

10. Place small tissue strips (1–2 cm), on a petri dish placed on a cooling block at 4 °C and dice using with a sterile scalpel until pulp-like (*see* **Note 7**).

11. Add 10 mL of prewarmed tissue digestion solution (37 °C in a water bath) to minced tissue through a 20 mL syringe and 0.2 μm syringe filter (*see* **Note 8**).

12. Transfer the media with the tissue fragments to a 50 mL conical tube using a 10 mL serological pipette.

13. Place 50 mL tubes in a shaking water bath for 1 h at 37 °C.

14. Add 10 mL of Hanks solution with 10% FBS to each tube to neutralize enzymatic activity.

15. Filter cell solution through a 40 μm Nylon mesh cell strainer. Flow-through will contain stromal cells but epithelial and endometrial gland clumps are retained in the strainer.

16. Turn the 40 μm cell strainer upside down and wash the epithelial cells into a clean 50 mL tube using 10 mL of neutralizing solution.

17. Centrifuge all tubes at room temperature for 7 min at $700 \times g$.

18. Carefully aspirate supernatant and resuspend pellet gently in 5 mL neutralizing solution.

19. Centrifuge at room temperature for 7 min at $700 \times g$.

20. Aspirate supernatant and resuspend the cell pellet in 1 mL warmed cell culture media.

21. Keep cell suspension in cell incubator at 37 °C in a 5% CO_2 incubator while determining the cell concentration by haemocytometer or electronic cell counter.

22. Seed cells at 1×10^5 cell/mL in 75 cm^2 culture flasks (*see* **Note 9**).

23. After 18 h, remove culture media from flasks containing stromal cells to remove any epithelial cells (*see* **Note 10**) [3]. Thereafter change media every 18 h until 80% cell confluency is reached.

24. *Do not change media from flasks containing epithelial cells for 72 h.* Thereafter change media every 48 h until 80% cell confluency is reached.

25. Detach cells to seed in their final environment or propagate cell cultures (*see* **Note 11**).

3.2 Vimentin/ Cytokeratin Double Immunofluorescence Staining for Validation of Pure Stromal and Epithelial Cell Cultures

At each cell passage, epithelial and stromal cells are seeded on sterile coverslips to check their purity and differentiation status. Carry out all procedures at room temperature, unless otherwise specified.

1. Seed cells from each culture flask on sterile coverslips place in culture plates (*see* **Note 12**).

2. Once cells seeded on coverslips have reached 80% of the confluence fix to detect cytoskeletal proteins.

3. Wash culture cells gently 3× with PBS 5 min each time.

4. Fix the cells with 4% PFA for 10 min (*see* **Note 4**).

5. Wash 3× with PBS 5 min each time on shaker (*see* **Note 13**).

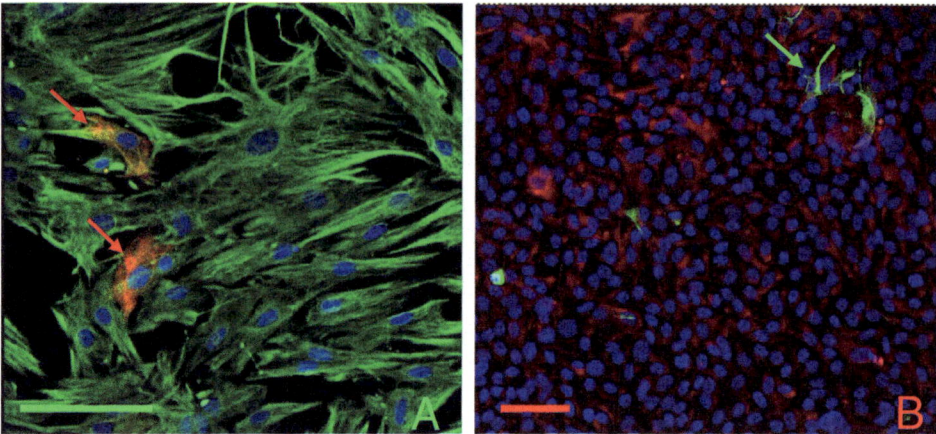

Fig. 2 (**a**) Stromal cell cultures and (**b**) epithelial cell cultures were processed for immunofluorescence using anti-cytokeratin and anti-vimentin primary antibodies. The secondary antibody used were Alexa Fluor® 488–conjugated anti-rabbit-IgG and Alexa Fluor® 555–conjugated anti-mouse-IgG. (**a**) Vimentin expression (in green) is expressed in the majority of cells present in stromal cell cultures whilst cytokeratin expressing cells (red) are rarely found (arrows). (**b**) Cytokeratin expression is detected (in red) in the majority of cells present in epithelial cell cultures although some vimentin expressing cell (in green) are also detected (arrows). (Scale bar = 100 μm)

6. Treat samples with permeabilization solution for 10 min.

7. Wash 3× with PBS 5 min each time on shaker.

8. To reduce nonspecific binding, incubate on shaker samples with blocking solution for 1 h.

9. Incubate samples overnight, at 4 °C in a humid chamber with a mixture of a rabbit anti-vimentin and a mouse anti-cytokeratin pan antibody diluted according to supplier indications, in the blocking solution.

10. Wash 3× with PBS 5 min each time on shaker.
    Hereinafter all steps should be performed in dim light/darkness to avoid fluorescence bleaching

11. Incubated samples on shaker at room temperature in a dark humid with goat Alexa Fluor 488–conjugated goat anti-rabbit IgG diluted according to supplier indications, in blocking solution for 1 h.

12. Wash 3× with PBS 5 min each time on shaker.

13. Incubated samples on shaker at room temperature in a dark humid chamber with a goat anti-mouse IgG (H+L) highly cross-absorbed Alexa 555 diluted according to supplier indications, in blocking solution for 1 h.

14. Wash 3× with PBS 5 min each time on shaker.

15. Incubate samples with DAPI at 0.2 mg/mL in PBS for 2 min at room temperature (*see* **Note 5**).

16. Rinse samples three times for 10 min each on shaker.

17. Mount coverslip up site down on slide with a drop of mounting medium.

18. Fluorescence was examined under a confocal microscope (ultraspectral Leica TCS-SP8-X; Leica Microsystems, Mannheim, Germany; Fig. 2).

    For control purposes representative sections were processed in the same way as described above using nonimmune rabbit or mouse sera instead of the primary antibodies, or omitting the primary antibodies in the incubation. Under these conditions no specific immunostaining was observed.

4   Notes

1. We use a commercially available antibiotic/antimycotic solution 100× solution suitable for cell culture.

2. We prepare the following stock solutions and keep them at −20 °C for 6 months at −20 °C.

    Collagenase (50 mg/mL): Dissolve 100 mg of collagenase in 2 mL Hanks solution. Sterilize them by passing through a 0.45-μm filter and store in aliquots at −20 °C.

    BSA (100 mg/mL): Dissolve 2 g of BSA in 20 mL of Hanks solution. Sterilize them by passing through a 0.45-μm filter and store for 6 months in aliquots at −20 °C.

    DNase I (10 mg/mL): Dissolve 1 mg of DNase I in 1 mL of 0.15 M NaCl. Sterilize them by passing through a 0.45-μm filter and store for 6 months in aliquots at −20 °C.

3. Endometrial cell culture is performed in a CO_2 incubator. Five percent CO_2 and 100% humidity at 37 °C are required.

4. Formaldehyde is toxic. Please read the MSDS before working with this chemical. Gloves and safety glasses should be worn and solutions are to be made inside a safety cabinet. Also use safety cabinet when fixing samples.

5. DAPI is a known mutagen and should be handled with care. The dye must be disposed of safely and in accordance with applicable local regulations.

6. We have found out that tracts in the in the early-luteal phase (days 1–4 of estrous cycle) provide a higher number of viable cells.

7. Do not allow the cells to dry out. Even while in the procedure of mincing, a small amount of PBS-Ab should be added to the tissue.

8. In our hands 1 mL of digestion solution for every 100 mg of tissue works nicely.

9. If the total number of cells is less than one million, plate all cells in 25 cm^2 flask with 5 mL of cell culture media. If the total number of cells is more than one million, plate all cells in 75 cm^2 flasks with 15 mL of cell culture media. Cell seeding density is important. If the seeding density is too low, the cells may not achieve a confluent monolayer, whereas higher seeding densities allow less cell spreading and can result in monolayers with lower transepithelial electrical resistances.

10. This step is very important as it will remove any epithelial cells that could be present in the stromal cell populations. Epithelial and stromal cells have different platting times.

11. We use Accutase® solution for detach cells. In our hands the recovery of cells more efficient than using trypsin-EDTA solution. We follow the instruction provided by supplier.

12. We seed 1.5×10^6 cells on 13 mm Nunc Thermanox® coverslip place on 4-well cell culture Nunc plates.

13. Shaker is set at the lowest speed possible, otherwise cells will detach.

Acknowledgments

The Authors are members of the COST Action CA16119 In vitro 3-D total cell guidance and fitness (CellFit).

References

1. Rijal G, Li W (2018) Native-mimicking in vitro microenvironment: an elusive and seductive future for tumor modeling and tissue engineering. J Biol Eng 12:20–42

2. Ireland JJ, Murphee RL, Coulson PB (1980) Accuracy of predicting stages of bovine estrous cycle by gross appearance of the corpus luteum. J Dairy Sci 63:155–160

3. Turner ML et al (2014) Epithelial and stromal cells of bovine endometrium have roles in innate immunity and initiate inflammatory responses to bacterial lipopeptides in vitro via toll-like receptors TLR2, TLR1, and TLR6. Endocrinology 155(4):1453–1465

Chapter 7

Using Decellularization/Recellularization Processes to Prepare Liver and Cardiac Engineered Tissues

Matteo Ghiringhelli, Yousef Abboud, Snizhanna V. Chorna, Irit Huber, Gil Arbel, Amira Gepstein, Georgia Pennarossa, Tiziana A.L. Brevini, and Lior Gepstein

Abstract

Tissue engineering provides unique opportunities for disease modeling, drug testing, and regenerative medicine applications. The use of cell-seeded scaffolds to promote tissue development is the hallmark of the tissue engineering. Among the different types of scaffolds (derived from either natural or synthetic polymers) used in the field, the use of decellularized tissues/organs is specifically attractive. The decellularization process involves the removal of native cells from the original tissue, allowing for the preservation of the three-dimensional (3D) macroscopic and microscopic structures of the tissue and extracellular matrix (ECM) composition. Following recellularization, the resulting scaffold provides the seeded cells with the appropriate biological signals and mechanical properties of the original tissue. Here, we describe different methods to create viable scaffolds from decellularized heart and liver as useful tools to study and exploit ECM biological key factors for the generation of engineered tissues with enhanced regenerative properties.

Key words Decellularization, Recellularization, Cardiomyocytes, Hepatocytes, Engineer slice, Patch, Transplantation

1 Introduction

A remarkable clinical need exists for the development of various methods to facilitate the regeneration of injured or diseased tissues and organs. The recent developments in stem cell biology, molecular interventions, biopolymers, and other related biological and engineering disciplines have paved the way to the emerging research and clinical discipline of "Regenerative Medicine." Regenerative medicine seeks to harness methods for the replacement or

Electronic supplementary material The online version of this chapter (https://doi.org/10.1007/978-1-0716-1246-0_7) contains supplementary material, which is available to authorized users.

Tiziana A.L. Brevini et al. (eds.), *Next Generation Culture Platforms for Reliable In Vitro Models: Methods and Protocols*,
Methods in Molecular Biology, vol. 2273, https://doi.org/10.1007/978-1-0716-1246-0_7,
© The Author(s), under exclusive license to Springer Science+Business Media, LLC, part of Springer Nature 2021

repair of dysfunctional cells, tissues, or organs, in an attempt to restore normal function [1].

As part of the emerging regenerative medicine discipline [2], tissue engineering is a multidisciplinary field that combines functional cells with three-dimensional scaffolds (made from synthetic or biological polymers) to create tissue substitutes. The use of cell-seeded scaffolds to promote tissue development is the hallmark of a tissue engineering strategy. The scaffold serves many purposes, including the delivery of biological signals to control and enhance tissue formation, to provide adequate biomechanical support for the cell graft, to control graft shape and size, to promote angiogenesis, and to protect the cells from physical damage.

Among the different tissue engineering strategies, the technique of whole-organ decellularization and recellularization has attracted increasing attention in the last decade [3]. When a decellularized extracellular matrix (ECM) is derived from an organ, it provides a unique noncytotoxic three-dimensional scaffold that allows for cell adhesion and proliferation [4, 5]. Cell removal techniques include utilization of mechanical, thermal, chemical, and enzymatic methods that simultaneously with cell detachment are preserving the micro- and macroanatomy of the ECM [6]. The patient-like organ shape is exploited and the goal is to construct a personalized neoorgan with a low potential for immunological rejection after transplantation [7].

Historically, one of the limitations of tissue engineering was the paucity of human cell sources for creation of the tissue-constructs, and specifically the ability to derive patient-specific cells. The advent of human embryonic stem cells and more recently of the human induced pluripotent stem cells (hiPSC) technology may provide a solution to this cell-sourcing problem. The latter technology allows to reprogram patient-specific somatic cells into hiPSC, which can then be coaxed to differentiate into any cell type. The desired cells can then be used for generation of functional tissue/organ like structures for in vitro and in vivo applications. Consequentially, hiPSC-derived cardiomyocytes and hepatocytes were combined with a variety of scaffolding polymer to generate functional liver and cardiac tissues using a variety of tissue engineering strategies [8–10]. Here, we describe two efficient methods for decellularization/recellularization to generate and utilize liver and heart tissue-constructs that resemble the 3D structural and functional characteristic of the native organ (Figs. 1 and 2). In addition to the detailed description of the methods used to generate such tissues, we provide two examples for each organ type in which the engineered tissues can be used as "models in a dish" or for regenerative medicine related in vivo transplantation projects.

In general, our approach can be used to generate relatively thin tissues termed engineered heart slices (EHSs) and engineered livers slices (ELSs) that are more suitable for in vitro modeling studies as

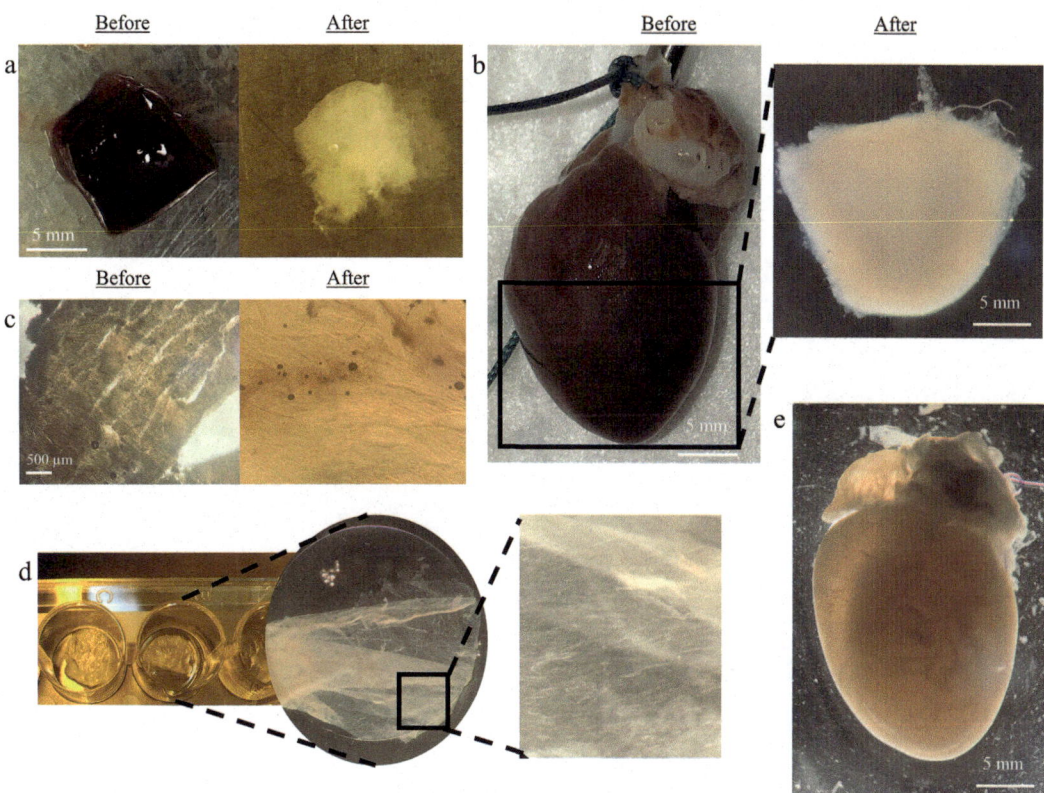

Fig. 1 Process of organ decellularized scaffolds creation. (**a**) Liver tile before and after decellularization. (**b**) Ventricle ECM patch obtained from the left wall of the rat heart. Note that the natural "V" shape of the ventricle is maintained. (**c**) EHS derived from the original heart slice. Note that the tissue before decellularization shows the typical myocardial morphological microarchitecture. (**d**) Engineered slices are hooked to the 12 mm glass coverslip to allow for fixation of the tissue at the bottom of the well and also to gently stretch the collagen fibers. (**e**) The decellularization process allows to keep the vascular tree of the original organ intact

well as very thick tissue patches that may be more suitable for in vivo transplantation applications (Fig. 3). We next describe two approaches that can be used to enhance the regenerative capacity of the tissue patches. These include supplementing the liver and heart tissue-constructs with chemical agents or extracellular vesicles (EVs) respectively. The first strategy includes the use of new culturing protocols to maintain hepatocytes viability and vitality for extensive time periods. To this end, different compound solutions are added to the culture medium, which demonstrate stimulatory effects and increase hepatic progenitor cells (HPC) replication capabilities [11]. For the liver tissue-constructs we report on the use of cocktail medium, which is based on branched chain amino acids, that seems to support the hepatocytes/liver organoid functionality in vitro (Fig. 2c and d).

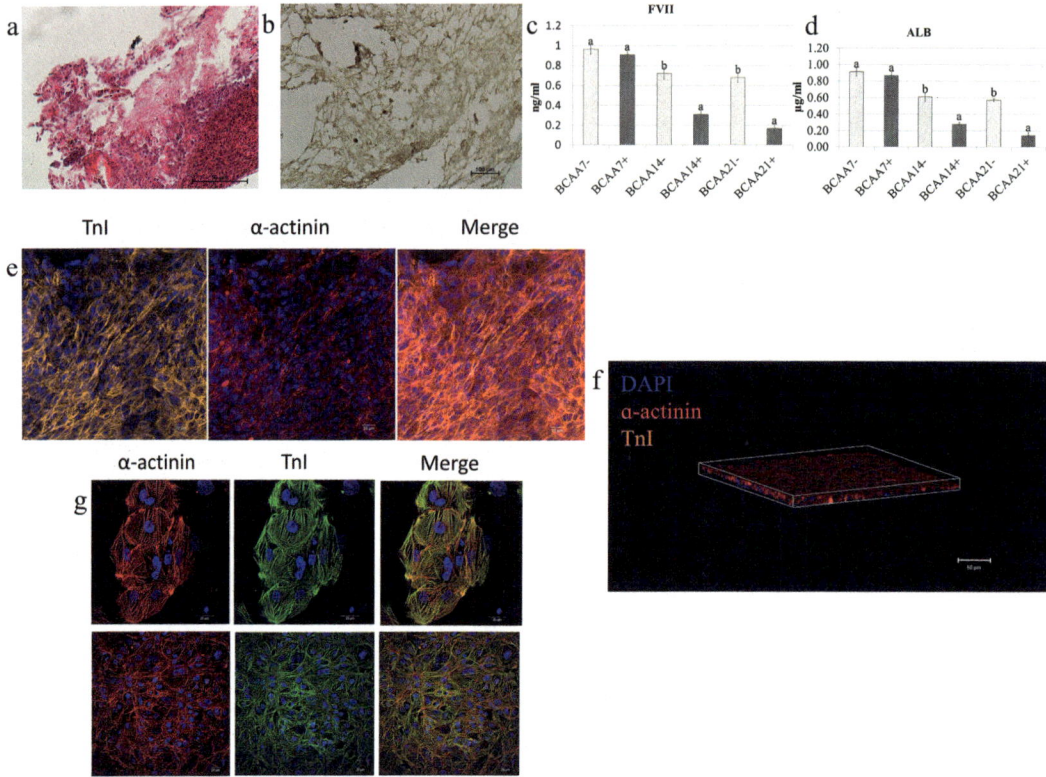

Fig. 2 Scaffold recellularization and tissue characterization. (**a**) Hematoxylin and Eosin (H&E) staining of the liver patch. (**b**) Immunohistochemical staining of the hepatic tile for albumin (ALB) in order to characterize the seeded cell population. (**c** and **d**) Tissue culturing of the liver patch with the addition of BCAAs (BCAA+) significantly decreases the release of the ALB and factor VII (FVII) proteins compared to the standard medium (BCAA−) at 14 and 21 day of culture. (**e** and **f**) Representative confocal images depicting the cardiomyocytes cultured within the EHS, which were immunostained for human α-actinin (red) human troponin I (TnI. orange) and DAPI (blue nuclei). (**g**) Morphology of the hiPSC-CMs cultured as monolayer as controls. Notice the different morphology of these cardiomyocytes (round) compared to the cells cultured within the scaffold that are markedly more elongated

For the cardiac patch model, in order to improve graft survival following tissue transplantation, we enrich the scafold's ECM with EVs to utilize their potential antiapoptotic, antifibrotic, and proangiogenic properties (Fig. 4). Extracellular vesicles (EVs) is a collective term for vesicles secreted or shed by cells, which are formed from the outer phospholipid bilayer and contain the specific cells' cytoplasmatic contents [12]. All examined prokaryotic and eukaryotic cells release EVs [13, 14] and they can be found in surrounding media in cell culture, or in body fluids such as blood [13]. EVs can carry different molecules in their payload: proteins, lipids, DNA, RNAs—most notably micro RNAs—and more [13, 15]. The process of vesicle formation and its cargo content is rather selective than passive [14]. Hence, It has been suggested that EVs can

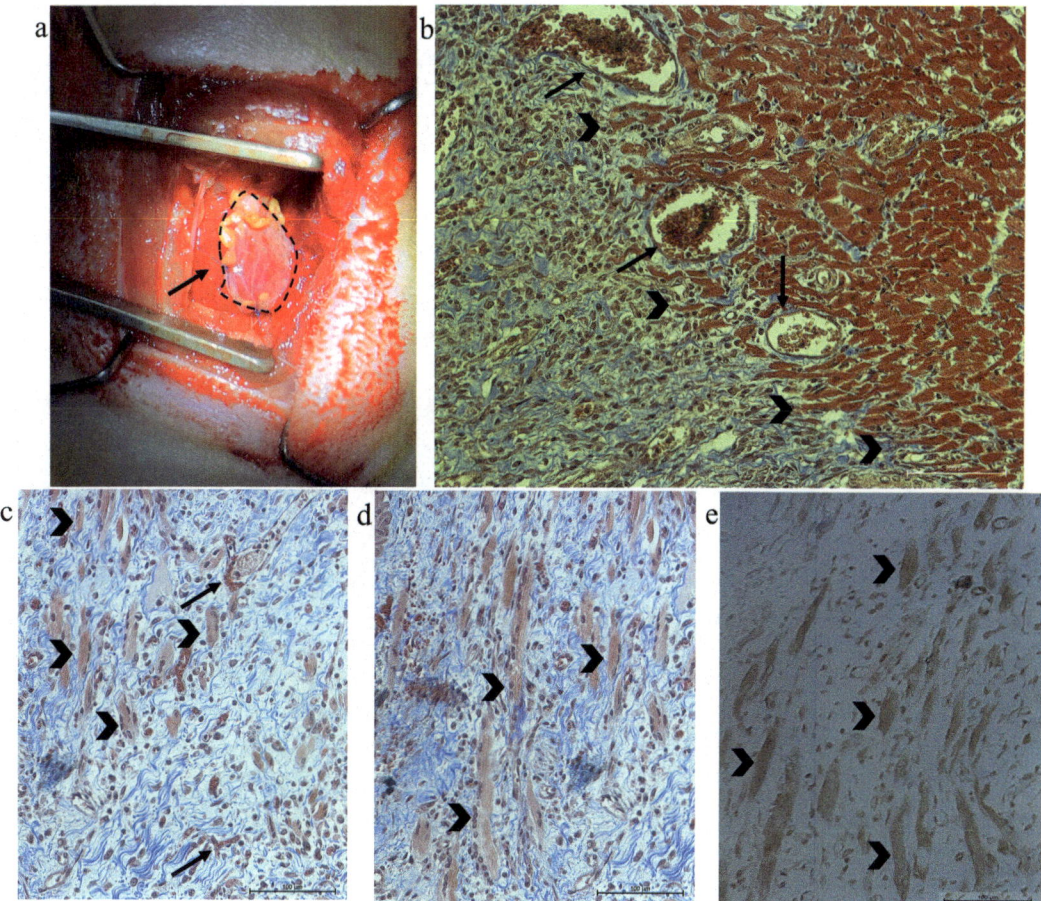

Fig. 3 Transplantation of the cardiac patch. (**a**) Representative photograph of an implanted 1 week old cardiopatch (arrow) sutured on the anterior ventricle wall of the rat heart. (**b**) Masson's trichrome staining of the transplanted recellularized patch after 4 weeks. This image focuses on the border between the patch and the heart where neovascularization (arrows) and rat cardiomyocytes' invasion (arrow heads) can be appreciated. (**c**) Masson's trichrome staining focusing on the center of the patch where relatively mature hiPSC-CMs (arrows) and blood vessels (arrow heads) are visible. (**d**) Immunohistochemical staining of the patch for human cardiac troponin I (TnI), positive cells are spotted (arrows)

mediate intercellular communication and serve as potential therapeutic agents [13]. Recent Studies demonstrated the therapeutic effect of EVs from different cell sources in an animal model of myocardial infarction [16]. Furthermore, the feasibility of vesicular entrapment in and slow release from engineered tissues was also demonstrated [17]. This may serve as a method to enrich engineered tissues, creating a fertile ground for recellularization. EVs are poorly immunogenic, a feature that may facilitate their use in regenerative therapies [18].

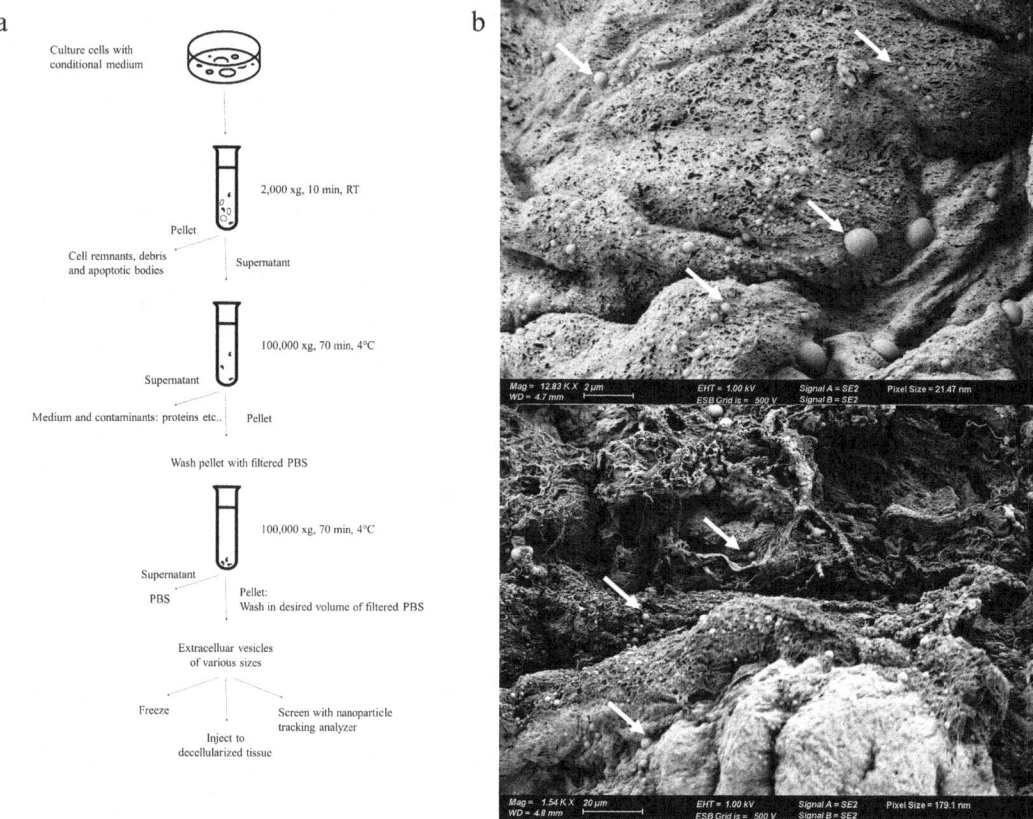

Fig. 4 EVs isolation and supplementation to the cardiac patch. (**a**) Scheme describing the process used to produce and isolate EVs. (**b**) SEM images of the ECM-EVs enriched cardiac patch where the EVs (arrows) can be identified as the spherical heterogeneous structures on the ECM surface

2   Materials

2.1   Liver Decellularization

2.1.1   Tissue Retrieval

1. Ice container.
2. Chlorhexidine skin disinfectant.
3. Surgical gaze.
4. Surgical blade handle # 3.
5. Surgical blades #15.
6. Metzenbaum scissor.
7. Surgical tweezers/Watchmaker's forceps.
8. 100 mm petri dish.

2.1.2   ELS

1. Optimal cutting temperature compound (OCT).
2. Cryomolds.
3. Cryostat.
4. 6 wells dish for cells culture.

5. Distilled water (DW).

6. 1% sodium dodecyl sulfate (SDS) solution in DW.

7. 1% Triton X-100 solution in DW.

8. Dulbecco's Phosphate Buffer Saline solution (PBS).

9. Orbital shaker.

10. Sterilized glass coverslip 12 mm.

11. 70% ethanol.

12. 1% penicillin/streptomycin solution (P/S).

2.1.3 Liver Patch

1. Watchmaker's forceps.

2. −80 °C freezer.

3. 0.05% ethylenediaminetetraacetic acid (EDTA) solution in DW.

4. 0.1% ammonium hydroxide in DW.

5. 3% Triton X-100 in DW.

6. PBS.

7. 1% P/S.

2.2 Liver Recellularization

2.2.1 ELS

1. Surgical blade # 15.

2. Trypsin–EDTA solution.

3. 40 μm cell strainer.

4. Centrifuge.

5. Pipette single channel, manually adjustable, 1000 μL.

6. Sterile tips for pipette single channel, manually adjustable, 1000 μL.

7. Culture medium for the liver organoid (CMLO): Dulbecco's Modified Eagle Medium/Nutrient Mixture F-12 (DMEM/12) supplemented with 20% fetal bovine serum (FBS), 2% antibiotic–antimycotic solution, 2 mM L-glutamine solution, 87.4 μg/mL Hepatic Growth Factor.

8. Albumin (ALB) ELISA kit.

9. Factor 7 (FVII) ELISA kit.

2.2.2 Liver Patch

1. 70% ethanol.

2. PBS.

3. Pipette single channel manual adjustable 200 μL.

4. Sterile tips for Pipette single channel manual adjustable 200 μL.

2.2.3 Liver Enriched Medium

1. Branched Chain Amino Acid (BCAA+) medium: CMLO supplemented with 25 mM Leucine, 12.5 mM Isoleucine, 12.5 mM Valine.

2.3 Heart Decellularization	1. Chlorhexidine skin disinfectant.
	2. Surgical gaze.
2.3.1 Tissue Retrieval	3. Surgical blade handle # 3.
	4. Surgical blades #15.
	5. Metzenbaum scissor.
	6. Surgical tweezers/Watchmaker's forceps.
	7. 100 mm petri dish.

2.3.2 EHS

1. Optimal cutting temperature compound (OCT).
2. Cryomolds.
3. Cryostat.
4. 6 wells dish for cells culture.
5. Distilled water (DW).
6. 1% sodium dodecyl sulfate (SDS) solution in DW.
7. 1% Triton X-100 solution in DW.
8. Dulbecco's Phosphate Buffer Saline solution (PBS).
9. Orbital shaker.
10. Sterilized glass coverslip 5 mm.
11. 70% ethanol.
12. 1% penicillin/streptomycin solution (P/S).

2.3.3 Cardiac Patch

1. Potts scissors.
2. −80 °C freezer.
3. 0.05% ethylenediaminetetraacetic acid (EDTA) solution in DW.
4. 0.1% ammonium hydroxide in DW.
5. 3% Triton X-100 in DW.
6. PBS.
7. 1% P/S.
8. Rotamix RM 1.

2.4 Heart Recellularization

2.4.1 EHS

1. TrypLE Express solution.
2. DNase I, Bovine pancreas 10 mg/mL in water.
3. Pipette single channel manual adjustable 1000 μL.
4. Sterile tips for pipette single channel, manually adjustable, 1000 μL.
5. 15 mL conical tube
6. RPMI 1640 Medium.
7. B27 supplement minus insulin.
8. 2 μM Thiazovivin.

2.4.2   Cardiac Patch

1. 70% ethanol.
2. PBS.
3. Sylgard® 184 Silicone Elastomer kit.
4. Stainless minutien pins 10 mm base × tip 0.
5. RPMI 1640 Medium.
6. B27 supplement.

2.4.3   EVs Enrichment of the Cardiac Patch

1. Centrifuge that can reach $100,000 \times g$ at 4 °C.
2. Filtered PBS (0.022 μm).
3. Nanoparticles tracking analysis system.
4. 30G needle.
5. 100 μL syringe.

2.5   In Vivo Transplantation

2.5.1   Liver Patch Transplantation

1. Cyclosporine.
2. Methylprednisolone.
3. Buprenorphine.
4. Carprofen.
5. Isoflurane.
6. Isoflurane vaporizer.
7. Piperacillin/tazobactam.
8. 0.9% Saline Solution.
9. 21 G needle.
10. 10 mL syringe.
11. Chlorhexidine skin disinfectant.
12. Surgical gaze.
13. Surgical blade handle # 3.
14. Surgical blades #15.
15. Metzenbaum scissor.
16. Surgical tweezers/Watchmaker's forceps.
17. PROLENE® 7-0 suture.
18. Nylon 4-0 suture.

2.5.2   Cardiac Patch Engraftment

1. Cyclosporine.
2. Methylprednisolone.
3. Buprenorphine.
4. Carprofen.
5. Isoflurane.
6. Isoflurane vaporizer.
7. Piperacillin/tazobactam.

8. 0.9% Saline Solution.

9. 21 G needle.

10. 10 mL syringe.

11. Chlorhexidine skin disinfectant.

12. Surgical gaze.

13. Surgical blade handle # 3.

14. Surgical blades #15.

15. Metzenbaum scissor.

16. Surgical tweezers/Watchmaker's forceps.

17. Electric shaver.

18. 18 G intravenous cannula.

19. Rodent ventilator.

20. Finocchietto rib spreader.

21. PROLENE® 8-0 suture.

22. PROLENE® 3-0 suture.

23. Nylon 5-0 suture.

3   Methods

All procedures are carried out according to the 3R rules of animal experimentation.

3.1   Liver Decellularization

3.1.1   Tissue Retrieval

1. Collect adult female New England white rabbit 3 kg not eviscerated carcasses from a local slaughterhouse, carrying it with a certified ice container.

2. Before the incision, treat the abdominal wall muscles with topical chlorhexidine skin disinfectant alternating with chlorhexidine-soaked gauze and dry gauze.

3. Perform a ventral midline laparotomy extending from the pubis to the xyphoid combined with a rooftop incision using the #3 handle with #15 blade.

4. Dissect the falciform ligament with a Metzenbaum scissor. Use a wet gauze bandage to hold the medial and left liver lobes cranially under the dome of the diaphragm.

5. Cut the bile duct 1.5 cm from the bile duct bifurcation.

6. Dissect the portal vein and hepatic artery from the surrounding tissue.

7. Transect the infrahepatic inferior vena cava.

8. Free the suprahepatic inferior vena cava.

9. Transfer the whole liver to a sterile 100 mm petri dish.

3.1.2 ELS

1. Select and cut a proper area from the liver parenchyma (*see* **Note 1**).

2. Place few drops of OCT onto the center of the bottom of cryomold. Be careful to select the proper size embedding mold according to the size of the tissues to be embedded.

3. Place the unfrozen tissue sample in the drops and oriented. Make sure that the side touching the bottom of the cryomold is the side you want sectioned first. Gently push the tissue with a forceps to ensure that the bottom surface of the tissue is placed properly, level with the container, touching the face of the bottom and the tissue is located in the center of the mold.

4. Carefully drop more OCT onto the specimen until it is completely covered. None of the tissue should remain exposed (*see* **Note 2**).

5. Let it settle for 15–30 s to allow the OCT to completely wet the surface of the tissue.

6. Place cryomold with OCT covered sample in it in the vapor phase right next to the liquid nitrogen with the flat side down using a long forceps.

7. After hardening of the OCT compound (it will happen in 0.5–1 min), store the samples at −80 °C (months) or directly transter it to the cryostat stage for cutting.

8. Slice the speciment with the cryostat ensuring that all the tissue is present and intact (*see* **Note 3**).

9. Transfer the slices in a 6 well dish filled with 2 mL DW for each well and let the OCT dissolve completely after that wash two times with DW.

10. Then treated with 1 mL of each of the following solutions (diluted in DW) while being rotated at 150 rpm on an orbital shaker: 2 washes of 1% SDS for 2 h each, 1 wash of DW for 15 min, 1 wash of 1% Triton X-100 for 10 min, and 3 washes of PBS for 15 min each.

11. Leave the decellularized slices in PBS to rotate 80 rpm on an orbital shaker overnight (*see* **Note 4**).

12. Sterilized glass coverslips 12 mm in diameter sterilized by rinsing in 70% ethanol and exposing to UV for 10 min.

13. Spread the ELS on the coverslips with the outer perimeter of the slice hooked around the edges of the coverslip and place them in a standard 24-well.

14. If required, the decellularized tissue could be store at 4 °C in PBS containing 1% P/S.

3.1.3 Liver Patch

1. Section a liver lobe with a scalpel and obtain a small tile with a volume of 125 cm^3 (*see* **Note 5**).

2. Gently grab with a micro dissecting Watchmaker's forceps the Glissonian capsule and detach it from the parenchymatous portion of the liver.

3. Store the tile at −80 °C for at least 24 h.

4. Thaw the liver specimen at room temperature (RT) avoiding any drying of the tissue.

5. Then start the chemical decellularization phase in the following succession of solution where the tissue is always place in a 50 mL tube on a laboratory shake: 0.05% EDTA in DW (24 h); 1% SDS/0.05% EDTA in DW (24 h); 0.1% ammonium hydroxide in DW (24 h); 3% Triton-X/0.05% EDTA in DW (24 h); DW (24 h); DW two time for 30 min; PBS overnight (*see* **Note 6**).

6. If required, the decellularized tissue could be store at 4 °C in PBS containing 1% penicillin/streptomicin.

3.2 Liver Recellularization

Authors describe the recellularization process for the liver with a primary cells line not purified as a rabbit cells allotransplantation.

3.2.1 ELS

1. Establish primary livers cells co-culture from fresh biopsies.

2. Cut liver fragments of approximately 200 mm^3 wiht a surgical blade.

3. Incubate the liver slivers in 30 mL of Trypsin-EDTA solution for 30 min at 37 °C with gentle shaking.

4. Filtered the digested tissue with 40 μm cell strainer and collect the cells throught centrifugation at 150 × g for 5 min.

5. Removed the supernatant and add the CMLO medium according to the size of the pellet (*see* **Note 7**).

6. Create the ELS slices plating at a density of 0.5–1 million cells/cm^2.

7. Add the specific culture medium.

8. Let the cells and medium solution dry/attach to the matrix for at least 6 h.

9. ELS were maintained in culture for at least 7 days before evaluation by ALB and factor 7 FVII production as markers of liver organoid fuctionality.

3.2.2 Liver Patch

1. Before any manipulation for cells seeding, sterilized the patch ECM in 70% EtOH for 30 min than wash it in PBS for 30 min and finally under UV for 10 min in a biosafety cabinet.

2. Put the liver ECM in a 35 mm petri dish let it dry for 2–5 min (*see* **Note 8**).

3. Follow the **steps 1–5** in Subheading 3.2.1.

4. After collecting the cells pellet, plate 1×10^6 cells directly onto the decellularized bioscaffold surface as a single drop (100–150 μL).

5. Let the cells and medium solution dry/attach to the matrix for at least 6 h.

6. Add slowly the culture medium avoiding cells washing.

7. Change the medium every 3 days.

3.2.3 Liver Enriched Medium

1. Add to the ELS or liver patch after the 6 h incubation with the isolated cells the BCCA+ medium.

2. Starve the ELS or liver patch for 12 h in PBS and measure in it the ALB and FVII productions.

3.3 Heart Decellularization

3.3.1 Tissue Retrieval

1. Lift the skin away from the abdominal cavity with forceps and then use scissors to incise the peritoneal cavity, following the curve of the diaphragm back to the posterior angle of the ribs.

2. Once the diaphragm is visible, using small scissors, cut along the anterior surface of the diaphragm following the direction of the prior cuts to allow for entry into the thorax. Extend each cut along the axillary line bilaterally to the axilla.

3. Retract the ribcage anteriorly from the xiphoid process using forceps. Incise the pericardium and pleura.

4. Identify the inferior vena cava (IVC) and aorta just above the diaphragm and retract them *en bloc* anteriorly using blunt forceps.

5. Using large, curved scissors rapidly make an incision across the IVC and the aorta, pulling the heart and lungs out of the chest *en bloc*. Cut the esophagus, trachea, brachiocephalic arteries and veins cephalad to remove the heart and lungs from the thorax. Excise the thymic tissue with this block of tissue.

3.3.2 EHS

1. Select and cut a proper area from the heart.

2. Follow the same steps reported in Subheading 3.1.2, **steps 2–13**.

3.3.3 Cardiac Patch

1. Isolate the ventricles from the atria trough a transversal section at the level of the heart base, then divide the left and right ventricles and keep the septum with the right portion.

2. Store the heart tissue for at least 24 h in a −80 °C freezer.

3. Thaw the heart tissue at RT avoind an excessive drying (*see* **Note 9**).

4. Wash three times 15 min each the ventricles in a 50 mL tube in DW with a laboratory shaker (*see* **Note 10**).

5. Then start the chemical decellularization phase in the following succession of solution where the tissue is always place in a 50 mL tube on a laboratory shake: 0.05% ethylenediaminetetraacetic acid (EDTA) in DW (24 h); 1% sodium dodecyl sulfate (SDS)/0.05% EDTA in DW (24 h); 0.1% ammonium hydroxide in DW (24 h); 3% Triton-X/0.05% EDTA in DW (24 h); DW (24 h); DW two time for 30 min; PBS overnight (*see* **Note 11**).

6. If required, the decellularized tissue could be store at 4 °C in PBS containing 1% penicillin/streptomicin.

3.4 Heart Recellularization

For the heart recellularization we used human induce pluripotent stem cells derived cardiomyocytes (hiPSCs-CM) differentiated using previously described protocols (*see* **Note 12**) [18].

3.4.1 EHS

1. On the day 10–12 of differentiation, dissociate the hiPSC-CM when the confluence and amount of hiPSC-CM positively beating reach 70% and 90% respectively.

2. Aspirate medium from the culture-well, and wash cell monolayer with 2 mL of PBS. Aspirate PBS and replace with 1 mL of TrypLE supplemented with 1 µL/mL DNase (1:1000 dilution). Ensure complete coverage of cell monolayer with TrypLE.

3. Place plate at 37 °C incubator for 3–10 min (*see* **Note 13**).

4. Triturate the cells with a P1000 tip to dislodge them from the plate.

5. Add the cells with the TrypLE to a 15-mL conical tube with 5 mL prewarmed complete medium for well (at least 1:2 dilution) supplemented with 1 µL/mL DNase.

6. Centrifuge the cells at 1100 rpm (150 × g) for 5 min.

7. Discard supernatant and resuspend cell pellet in 60 µL volume of RPMI/B27 medium, supplemented with 2 µM Thiazovivin (1:1000 dilution).

8. Plated on EHS slices at a density of 2 million cells/cm^2.

9. EHS were maintained in culture for 7–200 days before evaluation by optical mapping or contraction measurements (*see* Electronic Supplementary Material videos 1, 2, 3, available on this chapter's page on link.springer.com).

3.4.2 Cardiac Patch

1. Before any manipulation for cells seeding, sterilized the patch ECM in 70% EtOH for 30 min than wash it in PBS for 30 min and finally under UV for 10 min in a biosafety cabinet.

2. Localize the ventricle patch in a glass dish coated with silicone (*see* **Note 14**) and solidarized the ECM to the bottom with sterilizable thin pins (*see* **Note 15**).

3. Add the specific RPMI/B27 medium to the plate and let the patch absorb part of it in the incubator (T 37 °C, 5% CO_2) overnight.

4. Add to the plate the iPS-CM previously collected as cells aggregates and keep it in the incubator (T 37 °C, 5% CO_2) overnight (*see* **Note 16**).

5. Carefully and slowly, fulfill the plate with the RPMI/B27 plus insulin medium and keep the patch in the incubator (T 37 °C, 5% CO_2).

6. Change the medium every other day, according to the phenol red color deviation.

3.4.3 Patch EVs Enriched

1. Stimulate EVs production in cultured cells with a conditional medium.

2. Collect cells medium after 2 days of culture and centrifuge in 2000 × g for 10 min at room temperature to get rid of cell remnants and debris.

3. Collect supernatant, centrifuge in 100,000 × g for 70 min at 4 °C.

4. Get rid of supernatant, wash with filtered PBS and centrifuge in 100,000 × g for 70 min at 4 °C.

5. Get rid of supernatant, collect the pellet of EVs with the desired volume of filtered PBS.

6. Use freeze or screen sample with nanoparticles tracking analysis system to determine vesicle concentration.

7. Inject using 30 gauge needle into decellularized patch transplanted to the heart.

3.5 Transplantation

Animals received humane care in compliance with the Guide for the Principles of Laboratory Animals. The responsible local authority approved all animal protocols.

1. The days before the operation inject subcutaneous 15 mg/kg ciclosporin and 4 mg/kg methylprednisolone, then every 24 h 15 mg/kg ciclosporin and 2 mg/kg methylprednisolone.

2. Anesthetize the New Zealand White Rabbit weighing 4–4.5 kg with isoflurane on cone mask (3% for the induction, 2% during the operation, 1 l/min air flow, FiO_2 70%) and keep on a warm pad.

3. Inject 10 mL of 0.9% saline solution subcutaneously and 0.6 g of piperacillin/tazobactam intramuscular before laparotomy with a 21 G needle and a 10 mL syringe after induction of anesthesia.

4. Perform a midline xiphopubic laparotomy.

5. Reverse the xiphoid process with an autostatic forceps in order to better expose the liver.

6. Dislocate the left liver lobe and surrounding it with a wet surgical gauze.

7. Remove a segment from the left lateral lobe of the normal liver using a surgical trocar with inner diameter 1 cm.

8. Transplant the recellularized liver patch in the site of the detached part using PROLENE® 7-0 suture.

9. Close the animal by layers with Nylon 4-0 sutures. Allow free water and food from wakening.

10. Administer effective analgesia according to local institutional protocols.

3.5.1 Cardiac Patch Transplantation

1. The day before the operation inject subcutaneous 15 mg/kg ciclosporin and 4 mg/kg methylprednisolone. Then every 24 h 15 mg/kg ciclosporin and 2 mg/kg methylprednisolone.

2. Place the Wistar rat weighing 300–400 g in an induction chamber and anesthetize the animal with isoflurane check the depth of anesthesia by the lack of response to the toe-pinch.

3. Inject 0.05 mg/kg buprenorphine subcutaneously with a 27 G needle and a 10 mL syringe after induction of anesthesia.

4. Inject 10 mL of 0.9% saline solution subcutaneously and 0.03 g of piperacillin/tazobactam intramuscular before laparotomy with a 21 G needle and a 10 mL syringe after induction of anesthesia.

5. Place the rat on its back and keep anesthesia with a facemask covering mouth and nose.

6. Spread the rat's legs and fix the position using tape.

7. Shave the chest of the anesthetized animal with an electric shaver. Disinfect the area widely using iodine-based scrub, followed by 70% ethanol. Repeat this disinfection steps three times.

8. Perform an intubation with 18 G intravenous cannula and insert the flexible part of the cannula as a tracheal tube.

9. Connect the tracheal tube to an animal respirator to continuously ventilate the rat during the procedure.

10. Perform a 2 cm horizontal skin incision in the scar area of the left lateral side using scissors and tweezers.

11. Carefully open the pleural space with scissors. Insert a rib spreader to expose the heart.

12. Visually identify the region of interest where the heart patch will be suture over.

13. Secure it with four PROLENE® 8-0 sutures at all sides (ventral, dorsal, cranial and caudal). Inflate the lungs with pressure, to avoid atelectasis of the lung. Remove the retractor from the intercostal space.

14. Close the ribs with two 3-0 sutures. Close the muscles over the ribs with a 4-0 running suture. For closure of the skin use 5-0 suture single stitches.

15. Reduce the isoflurane to 1%. When the animal is breathing spontaneously, remove the tracheal tube, and continue to administer 100% O_2.

16. Use buprenorphine (0.05 mg/kg per 12 h) for pain medication for the following 5 days.

17. For Evs injection is preferable to use a Hamilton's needle 30G where a proper low volume.

4   Notes

1. For the ELS creation, the cryostat sectioning will control the thickness. The area and extension of the specimen is chose according by the type of the experiment and the size of the cryomold.

2. Try to avoid the formation of air bubbles. Remove any bubbles inside the OCT. This is important because the air bubbles will create problems when cutting sections.

3. A 100 μm thickness is suggested to obtain a durable and flexible ECM (see **Note 1**).

4. If the ELS (or EHS) are stored for up to 2 months, place the slides in a 1% penicillin/streptomycin and PBS solution.

5. To obtain the tile also a microtome blade or a hollow hole square punch 5 mm cutter steel tool leather belt watch band could be utilized.

6. To prepare the decellularization solutions mix with a magnetic stirrer at RT all the detergent in DW and The laboratory shaker used by the authors is Rotamix RM 1, ELMI, and the set up suggested is 50 RPM function #5.

7. Consider the proper volume of seeding, for one ELS with $0.5–1 \times 10^6$ cells/cm^2 not exceed 150 μL of culture medium for the liver organoid.

8. This step could be critical, pay attention not to dry to much the tissue. Using a plastic culture dishes with any coating and let the bottom of the scaffold dry for few minutes will help the patch to adhere to the plate.

9. Check the tissue every 15 min to be sure that the thawing phase is not too long.

10. The laboratory shaker used by the authors is Rotamix RM 1, ELMI, and the set up suggested is 50 RPM #five.

11. To prepare the decellularization solutions mix with a magnetic stirrer at RT all the detergent in DW.

12. In the present manuscript, the procedure to proliferate and differentiate the hiPS-CM lines are not describe because is not concerning the aim of the methodology paper.

13. Depending on the age of the cell culture.

14. Glass petri dishes 60 mm × 20 mm.

15. Stainless steel minutien pins 10 mm base × tip 0.

16. Each cell aggregates could be formed by 0.5–3 × 10^6 cells. At this step, the cells could be added directly in the empty medium plate or proceed in a double phase seeding. In the former case, the aggregates are lain in the plate medium, push them on the top of the matrix and then the medium is gently sucked from the plate.

Acknowledgments

GM, PG, and BTAL are members of the COST Action CA16119 In vitro 3-D total cell guidance and fitness (CellFit).

References

1. Terzic A, Pfenning MA, Gores GJ et al (2015) Regenerative medicine build-out. Stem Cells Transl Med 4(12):1373–1379

2. Ginsburg GS, Phillips KA (2018) Precision medicine: from science to value. Health Aff (Millwood) 37(5):694–701

3. Gilbert TW, Sellaro TL, Badylak SF (2006) Decellularization of tissues and organs. Biomaterials 27(19):3675–3683

4. Tong C, Li C, Xie B et al (2019) Generation of bioartificial hearts using decellularized scaffolds and mixed cells. Biomed Eng Online 18 (1):71

5. Modulevsky DJ, Lefebvre C, Haase K et al (2014) Apple derived cellulose scaffolds for 3D mammalian cell culture. PLoS One 9(5): e97835

6. Hillebrandt KH, Everwien H, Haep N et al (2019) Strategies based on organ decellularization and recellularization. Transpl Int 32 (6):571–585

7. Moroni F, Mirabella T (2014) Decellularized matrices for cardiovascular tissue engineering. Am J Stem Cells 3(1):1–20

8. Lu TY, Lin B, Kim J et al (2013) Repopulation of decellularized mouse heart with human induced pluripotent stem cell-derived cardiovascular progenitor cells. Nat Commun 4:2307

9. Palakkan AA, Nanda J, Ross JA (2017) Pluripotent stem cells to hepatocytes, the journey so far. Biomed Rep 6(4):367–373

10. Jaramillo M, Yeh H, Yarmush ML et al (2018) Decellularized human liver extracellular matrix (hDLM)-mediated hepatic differentiation of human induced pluripotent stem cells (hIPSCs). J Tissue Eng Regen Med 12(4): e1962–e1973

11. Chen L, Zhang J, Yang L et al (2018) The effects of conditioned medium derived from mesenchymal stem cells Cocultured with hepatocytes on damaged hepatocytes and acute liver failure in rats. Stem Cells Int 2018:9156560

12. van der Pol E, Coumans F, Varga Z et al (2013) Innovation in detection of microparticles and exosomes. J Thromb Haemost 11(1):36–45

13. Ibrahim A, Marb E, Marbán E (2016) Exosomes: fundamental biology and roles in cardiovascular physiology. Annu Rev Physiol 78:1–17

14. Fleury A, Martinez MC, Le Lay S (2014) Extracellular vesicles as therapeutic tools in cardiovascular diseases. Front Immunol 5:1–8

15. Chistiakov DA, Orekhov AN, Bobryshevy YV (2016) Cardiac extracellular vesicles in normal and infarcted heart. Int J Mol Sci 17(1):1–18

16. El Harane N, Kervadec A, Bellamy V et al (2018) Acellular therapeutic approach for heart failure: in vitro production of extracellular vesicles from human cardiovascular progenitors. Eur Heart J 39(20):1835–1847

17. Liu B, Lee BW, Nakanishi K et al (2018) Cardiac recovery via extended cell-free delivery of extracellular vesicles secreted by cardiomyocytes derived from induced pluripotent stem cells. Nat Biomed Eng 2(5):293–303

18. Burridge PW, Matsa E, Shukla P et al (2014) Chemically defined generation of human cardiomyocytes. Nat Methods 11(8):855–860

Chapter 8

Use of Virus-Mimicking Nanoparticles to Investigate Early Infection Events in Upper Airway 3D Models

Georgia Pennarossa, Alireza Fazeli, Sergio Ledda, Fulvio Gandolfi, and Tiziana A.L. Brevini

Abstract

The current coronavirus disease-19 (COVID-19) pandemic, caused by "severe acute respiratory syndrome coronavirus 2" (SARS-CoV-2), underscores the threat posed by newly emerging viruses. The understanding of the mechanisms driving early infection events, that are crucial for the exponential spread of the disease, is mandatory and can be significantly implemented generating 3D in vitro models as experimental platforms to investigate the infection substrates and how the virus invades and ravages the tissues.

We here describe a protocol for the creation of a synthetic hydrogel-based 3D culture system that mimics in vitro the complex architectures and mechanical cues distinctive of the upper airway epithelia. We then expose the in vitro generated 3D nasal and tracheal epithelia to gold nanoparticles (AuNPs) that display the typical shape and size distinctive of SARS-CoV-2 and of the majority of Coronaviridae presently known.

The infection platform here described provides an efficient and highly physiological in vitro model that reproduces the host–pathogen early interactions, using virus-mimicking nanoparticles, and offers a flexible tool to study virus entry into the cell. At the same time, it reduces the risk of accidental infection/spillovers for researchers, which represents a crucial aspect when dealing with a virus that is highly contagious, virulent, and even deadly.

Key words Coronaviridae, Covid-19 pandemic, Hydrogel-based culture system, Gold nanoparticles, SARS-CoV-2, Upper respiratory tract

1 Introduction

The current coronavirus outbreak caused by "severe acute respiratory syndrome coronavirus 2" (SARS-CoV-2) is the gravest health crisis the World has seen in a century, with an unprecedent death toll and number of patients in urgent need of hospital treatment [1]. According to the World Health Organization, the situation report update at 08 July 2020 is 11,635,939 of total confirmed cases and 539,026 the deaths, with the USA and Europe as epicenters of the global pandemic, and 6,004,685 and 2,809,848 confirmed cases respectively [2].

Tiziana A.L. Brevini et al. (eds.), *Next Generation Culture Platforms for Reliable In Vitro Models: Methods and Protocols*, Methods in Molecular Biology, vol. 2273, https://doi.org/10.1007/978-1-0716-1246-0_8,
© The Author(s), under exclusive license to Springer Science+Business Media, LLC, part of Springer Nature 2021

Although, SARS-CoV-2 has a relatively low/medium mortality rate, it is highly contagious, leading to an explosive and exponential spread of the infection [3]. The understanding of the first events controlling the mechanisms that lead to the rapid spread of the disease are therefore crucial for its containment and the early detection [4]. Typical in vitro experiments carried out by virologists use cell lines cultured into 2D culture systems [5], that are only poorly resemble real tissue behavior, especially in terms of structural and functional proprieties, nor its interactions with SARS-CoV-2. 3D in vitro models have been currently developed to better investigated the in vivo infection substrates and allow for studying how the new coronavirus invades and ravages the body.

We here describe a protocol for the generation of reliable 3D in vitro models of upper respiratory tract, namely, nose and trachea. We propose the use of a synthetic hydrogel-based 3D culture system that is able to reproduce in vitro the complex biochemical and mechanical cues deriving from the native extracellular matrix [6, 7] as well as several aspects of the native cellular microenvironment, mimicking the original conditions of the airway epithelia [8].

To stimulate the mechanisms active at early stages of infection—prior to viral replication—and allows to study how the virus interacts and invades cells of the upper respiratory tract, we implement a nanomedicine-based approach, exposing the 3D in vitro generated nasal and tracheal epithelia to gold nanoparticles (AuNPs) (Fig. 1). In particular, we select AuNPs spherical in shape and with size ranging from 60 to 140 nm, which are distinctive features of SARS-CoV-2 and common to the majority of Coronaviridae presently known [9–10]. In addition, in order to address the different mechanisms used by Coronaviridae to enter cells, the same protocol can be applied using nanoparticles with surface ad hoc activated or caged with coronavirus envelope spike proteins [11, 12].

Overall, the bioengineered model of infection here described provides an efficient and highly physiological in vitro platform that mimics the host–pathogen early interaction and offers a powerful insight on the possible strategies to prevent virus entry into the cell (Fig. 2). In addition, beside its predictive power and flexibility, it also ensures safety. This latter represents a crucial aspect, since dealing with a virus that is highly contagious, virulent and even deadly, inevitably reduces the manpower and the number of laboratories that have high protection facilities, significantly affecting in a way the intellectual and scientific resources to fight this as well as other future pandemics. Finally, the 3D cell-based assay here proposed could contribute to reduce the use of experimental animal studies, the uncertainties arising from monolayer cultures and hence the cost of subsequent screening processes.

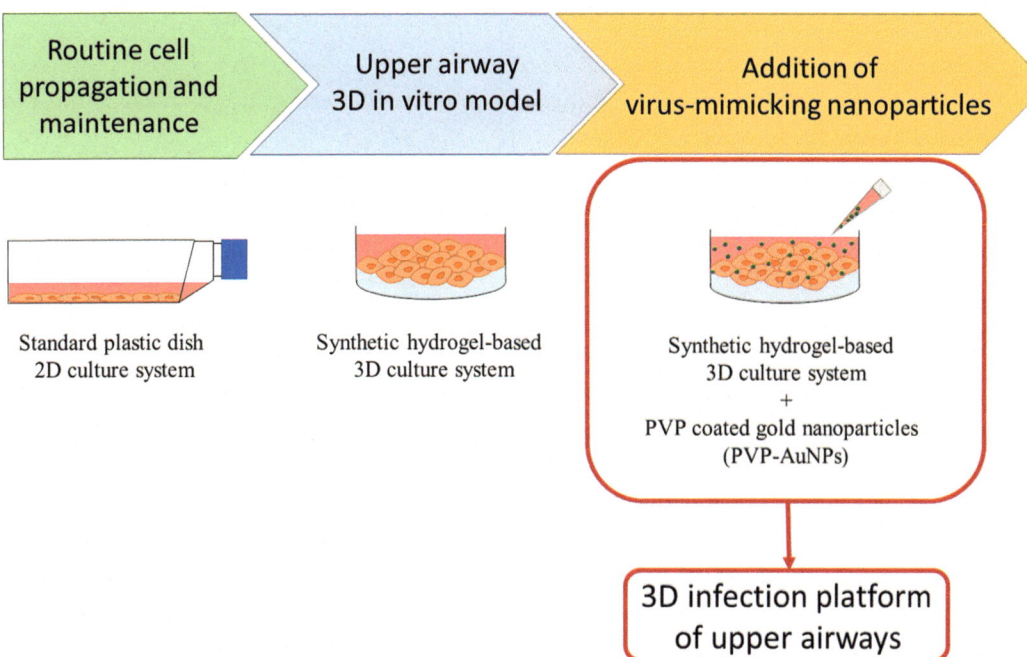

Fig. 1 Schematic representation of the protocol used to generate a 3D infection platform of the upper airways

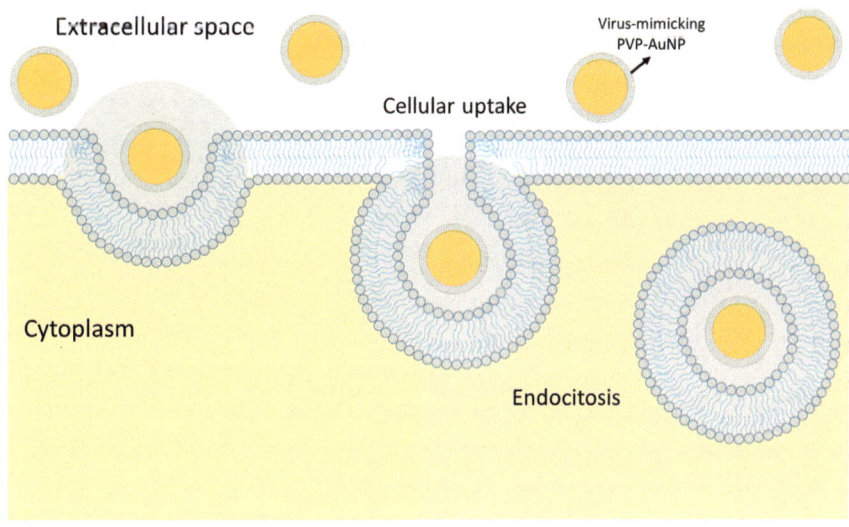

Fig. 2 An overview of virus-mimicking nanoparticle uptake by target cells

2   Materials

Prepare all solutions immediately before use (unless indicated otherwise).

2.1   Cell Lines

1. Human nasal epithelial cells (HNEpCs).
2. Human tracheal epithelial cells (HTEpCs).

2.2   HNEpC and HTEpC 2D Culture

1. Water bath.
2. Culture dish.
3. CO_2 incubator.
4. Inverted microscope.
5. Dulbecco's phosphate-buffered saline (PBS): dissolve 8 g of NaCl (137 mM), 200 mg of KCl (2.7 mM), 1.44 g of Na_2HPO_4 (8 mM), and 240 mg of KH_2PO_4 (2 mM) in 800 mL of distilled water. Adjust pH to 7.4. Add distilled water until volume is 1 L. Sterilize solution with autoclave and store at +4 °C.
6. Antibiotic–Antimycotic Solution.
7. Trypsin–EDTA solution: dissolve 0.5 g of porcine trypsin and 0.2 g of EDTA 4Na in 1 L of HBSS with phenol red.
8. Epithelial cells (EC) culture medium: Dulbecco's Modified Eagle Medium (DMEM) high glucose, 10% (v/v) fetal bovine serum (FBS), 1% (v/v) L-glutamine solution, 5 ng/mL of recombinant epidermal growth factor (EGF), 30μg/mL bovine pituitary extract (BPE), 0.05μM Retinoic acid, and 1% (v/v) antibiotic–antimycotic solution.

2.3   Hydrogel-Based 3D Culture System

1. Water bath.
2. 15 mL centrifuge polystyrene tube.
3. Centrifuge.
4. CO_2 incubator.
5. Cell counting chamber.
6. Inverted microscope.
7. Collagen precoated hydrogel with Young's modulus ranging from 20 kPa to 30 kPa (*see* **Note 1**).
8. Dulbecco's phosphate-buffered saline (PBS): dissolve 8 g of NaCl (137 mM), 200 mg of KCl (2.7 mM), 1.44 g of Na_2HPO_4 (8 mM), and 240 mg of KH_2PO_4 (2 mM) in 800 mL of distilled water. Adjust pH to 7.4. Add distilled water until volume is 1 L. Sterilize solution with autoclave and store at +4 °C.
9. Antibiotic–antimycotic solution.

10. Trypsin–EDTA solution: dissolve 0.5 g of porcine trypsin and 0.2 g of EDTA 4Na in 1 L of HBSS with phenol red.

11. Epithelial cells (EC) culture medium: Dulbecco's Modified Eagle Medium (DMEM) high glucose, 10% (v/v) fetal bovine serum (FBS), 1% (v/v) L-glutamine solution, 5 ng/mL of recombinant epidermal growth factor (EGF), 30μg/mL bovine pituitary extract (BPE), 0.05μM retinoic acid, and 1% (v/v) antibiotic–antimycotic solution.

2.4 Cellular Uptake of Nanoparticles

1. Water bath.

2. 15 mL centrifuge polystyrene tube.

3. CO_2 incubator.

4. Inverted microscope.

5. Polyvinylpyrrolidone-coated gold nanoparticle (AuNP-PVP) (*see* **Note 2**).

6. Dulbecco's phosphate-buffered saline (PBS): dissolve 8 g of NaCl (137 mM), 200 mg of KCl (2.7 mM), 1.44 g of Na_2HPO_4 (8 mM), and 240 mg of KH_2PO_4 (2 mM) in 800 mL of distilled water. Adjust pH to 7.4. Add distilled water until volume is 1 L. Sterilize solution with autoclave and store at +4 °C.

7. Antibiotic–antimycotic solution.

8. Epithelial cells (EC) culture medium: Dulbecco's Modified Eagle Medium (DMEM) high glucose, 10% (v/v) fetal bovine serum (FBS), 1% (v/v) L-glutamine solution, 5 ng/mL of recombinant epidermal growth factor (EGF), 30μg/mL bovine pituitary extract (BPE), 0.05μM retinoic acid, and 1% (v/v) antibiotic–antimycotic solution.

3   Methods

All the procedures described below must be performed under sterile conditions. Cell manipulation must be carried out under laminar flow hood and cell cultures have to be maintained at 37 °C during their handling using thermostatically controlled stages.

3.1 HNEpC and HTEpC Propagation and Maintenance

1. Monitor cells daily.

2. Once cells have reached 80% confluency, carefully aspirate the culture medium (*see* **Note 3**).

3. Wash cells three time with 4 mL of PBS supplemented with 1% antibiotic–antimycotic solution.

4. Add 600μL of trypsin–EDTA solution (*see* **Note 4**) and incubate at 37 °C until cell monolayer begins to detach from the bottom of the tissue culture dish and cells dissociate (*see* **Note 5**).

5. Dilute cell suspension with 9 parts of EC culture medium to neutralize trypsin action.

6. Dislodge cells by repeatedly and gently pipetting.

7. Plate cells in a new culture dish and culture at 37 °C in 5% CO_2 incubator. Keep the passage ratio between 1:4 and 1:6, depending on growth rate (*see* **Note 6**).

8. Change medium every 2–3 days.

9. Maintain cells in culture until they have reached 80% confluency and passage them.

3.2 Establishment of a Hydrogel-Based 3D Culture System

1. Remove the sterile water or buffer used for hydrogel maintenance.

2. Immerse hydrogels in EC culture medium and incubate at room temperature for a minimum of 30 min.

3. Trypsinize cells (*see* Subheading 3.1, **steps 3–6**).

4. Collect cell suspension and transfer it in a 15 mL centrifuge polystyrene tube.

5. Count cells using a counting chamber under an optical microscope at room temperature. Calculate the volume of medium needed to resuspend cells in order to obtain a concentration of 6.5×10^4 cells/cm^2 (*see* **Note 7**).

6. Centrifuge at $300 \times g$ for 5 min.

7. Remove supernatant carefully and resuspend the cell pellet in the previously calculated volume (*see* **step 3**) of EC culture medium by pipetting.

8. Plate cells onto hydrogels.

9. Transfer into CO_2 incubator.

10. Incubated overnight for cell attachment.

3.3 Cellular Uptake of Nanoparticles

1. Prepare PVP-AuNP solution by diluting PVP-AuNPs to 70μg/mL in EC culture medium (*see* **Note 8**).

2. Mix solution by gently pipetting three times.

3. Carefully aspirate the culture medium for culture dish.

4. Wash cells three times with PBS supplemented with 1% antibiotic–antimycotic solution.

5. Add of PVP-AuNPs solution to cells.

6. Transfer into CO_2 incubator.

7. At the end of the desired culture period, wash cells twice with PBS and analyze PVP-AuNPs absorption of by transmission electron microscopy.

4   Notes

1. Hydrogels must be stored at 4 °C and used within 6 months of the printed manufacture date.

2. PVP-AuNPs must be stored at 4 °C. As an alternative AuNPs can be replaced with polyvinylpyrrolidone coated silver NPs, which, however, must be protected from light, beside storing at 4 °C.

3. Confluency normally takes between 7 and 10 days. If cells are not confluent after 10 days, they are not successfully growing.

4. The trypsin volume here reported is necessary for detaching cell cultured in a T25 flask. When working with bigger dish, scale up the volumes accordingly.

5. It usually takes 2–3 min.

6. The recommended seeding density is 10,000–15,000 cells per cm^2.

7. The formula to be used depends on the specific type of chamber. Cells/μL = Average number of cells per small grid × chamber multiplication factor × dilution.

8. PVP-AuNPs tend to rapidly cluster, prepare PVP-AuNP solution immediately prior to the use.

Acknowledgments

This work was funded by Carraresi Foundation. Authors are member of the COST Action CA16119 In vitro 3-D total cell guidance and fitness (CellFit) and of the Inter-COST Actions Task-Force on Covid-19.

References

1. Vaillant L, La RG, Tarantola A et al (2009) Epidemiology of fatal cases associated with pandemic H1N1 influenza. Euro Surveill 14:19309

2. WHO Coronavirus Disease (COVID-19) Dashboard|WHO Coronavirus Disease (COVID-19) Dashboard. https://covid19.who.int/

3. How Coronavirus Spreads|CDC. https://www.cdc.gov/coronavirus/2019-ncov/

prevent-getting-sick/how-covid-spreads.html? CDC_AA_refVal=https%3A%2F%2Fwww.cdc.gov%2Fcoronavirus%2F2019-ncov%2Fprepare%2Ftransmission.html

4. Vasconcelos MH, Alcaro S, Arechavala-Gomeza V et al (2020) Joining European Scientific Forces to Face Pandemics Trends in Microbiology (in press)

5. Leland DS, Ginocchio CC (2007) Role of cell culture for virus detection in the age of technology. Clin Microbiol Rev 20:49–78

6. Pennarossa G, Santoro R, Manzoni EFM et al (2018) Epigenetic erasing and pancreatic differentiation of dermal fibroblasts into insulin-producing cells are boosted by the use of low-stiffness substrate. Stem Cell Rev Rep 14 (3):398–411

7. Hussey GS, Dziki JL, Badylak SF (2018) Extracellular matrix-based materials for regenerative medicine. Nat Rev Mater 3:159–173

8. Luengen AE, Kniebs C, Buhl EM et al (2020) Choosing the right differentiation medium to develop Mucociliary phenotype of primary nasal epithelial cells in vitro. Sci Rep 10:6963

9. Zhu N, Zhang D, Wang W et al (2020) A novel Coronavirus from patients with pneumonia in China, 2019. N Engl J Med 382:727–733

10. Wolfram J, Ferrari M (2019) Clinical cancer nanomedicine. Nano Today 25:85–98

11. Hu TY, Frieman M, Wolfram J (2020) Insights from nanomedicine into chloroquine efficacy against COVID-19. Nat Nanotechnol 15:247–249

12. Verdecchia P, Cavallini C, Spanevello A et al (2020) The pivotal link between ACE2 deficiency and SARS-CoV-2 infection. Eur J Intern Med 76:14–20

Chapter 9

Creation of a Bioengineered Ovary: Isolation of Female Germline Stem Cells for the Repopulation of a Decellularized Ovarian Bioscaffold

Georgia Pennarossa, Matteo Ghiringhelli, Fulvio Gandolfi, and Tiziana A.L. Brevini

Abstract

Ovarian failure is the most common cause of infertility and affects about 1% of young women. One innovative strategy to restore ovarian function may be represented by the development of a bioprosthetic ovary, obtained through the combination of tissue engineering and regenerative medicine.

We here describe the two main steps required for bioengineering the ovary and for its ex vivo functional reassembling. The first step aims at producing a 3D bioscaffold, which mimics the natural ovarian milieu in vitro. This is obtained with a whole organ decellularization technique that allows the maintenance of microarchitecture and biological signals of the original tissue. The second step involves the use of magnetic activated cell sorting (MACS) to isolate purified female germline stem cells (FGSCs). These cells are able to differentiate in ovarian adult mature cells, when subjected to specific stimuli, and can be used them to repopulate ovarian decellularized bioscaffolds. The combination of the two techniques represents a powerful tool for in vitro recreation of a bioengineered ovary that may constitute a promising solution for hormone and fertility function restoring. In addition, the procedures here described allow for the creation of a suitable 3D platform with useful applications both in toxicological and transplantation studies.

Key words 3D bioscaffold, Bioprosthetic ovary, Extracellular matrix, Female germline stem cells, Ovary bioengineering and repopulation, Whole-ovary decellularization

1 Introduction

Infertility is a growing issue in modern society. According to the World Health Organization, it represents the fifth highest serious global disability, with an alarming incidence of one out of 1000 women, under the age of 30, rising to 1.0–1.5% in women younger than 40 years [1, 2]. Ovary dysfunction and premature ovarian insufficiency (POI) represent the main causes of infertility and can occur as a result of an inherited condition, de novo mutation or from insults to the ovarian tissue, including viral infections and

Tiziana A.L. Brevini et al. (eds.), *Next Generation Culture Platforms for Reliable In Vitro Models: Methods and Protocols*, Methods in Molecular Biology, vol. 2273, https://doi.org/10.1007/978-1-0716-1246-0_9,
© The Author(s), under exclusive license to Springer Science+Business Media, LLC, part of Springer Nature 2021

environmental factors [2–4]. Furthermore, therapy-induced ovarian failure due to chemotherapy and radiation treatments, can cause oocyte and/or surrounding support cell apoptosis in cancer survivors [5–9]. To date, several approaches have been developed and used in clinics to restore ovarian functions, including oocyte, embryo and ovarian tissue cryopreservation [10–18]. However, since these procedures are largely devoted to cancer patients, the high risk of malignant cell reintroduction pose a severe limit to their use in clinical practices [19, 20]. Development of a bioengineered ovary may provide a safe option in fertility restoration for all patients, including cancer survivors.

We here describe the two main steps required for the ex vivo creation of a bioprosthetic ovary and its functional assembling. The first step is based on a decellularization technique that produces an extracellular matrix (ECM)-based3D-scaffold. The second one allows the repopulation of the decellularized bioscaffold in order to create a functional in vitro bioengineered ovary. To date, many reports in the literature describe the regeneration in vitro of different organs using decellularized scaffolds [21–30]. However, limited studies have been performed in the reproductive system, and, more specifically, in the ovarian tissue [31, 32]. Indeed, the majority of decellularization protocols was specifically developed for ovarian tissue fragments and cortical slides [33–35], while the use of an entire ovary was limited to the bovine [32], the mouse [36, 37], and the porcine [38]. The decellularization protocol here described leads to the creation of a whole-ovary 3D bioscaffold preserves intact microarchitecture as well as ECM structures and components [38] (Fig. 1). It combines the use of physical and chemical methods to remove cellular components and generate 3D ovarian bioscaffold that recreates in vitro the complex in vivo ovarian milieu, facilitating the necessary interactions between cells and their surroundings and ensuring a correct cell growth, differentiation and function [39].

The obtained whole-ovary bioscaffold can be repopulated with a single cell type or with different ovarian cell populations, including fibroblasts, stromal and granulosa cells, or follicles. Among the several cell types present in the ovary, female germline stem cells (FGSCs) can be a promising candidate. Indeed, when subjected to specific stimuli, these cells are able to differentiate in ovarian adult mature cells and generate fully functional oocytes [40, 41]. We here describe the isolation of FGSCs by using magnetic activated cell sorting (MACS). Obtained cells can be stably maintained in culture, undergo mitotic division and steadily express germline and pluripotency-related genes (Fig. 2). Moreover, when used for the recellularization of decellularized ovary, FGSCs are able to rapidly migrate into the bioscaffold, adhering and colonizing the ECM within 24 h, and, during the subsequent days of culture, they increase in number and form cluster-like structures (Fig. 3).

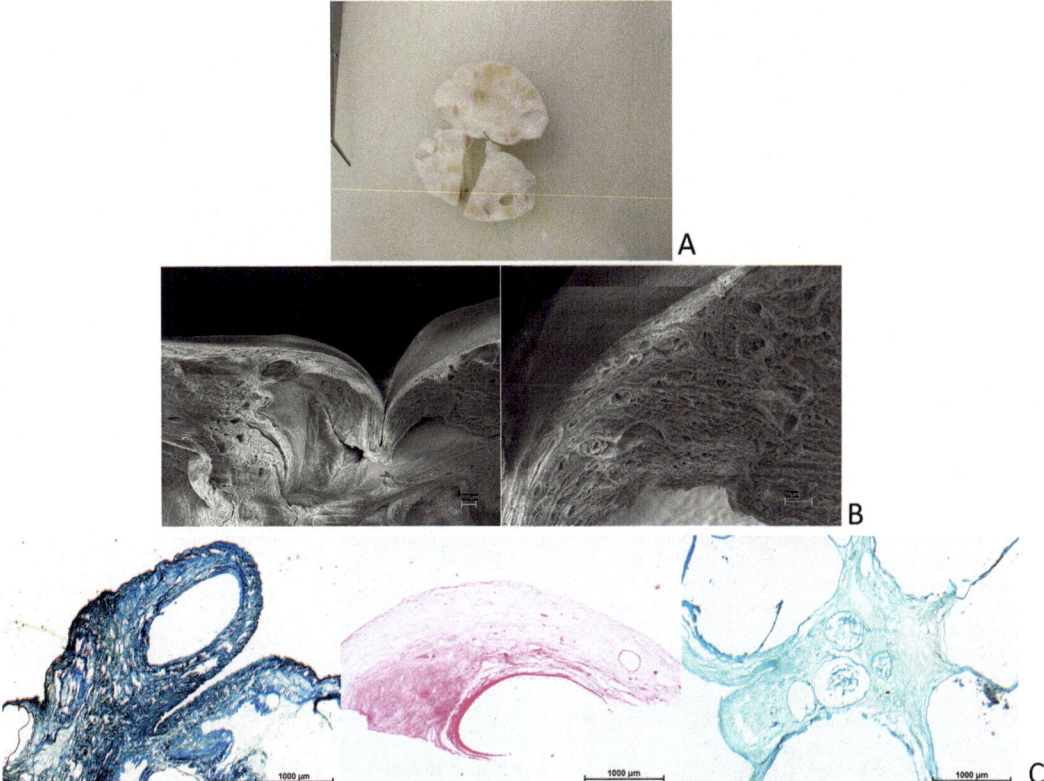

Fig. 1 (**a**) Whole-ovary decellularized bioscaffold preserves an intact macrostructure, maintaining the original shape and tissue homogenoity. (**b**) Decellularized ovary displays three-dimensional microarchitecture and ECM integrity with complex and well-organized fiber network. (**c**) Masson's trichrome (left panel), Gomori's aldehyde-fuchsin (middle panel), and Alcian Blue (right panel) staining show the persistence of collagen fibers (dark blue), elastic fibers (magenta), and the retention of glycosaminoglycans (GAG, light blue) after decellularization process

Overall, the method here reported is simple, fast and highly efficient and paves the way for a possible in vitro ovarian tissue reconstruction that may result advantageous for a general improvement of reproductive technologies, and possible future application to organ transplantation for hormone and fertility function restoring.

2   Materials

Prepare all solutions immediately before use (unless indicated otherwise).

2.1   Ovary Collection

1. Porcine ovaries collected from a local slaughterhouse.

2. 500 mL plastic bottle.

3. Ice container.

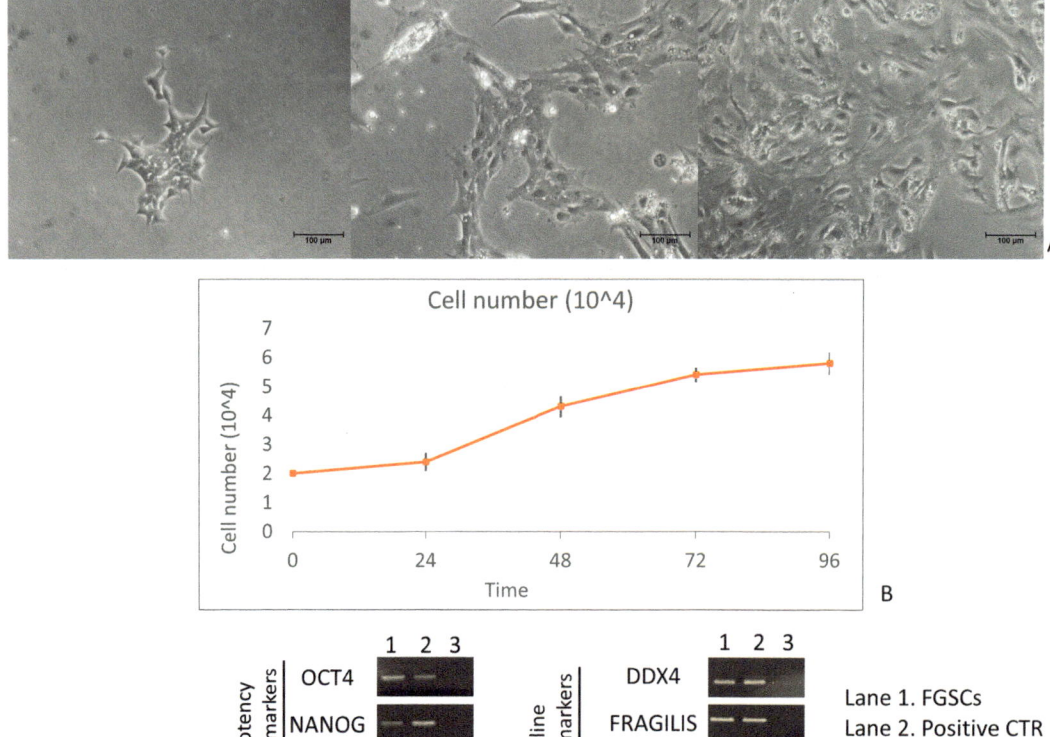

Fig. 2 (**a**) FGSCs isolated using MACS in vitro cultured. (**b**) FGSCs could be expanded in vitro with an estimated cell doubling time of 48–72 h. (**c**) FGSCs express pluripotency-related genes (OCT4, NANOG, REX1, and SOX2) and (**d**) germline specific markers (DDX4/VASA, FRAGILIS, BLIMP1, and DAZL)

4. Surgical scissor.

5. Dulbecco's phosphate-buffered saline (PBS): dissolve 8 g of NaCl (137 mM), 200 mg of KCl (2.7 mM), 1.44 g of Na_2HPO_4 (8 mM), and 240 mg of KH_2PO_4 (2 mM) in 800 mL of distilled water. Adjust pH to 7.4. Add distilled water until volume is 1 L. Sterilize solution with autoclave and store at +4 °C.

6. Antibiotic/Antimycotic Solution.

2.2 Whole-Ovary Decellularization

1. 50 mL centrifuge polypropylene tubes.

2. Water bath.

3. Orbital shaker.

4. 500 mL plastic or glass bottle.

5. Deionized water (DI-H_2O).

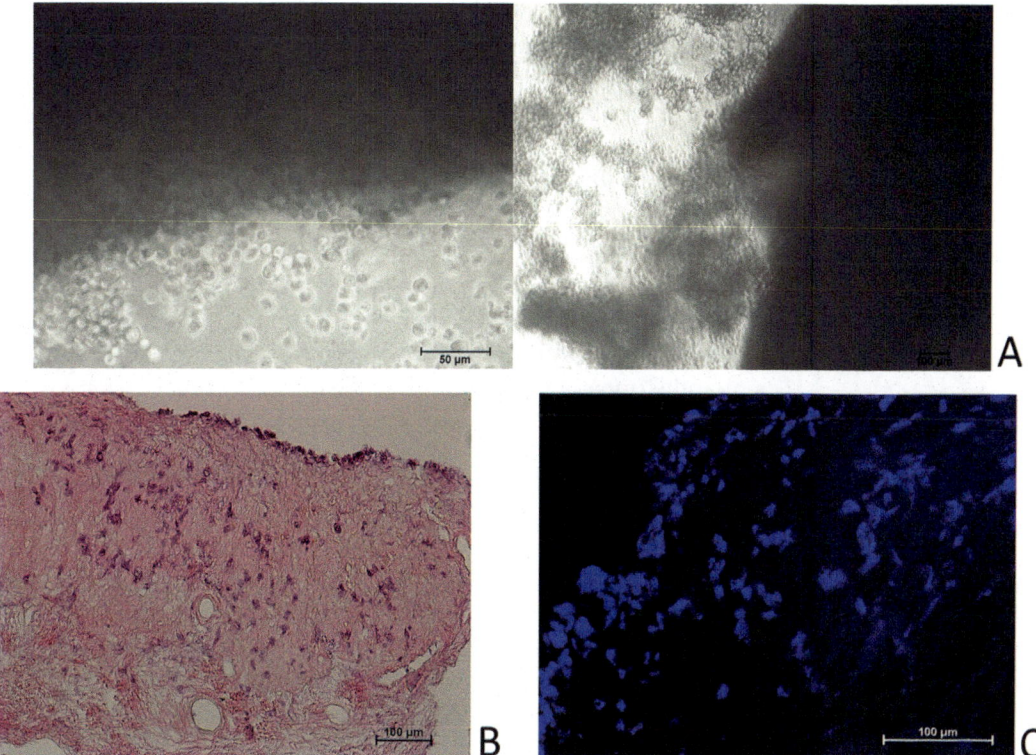

Fig. 3 (**a**) Reseeded FGSCs rapidly migrate into the bioscaffolds, adhering and colonizing the ECM within 24 h (left panel) and, during the subsequent days of culture, the cells form cluster-like structures (right panel). (**b**) H&E staining demonstrates the presence of cells into the bioscaffolds after recellularization. (**c**) DAPI staining confirms the positivity for nuclei

6. 0.5% sodium dodecyl sulfate (SDS): dissolve 2.5 g of SDS in 500 mL of DI-H_2O.

7. 1% Triton X-100: add 5 mL in 495 mL of DI-H_2O.

8. 2% deoxycholate: dissolve 10 g of deoxycholate in 500 mL of DI-H_2O.

2.3 FGSC Isolation

1. 4-well dish.

2. 100 mm petri dish.

3. Surgical scalpels.

4. 15 mL centrifuge polystyrene tube.

5. 50 mL centrifuge polypropylene tubes.

6. Centrifuge.

7. CO_2 incubator.

8. MACS cell separator.

9. 30-μm nylon mesh cell strainer.

10. Cell counting chamber.

11. Anti-SSEA-4 MicroBeads.

12. LS Column.

13. Inverted microscope.

14. 0.5% porcine gelatin: dissolve 0.5 g of porcine gelatin in 100 mL of distilled water. Sterilize solution with autoclave.

15. Dulbecco's phosphate-buffered saline (PBS): dissolve 8 g of NaCl (137 mM), 200 mg of KCl (2.7 mM), 1.44 g of Na_2HPO_4 (8 mM), and 240 mg of KH_2PO_4 (2 mM) in 800 mL of distilled water. Adjust pH to 7.4. Add distilled water until volume is 1 L. Sterilize solution with autoclave and store at +4 °C.

16. Hank's Balanced Salt Solution (HBSS) with phenol red.

17. 1 mg/mL collagenase (type IV): dissolve 5 mg of collagenase in 5 mL of HBSS with phenol red. Sterilize solution with 0.22μm filter.

18. Trypsin-EDTA solution: dissolve 0.5 g of porcine trypsin and 0.2 g of EDTA 4Na in 1 L of HBSS with phenol red.

19. Fetal bovine serum (FBS).

20. FGSC culture medium: Dulbecco's Modified Eagle Medium: Nutrient Mixture F-12 (DMEM/F12), 40 ng/mL Human Stem Cell Factor (SCF) Recombinant Protein, 1% B27, 1 mM MEM nonessential amino acids, 0.1 mM β-mercaptoethanol, 10% KnockOut serum replacement (KO serum), 2 mM L-glutamine, 1% antibiotic–antimycotic solution (*see* **Note 1**).

---

3   Methods

All the procedures described below must be performed under sterile conditions. Instruments touching or in connection to the ovary have to be sterilized. Cell isolation must be carried out under laminar a flow hood and cell cultures have to be maintained at 37 °C during their handling using thermostatically controlled stages.

3.1   *Ovary Collection*

1. Collect ovaries from gilts weighing approximately 120 kg.

2. Separate ovaries from fallopian tubes by cutting them with surgical scissor.

3. Transfer ovaries in cold sterile PBS containing antibiotic–antimycotic solution (5 mL/500 mL) and transport them to the laboratory using ice container.

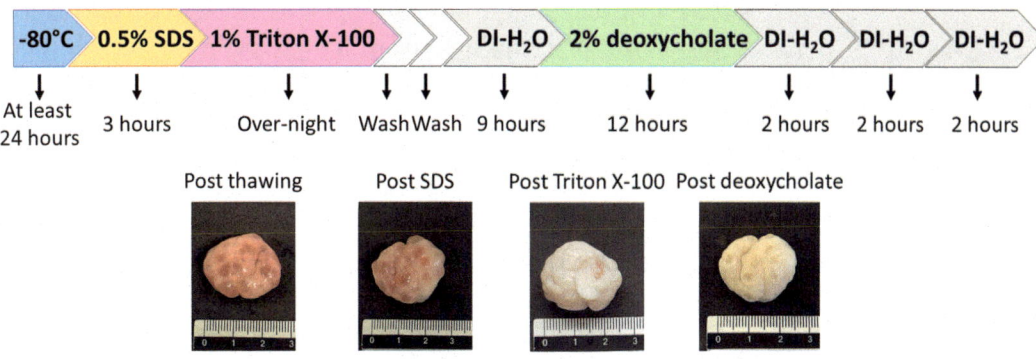

| -80°C | 0.5% SDS | 1% Triton X-100 | | | DI-H$_2$O | 2% deoxycholate | DI-H$_2$O | DI-H$_2$O | DI-H$_2$O |

| At least 24 hours | 3 hours | Over-night | Wash | Wash | 9 hours | 12 hours | 2 hours | 2 hours | 2 hours |

Post thawing   Post SDS   Post Triton X-100   Post deoxycholate

Fig. 4 Schematic representation of decellularization protocol and chronological macroscopic images illustrating changes in ovary color, turning from red to white, while maintaining original shape

3.2 Whole-Ovary Decellularization

1. Wash ovary in fresh PBS, completely remove the PBS, place ovary in 50 mL tube and store organ at −80 °C for at least 24 h (*see* **Note 2**, Fig. 4).

2. Thaw whole-ovary at 37 °C in a water bath for 30 min.

3. Transfer whole-ovary in a bottle containing 500 mL of 0.5% SDS. Place the bottle onto an orbital shaker at 200 rpm and incubate for 3 h at room temperature.

4. Remove SDS solution from the bottle containing the whole-ovary and add 500 mL of 1% Triton X-100. Incubate whole-ovary over-night at room temperature in 1% Triton X-100, using an orbital shaker at 200 rpm.

5. Remove Triton X-10 solution from the bottle containing the whole-ovary and wash ovary with 500 mL of DI-H$_2$O twice. Add for a third time 500 mL of DI-H$_2$O and extensively wash whole-ovary for 9 h at room temperature, using an orbital shaker at 200 rpm.

6. Remove DI-H$_2$O from the bottle and add 500 mL of 2% deoxycholate for 12 h at room temperature, using an orbital shaker at 200 rpm.

7. Remove deoxycholate and wash whole-ovary in DI-H$_2$O for 6 h at room temperature, using an orbital shaker at 200 rpm. Changes DI-H$_2$O every 2 h (*see* **Note 3**).

3.3 FGSC Isolation

1. Add 500 μL of sterile 0.5% porcine gelatin to 4-well dishes. Wait 2 h to polymerize, maintaining them at room temperature.

2. Wash ovary twice in 50 mL of fresh sterile PBS at room temperature.

3. Place ovary into a 100 mm sterile petri dish and cut ovarian cortex into small pieces (approximately cubes of 1–2 mm^2) using a surgical scalpel.

4. Wash fragments four times in sterile PBS at room temperature.

5. Enzymatically digest 20–30 fragments by 30 min incubation with 5 mL of 1 mg/mL collagenase (type IV) in 15 mL tube, with gentle shaking every 5 min.

6. Centrifuge digested tissue at 300 × g for 5 min. Remove supernatant, add 5 mL of HBSS and resuspend digested tissue.

7. Centrifuge at 300 × g for 5 min. Remove supernatant, add 5 mL of trypsin–EDTA solution and resuspend digested tissue. Incubate for 15 min at 37 °C.

8. Remove the 0.5% porcine gelatin excess for the 4-well dish and let the dish open under laminar flow hood to dry (*see* Subheading 3.3, **step 1**).

9. Neutralize trypsin by adding 500µL FBS. Disperse digested tissues into single cells by gentle pipetting and centrifuge at 300 × g for 5 min. Remove supernatant and resuspend pellet in FGSC culture medium.

10. Dissociate to single-cell suspension by pipetting up and down using a 10 mL serological pipette.

11. Pass cell suspension through a 30-µm nylon mesh filter to remove cell clumps which may clog the column (*see* **Note 4**).

12. Count cells using a counting chamber under an optical microscope at room temperature. Calculate the volume of medium needed to resuspend cells in order to obtain 10^7 cells in 80µL (*see* **Note 5**).

13. Centrifuge cell suspension at 300 × g for 5 min. Aspirate supernatant completely and resuspend cell pellet in 80µL of precooled FGSC culture medium per 10^7 total cells (*see* **Note 6**).

14. Add 20µL of Anti-SSEA-4 MicroBeads per 10^7 total cells (*see* **Note 7**). Mix well and incubate for 15 min at +4 °C (*see* **Note 8**).

15. Wash cells by adding 2 mL of FGSC culture medium and centrifuge at 300 × g for 5 min. Aspirate supernatant and resuspend in 500µL of fresh precooled FGSC culture medium.

16. Place column in the magnetic field of a MACS Separator and prepare column by rinsing 3 mL of precooled FGSC culture medium.

17. Apply cell suspension onto the column and collect flow-through containing unlabeled cells (*see* **Note 9**). Wash column tree time with 3 mL of FGSC culture medium (*see* **Note 10**).

18. Remove column from the separator and place it on a 50 mL collection tube. Pipette 5 mL of FGCS culture medium onto the column. Immediately flush out the magnetically labeled cells by firmly pushing the plunger into the column (*see* **Note 11**).

19. Centrifuge cell suspension at 300 × g for 5 min. Aspirate supernatant completely and resuspend cell pellet in of FGSC

culture medium and plate cells in gelatin precoated 4-well dish (*see* Subheading 3.3, **steps 1** and **8**) and culture the incubator under aseptic conditions with 5% CO_2 at 37 °C.

4 Notes

1. FGCS culture medium can be stored at +4 °C a maximum of 10 days.

2. Intact ovaries can be stored at −80 °C for long time periods without causing matrix alteration.

3. The obtained bioscaffold can be either directly used for histological analysis or sterilized for cell repopulation. Its sterilization procedure can be performed using 70% ethanol and 2% antibiotic in sterile H_2O for 30 min at room temperature. Before cell repopulation, wash bioscaffold extensively with PBS and 4% Antibiotic at room temperature using an orbital shaker at 200 rpm.

4. Moisten filter with culture medium before use.

5. The formula to be used depends on the specific type of chamber. Cells/µL = Average number of cells per small grid × chamber multiplication factor × dilution.

6. Work fast, keep cells cold, and use precooled solutions. This will prevent capping of antibodies on the cell surface and nonspecific cell labelling.

7. The microbead volume here reported is necessary for up to 10^7 cells. When working with fewer cells, use the same quantity. When working with higher cell numbers, scale up the volumes accordingly.

8. The incubation temperature and period are fundamental for specific cell labeling. Higher temperatures and/or longer incubation times may lead to nonspecific cell labelling.

9. FGCS are labelled cells. The unlabeled ones, collected in this step, can be discarded or cultured for negative control.

10. Add medium only when the column reservoir is empty.

11. To increase the purity of SSEA-4+ cells, the eluted fraction can be enriched with a new second column, repeating the magnetic separation procedure.

Acknowledgments

This work was funded by Carraresi Foundation and PSR2017. The Laboratory of Biomedical Embryology is member of the COST Action CA16119 In vitro 3-D total cell guidance and fitness (Cell-Fit) and of the Inter-COST Actions Task-Force on Covid-19.

References

1. Hewlett M, Mahalingaiah S (2015) Update on primary ovarian insufficiency. Curr Opin Endocrinol Diabetes Obes 22:483–489

2. Qin Y, Jiao X, Simpson JL et al (2015) Genetics of primary ovarian insufficiency: new developments and opportunities. Hum Reprod Update 21:787–808

3. Yalcinkaya TM, Sittadjody S, Opara EC (2014) Scientific principles of regenerative medicine and their application in the female reproductive system. Maturitas 77:12–19

4. Sadri-Ardekani H, Atala A (2015) Regenerative medicine for the treatment of reproductive system disorders: current and potential options. Adv Drug Deliv Rev 82–83:145–152

5. Siegel RL, Miller KD, Jemal A (2019) Cancer statistics, 2019. CA Cancer J Clin 69:7–34

6. Wallace WHB, Thomson AB, Saran F, Kelsey TW (2005) Predicting age of ovarian failure after radiation to a field that includes the ovaries. Int J Radiat Oncol Biol Phys 62:738–744

7. Wallace WHB, Critchley HOD, Anderson RA (2012) Optimizing reproductive outcome in children and young people with cancer. J Clin Oncol 30:3–5

8. Duffy C, Allen S (2009) Medical and psychosocial aspects of fertility after cancer. Cancer J 15:27–33

9. Jadoul P, Dolmans M-M, Donnez J (2010) Fertility preservation in girls during childhood: is it feasible, efficient and safe and to whom should it be proposed? Hum Reprod Update 16:617–630

10. Vajta G, Kuwayama M (2006) Improving cryopreservation systems. Theriogenology 65:236–244

11. Wong KM, Mastenbroek S, Repping S (2014) Cryopreservation of human embryos and its contribution to in vitro fertilization success rates. Fertil Steril 102:19–26

12. Sparks A (2015) Human embryo cryopreservation—methods, timing, and other considerations for optimizing an embryo cryopreservation program. Semin Reprod Med 33:128–144

13. Chen C (1986) Pregnancy after human oocyte cryopreservation. Lancet 327:884–886

14. Westphal LM, Massie JAM, Lentscher JA (2019) Embryo and oocyte banking. In: Textbook of Oncofertility research and practice. Springer International Publishing, Cham, pp 71–79

15. Porcu E (2001) Oocyte freezing. Semin Reprod Med 19:221–230

16. Davis VJ (2006) Female gamete preservation. Cancer 107:1690–1694

17. Gook DA, Edgar DH (2019) Cryopreservation of female reproductive potential. Best Pract Res Clin Obstet Gynaecol 55:23–36

18. Laronda MM (2020) Engineering a bioprosthetic ovary for fertility and hormone restoration. Theriogenology 150:8–14

19. Meirow D, Hardan I, Dor J et al (2008) Searching for evidence of disease and malignant cell contamination in ovarian tissue stored from hematologic cancer patients. Hum Reprod 23:1007–1013

20. Dolmans M-MM, Marinescu C, Saussoy P et al (2010) Reimplantation of cryopreserved ovarian tissue from patients with acute lymphoblastic leukemia is potentially unsafe. Blood 116:2908–2914

21. Rajabi-Zeleti S, Jalili-Firoozinezhad S, Azarnia M et al (2014) The behavior of cardiac progenitor cells on macroporous pericardium-derived scaffolds. Biomaterials 35:970–982

22. Lecht S, Stabler CT, Rylander AL et al (2014) Enhanced reseeding of decellularized rodent lungs with mouse embryonic stem cells. Biomaterials 35:3252–3262

23. Lee H, Han W, Kim H et al (2017) Development of liver Decellularized extracellular matrix bioink for three-dimensional cell printing-based liver tissue engineering. Biomacromolecules 18:1229–1237

24. Yu YL, Shao YK, Ding YQ et al (2014) Decellularized kidney scaffold-mediated renal regeneration. Biomaterials 35:6822–6828

25. Aulino P, Costa A, Chiaravalloti E et al (2015) Muscle extracellular matrix scaffold is a multipotent environment. Int J Med Sci 12:336–340

26. Baiguera S, Del GC, Kuevda E et al (2014) Dynamic decellularization and cross-linking of rat tracheal matrix. Biomaterials 35:6344–6350

27. Sjöqvist S, Jungebluth P, Lim ML et al (2014) Experimental orthotopic transplantation of a tissue-engineered oesophagus in rats. Nat Commun 5:3562

28. Singh A, Bivalacqua TJ, Sopko N (2018) Urinary tissue engineering: challenges and opportunities. Sex Med Rev 6:35–44

29. Kajbafzadeh A-M, Khorramirouz R, Kameli SM et al (2017) Decellularization of human internal mammary artery: biomechanical properties and histopathological evaluation. Biores Open Access 6:74–84

30. Zhang J-K, Du R-X, Zhang L et al (2017) A new material for tissue engineered vagina reconstruction: acellular porcine vagina matrix. J Biomed Mater Res A 105:1949–1959

31. Laronda MM, Rutz AL, Xiao S et al (2017) A bioprosthetic ovary created using 3D printed microporous scaffolds restores ovarian function in sterilized mice. Nat Commun 8:15261

32. Laronda MM, Jakus AE, Whelan KA et al (2015) Initiation of puberty in mice following decellularized ovary transplant. Biomaterials 50:20–29

33. Pors SE, Ramløse M, Nikiforov D et al (2019) Initial steps in reconstruction of the human ovary: survival of pre-antral stage follicles in a decellularized human ovarian scaffold. Hum Reprod 34:1523–1535

34. Liu W-Y, Lin S-G, Zhuo R-Y et al (2017) Xenogeneic Decellularized scaffold: a novel platform for ovary regeneration. Tissue Eng Part C Methods 23:61–71

35. Hassanpour A, Talaei-Khozani T, Kargar-Abarghouei E et al (2018) Decellularized human ovarian scaffold based on a sodium lauryl ester sulfate (SLES)-treated protocol, as a natural three-dimensional scaffold for construction of bioengineered ovaries. Stem Cell Res Ther 9:252

36. Eivazkhani F, Abtahi NS, Tavana S et al (2019) Evaluating two ovarian decellularization methods in three species. Mater Sci Eng C 102:670–682

37. Alshaikh AB, Padma AM, Dehlin M et al (2019) Decellularization of the mouse ovary: comparison of different scaffold generation protocols for future ovarian bioengineering. J Ovarian Res 12:58

38. Pennarossa G, Ghiringhelli M, Gandolfi F et al (2020) Whole-ovary decellularization generates an effective 3D bioscaffold for ovarian bioengineering. J Assist Reprod Genet 37 (6):1329–1339

39. Gilpin A, Yang Y (2017) Decellularization strategies for regenerative medicine: from processing techniques to applications. Biomed Res Int 2017:9831534

40. Telfer EE, Anderson RA (2019) The existence and potential of germline stem cells in the adult mammalian ovary. Climacteric 22:22–26

41. Clarkson YL, McLaughlin M, Waterfall M et al (2018) Initial characterisation of adult human ovarian cell populations isolated by DDX4 expression and aldehyde dehydrogenase activity. Sci Rep 8:6953

Chapter 10

A Two-Step Protocol to Erase Human Skin Fibroblasts and Convert Them into Trophoblast-like Cells

Sharon Arcuri, Fulvio Gandolfi, Edgardo Somigliana, and Tiziana A.L. Brevini

Abstract

The first differentiation event in mammalian embryos is the formation of the trophectoderm, which is the progenitor of the outer epithelial component of the placenta and supports the fetus during intrauterine life. Our understanding of these events is limited, particularly in human, because of ethical and legal restrictions and availability of adequate in vitro models would be very advantageous. Here we describe a method that converts human fibroblasts into trophoblast-like cells, combining the use of 5-azacytidine-CR (5-aza-CR) to erase the original cell phenotype and a cocktail containing bone morphogenetic protein 4 (BMP4) with inhibitors of the Activin/Nodal/ERK signaling pathways, to drive erased fibroblasts into the trophoblastic differentiation. This innovative method uses very easily accessible cells to derive trophoblast-like cells and it can be useful to study embryo implantation disorders related to aging.

Key words 5-aza-CR, Activin/Nodal/ERK inhibitors, BMP4, Epigenetics, Fibroblast, Trophoblast

1 Introduction

In mammals, trophectoderm is the progenitor tissue of the entire outer epithelial component of the placenta that supports the fetus during intrauterine life. Its primary function is to allow implant of the embryo to the uterine wall, to protect the fetus from immunological response, to secrete hormones for pregnancy maintenance and to allow gasses and nutrient exchange [1]. Trophectoderm is composed by two subpopulations: villous cytotrophoblast that layers the basement membrane surrounding placental villi and syncytiotrophoblast, a cell population that makes direct contact with maternal blood and is characterized by production of chorionic gonadotropin (CG) and other placental hormones [2]. Although placental dysfunction seems to be the main disorder of pregnancy, with immediate consequences for the mother and child, aging is

Tiziana A.L. Brevini et al. (eds.), *Next Generation Culture Platforms for Reliable In Vitro Models: Methods and Protocols*, Methods in Molecular Biology, vol. 2273, https://doi.org/10.1007/978-1-0716-1246-0_10,

also one of the main causes of infertility. In this perspective, it comes as no surprise that the percentage of infertility has increased dramatically in recent years, given the fact that women's empowerment causes the postponement of motherhood, with an increase in placental and embryonic implantation defect. Recent evidence has shown that aged uterine environment is decisively involved, causing a reduced uterus decidualization and increased severe placentation defects resulting in abnormal embryonic development [3].

At the moment, only few in vitro models are available to study the main mechanisms involved in reproductive disorders and, all of them, are obtained from embryonic stem cells (ESCs). In particular, Roberts et al. have demonstrated that human pluripotent stem cells (hPSCs), bone morphogenetic protein 4 (BMP4) and hypoxic environment allow to derive trophoblast cells [2]; Wang et al. have described that hESC seems to be a robust model to acquire trophoblast [4] and Turco et al. have highlighted how hESC together with BMP4, 3D matrix and hypoxic environment permits to obtain trophoblast organoids [5]. Here we describe a model that uses a two-step protocol to derive trophoblast-like cells (Fig. 1). Step 1 involves the use of epigenetic erasing to drive human adult somatic cells into, into a high plasticity state [6, 7]. Step 2 takes advantage of this induced high plasticity window and uses a cocktail of trophoblast inducers to encourage the acquisition of the trophoblastic phenotype [8] (Fig. 2). More in details, Step 2 is based on the use of mouse embryonic fibroblasts (MEF) conditioned medium combined with a cocktail of BMP4 and inhibitors of the

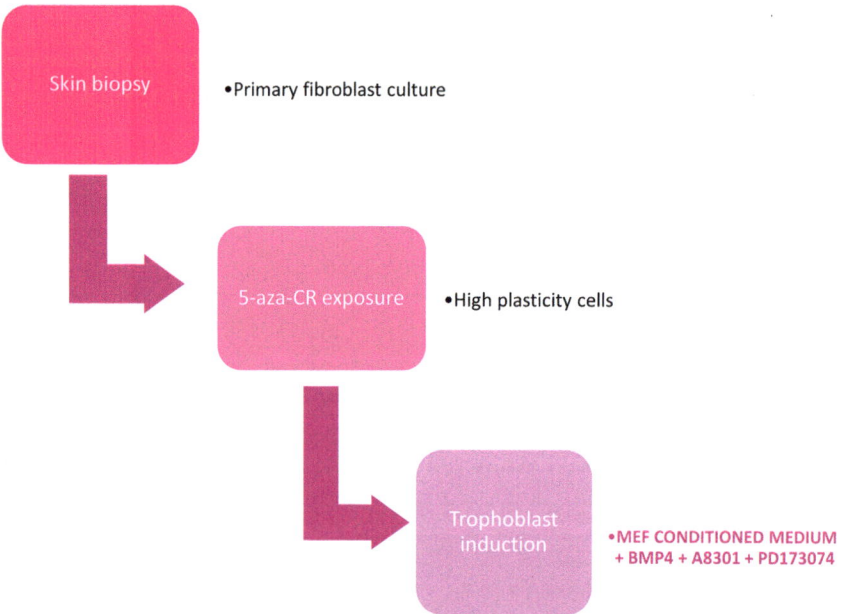

Fig. 1 An overview of the steps involved in the protocol

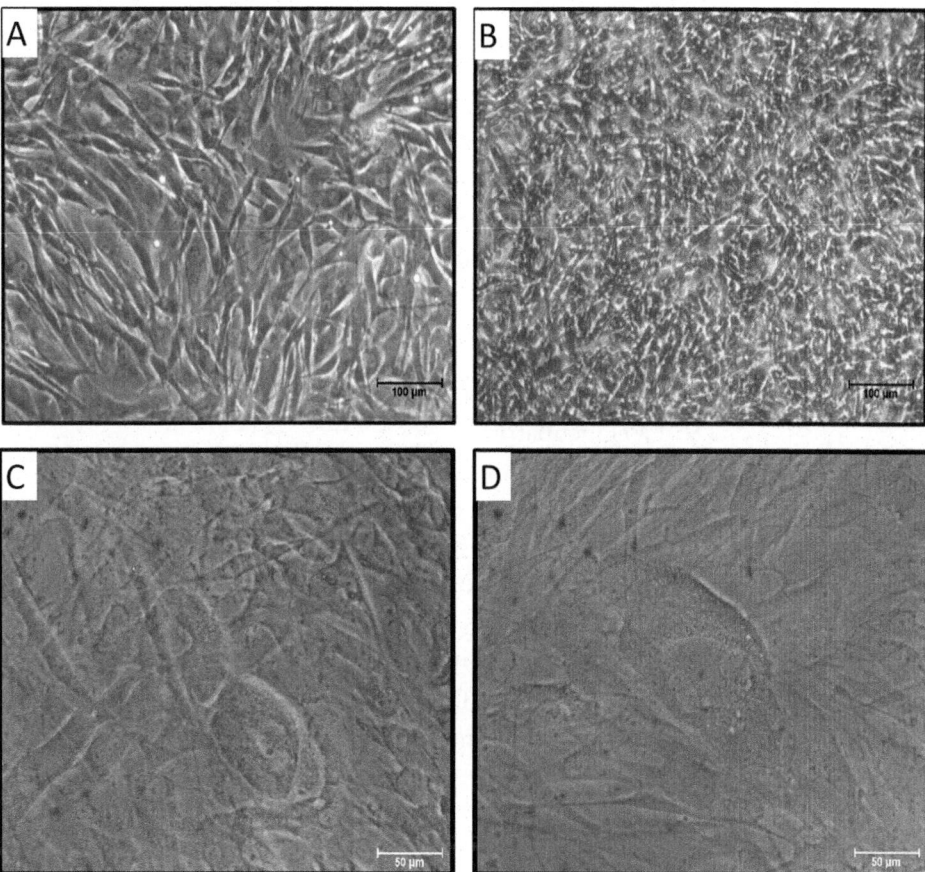

Fig. 2 Morphological changes of fibroblasts (**a**) after 5-azacytidine treatment (**b**). Trophoblast-like cells obtained after 11 days of chemical induction in hypoxic environment: cytotrophoblast cells (**c**) and syncytiotrophoblast cells (**d**)

Activin/Nodal/ERK signaling pathways as previously described by Roberts et al., Schulz et al., and Wang et al. to drive cells into the new phenotype [2, 4]. This method allows the generation of trophoblast cells from easily accessible cells, such as dermal fibroblasts that can be simply propagated in vitro. It is free of any genetic modifications that make cells prone to instability and transformation. Because of their stable phenotype, the cells generated with this procedure are more easily applied in regenerative medicine.

The in vitro model generated is efficient and reproducible. It can be used for the acquisition of useful information on the pathogenesis of developmental disorders based on trophoblast defects and aging, as well as drug discovery and regenerative medicine.

2   Materials

2.1   Mouse Embryonic Fibroblast (MEF) Thawing and Inactivation

1. MEF cells.
2. Water bath.
3. 15 mL sterile tubes.
4. Centrifuge.
5. 35 mm petri dishes.
6. T25 flasks.
7. 0.20 μm filter.
8. Inverted microscope.
9. CO_2 incubator.
10. Trypsin–EDTA solution.
11. MEF culture medium: 88% (v/v) Dulbecco's modified Eagle medium, 10% (v/v) fetal bovine serum (FBS), 2 mM (v/v) L-glutamine solution, 1% antibiotics.
12. MEF inactivating medium: 6 mL of fibroblast culture medium with 60 μL of Mitomycin C.
13. ESC culture medium without bFGF: 40% Ham's F-10 Nutrient mix, 40% DMEM Low glucose, 10% KnockOut Serum Replacement, 5% FCS, 1% antibiotics, 2 mM L-glutamine, 1% Nucleoside Mix, 1% MEM nonessential amino acids solution, 0.1 mM 2-Mercaptoethanol, 1 unit/mL LIF.

2.2   Isolation of Human Skin Fibroblasts

1. Skin biopsy collected from adult women.
2. 15 mL sterile tubes.
3. Sterile surgical instruments.
4. 35 mm petri dishes.
5. T25 flasks.
6. CO_2 incubator.
7. Inverted microscope.
8. Dulbecco's phosphate-buffered saline (PBS) containing 2% antibiotics.
9. 0.1% porcine gelatin: dissolve 0.1 g of porcine gelatin in 100 mL of sterile water. Autoclave and stock at +4 °C.
10. Fibroblast culture medium: 77% (v/v) Dulbecco's modified Eagle medium, 20% (v/v) fetal bovine serum (FBS), 2 mM (v/v) L-glutamine solution, 1% antibiotics.

2.3   Seeding of Human Skin Fibroblasts

1. 4-well multidishes.
2. 15 mL sterile tubes.
3. Cell counter.

4. Centrifuge.

5. Inverted microscope.

6. CO_2 incubator.

7. Water bath.

8. 0.1% porcine gelatin: dissolve 0.1 g of porcine gelatin in 100 mL of sterile water. Autoclave and stock at +4 °C.

9. Fibroblast culture medium: 88% (v/v) Dulbecco's modified Eagle medium, 10% (v/v) fetal bovine serum (FBS), 2 mM (v/v) L-glutamine solution, 1% antibiotics.

2.4 5-Azacytidine-CR (5-aza-CR) Treatment

1. 15 mL sterile tubes.

2. Water bath.

3. CO_2 incubator.

4. 1 mM 5-aza-CR: dissolve 0.0024 g of 5-aza-CR in 10 mL of warm DMEM; vortex the solution and sterilize using 0.22 μm filter.

5. Fibroblast culture medium: fresh culture medium: 88% (v/v) Dulbecco's modified Eagle medium, 10% (v/v) fetal bovine serum (FBS), 2 mM (v/v) L-glutamine solution, 1% antibiotics.

6. ESC culture medium: 40% Ham's F-10 Nutrient mix, 40% DMEM Low glucose, 10% KnockOut Serum Replacement, 5% FCS, 1% antibiotics, 2 mM L-glutamine, 1% Nucleoside Mix, 1% MEM nonessential amino acids solution, 0.1 mM 2-mercaptoethanol, 1 unit/mL LIF, 5 ng/mL human basic fibroblast growth factor (bFGF).

2.5 Trophoblastic Induction

1. Water bath.

2. Inverted microscope.

3. Tri-gas incubator.

4. Trophoblast induction medium: 10 mL of conditioned MEF inactivated medium, 10 μL of BMP4 (50 ng/mL), 10 μL A83-01 (1 μM), and 10 μL PD173074 (0.1 μM).

3 Methods

3.1 MEF Thawing and Inactivation

1. Thaw a cryo-vial containing MEFs using a 37 °C water bath.

2. Add 2 mL of fibroblast culture medium in a 15 mL sterile tube.

3. Transfer MEFs in the 15 mL sterile tube containing the 2 mL of fibroblast culture medium.

4. Centrifuge at $300 \times g$ for 5 min.

5. Resuspend the pellet in 2 mL of fibroblast culture medium.

6. Seed in a 35 mm petri dish and incubate at 37 °C in CO_2 incubator.

7. Change fibroblast culture medium every day.

8. When they reach confluency, remove culture medium from the dish.

9. Wash three times in sterile PBS supplemented with 1% antibiotic–antimycotic solution.

10. Add 500 μL of trypsin–EDTA solution and incubate at 37 °C until cell monolayer begins to detach from the bottom of the tissue culture dish and cells dissociate.

11. Neutralize trypsin–EDTA using 1.5 mL of MEF culture medium.

12. Dislodge cells by repeatedly and gently pipetting.

13. Plate cells in a new culture dish using a 1:2 passage ratio and culture at 37 °C in 5% CO_2 incubator.

14. Change fibroblast culture medium every day and passage every 2 days.

15. Seed MEFs at a density of 2×10^6 in T25 flask and incubate for 24 h in CO_2 incubator.

16. Rinse cells three times in sterile PBS and expose subconfluent monolayers to the MEF inactivating medium for 3 h.

17. Eliminate MEF inactivating medium and incubate cells with ESC medium without b-FGF for 24 h.

18. Collect conditioned medium, filter using 0.20 μm filter and storage at −20 °C until use.

3.2 Isolation of Human Skin Fibroblasts

1. Isolate skin biopsy and transfer it in a 50 mL sterile tube containing sterile PBS with 2% antibiotics.

2. Prepare 2 mL of 0.1% of porcine gelatin in a 35 mm petri dish. Wait 45 min to coat the surface at room temperature.

3. Wash the biopsy in sterile PBS and cut it in small fragments of approximately 2 mm^3 (*see* **Note 1**).

4. Transfer 5 skin fragments in a petri dish.

5. Add 20 μL of fibroblast culture medium onto each fragment and incubate at 37 °C in CO_2 incubator.

6. Add around 20–50 μL of fibroblast culture medium daily for the first 5 days.

7. Add 1 mL in each petri dish and culture until cells reach confluency.

8. When cells reach confluency, trypsinize (*see* Subheading 3.1, **steps 9–12**) and transfer cells in a new T25 flask.

9. Incubate at 37 °C.

10. Passage every 3 days with a ratio of 1:2.

3.3 Seeding of Human Skin Fibroblasts

1. Add 0.5 mL of 0.1% of porcine gelatin in each well of 4-well multidish and wait 20 min to coat the surface at room temperature.

2. Eliminate gelatin and let it dry for 45 min.

3. Remove the medium from T25 flask, and ten trypsinized using 0.6 mL of Trypsin-EDTA (*see* Subheading 3.1, **steps 9–12**) (*see* **Note 2**).

4. Collect cell suspension in a 15 mL sterile tube.

5. Count cells using a counting chamber under an optical microscope at room temperature. Calculate the volume of medium needed to resuspend cells in order to obtain 7.8×10^4 cells/cm^2.

6. Centrifuge cells at $300 \times g$ for 5 min.

7. Resuspend the pellet in a calculated volume by pipetting carefully.

8. Plate cells in 4-well multidish.

9. Incubate at 37 °C for 24 h.

3.4 5-aza-CR Treatment

1. Rinse cells in a 4-well multidish using 0.5 mL of sterile PBS/well.

2. Dilute 1 μL of 5-aza-CR stock solution in 1 mL of fibroblast culture medium (*see* **Note 3**).

3. Remove the medium from the dish.

4. Add 0.5 mL of fibroblast culture medium containing 5-aza-CR in three wells. Use one as a negative control.

5. Incubate at 37 °C for 18 h in CO_2 incubator.

6. Remove 5-aza-CR.

7. Rinse cells three times in sterile PBS and add 0.5 mL of ESC medium for 3 h at 37 °C in CO_2 incubator.

3.5 Trophoblastic Induction

1. Rinse cells three times in sterile PBS.

2. Expose cells to the trophoblast induction medium (*see* **Notes 4–5**) and incubate in a hypoxic environment (5% O_2) in the tri-gas incubator for 11 days.

3. Refresh culture medium every other day.

4 Notes

1. It is crucial to cut very small pieces of skin tissue in order to facilitate fibroblasts to grow out of the tissue fragments.

2. Before seeding cells for 5-aza-CR treatment, it is important that they are in subconfluency.

3. Prepare 5-aza-CR solution just before use.

4. Aliquot differentiation factors in small volumes.

5. It is very important to add differentiation factors just before use.

Acknowledgments

This work was supported by Carraresi Foundation. T.A.L.B., F.G., and S.A are members of the COST Actions CA16119 and of the Inter-COST Actions Task-Force on Covid-19.

References

1. Knöfler M, Haider S, Saleh L et al (2019) Human placenta and trophoblast development: key molecular mechanisms and model systems. Cell Mol Life Sci 76:3479–3496

2. Schulz L, Ezashi T, Das P et al (2009) Human embryonic stem cells as models for trophoblast differentiation Introduction: the trophoblast lineage and its emergence. Semin Reprod Med 29:1–12

3. Inhorn MC, Patrizio P (2014) Infertility around the globe: new thinking on gender, reproductive technologies and global movements in the 21st century. Hum Reprod Update 21:411–426

4. Szaraz P, Gratch YS, Iqbal F et al (2017) In vitro differentiation of human mesenchymal stem cells into functional cardiomyocyte-like cells. J Vis Exp 01:2–10

5. Turco MY, Gardner L, Kay RG, Hamilton RS, Prater M, Hollinshead M, McWhinnie A, Esposito L, Fernando R, Skelton H, Reimann F, Gribble F, Sharkey A, Marsh SGE, O'Rahilly S, Hemberger M, Burton GJ, Moffett A (2018) Trophoblast organoids as a model for maternal-fetal interactions during human placentation. Nature 564(7735):263–267

6. Pennarossa G, Santoro R, Manzoni EFM et al (2018) Epigenetic erasing and pancreatic differentiation of dermal fibroblasts into insulin-producing cells are boosted by the use of low-stiffness substrate. Stem Cell Rev Rep 14 (3):398–411

7. Manzoni EFM, Pennarossa G, Deeguileor M et al (2016) 5-azacytidine affects TET2 and histone transcription and reshapes morphology of human skin fibroblasts. Sci Rep 6:37017

8. Erb TM, Schneider C, Mucko SE et al (2011) Paracrine and epigenetic control of Trophectoderm differentiation from human embryonic stem cells: the role of bone Morphogenic protein 4 and histone deacetylases. Stem Cells Dev 20:1601–1614

Chapter 11

3D-ViaFlow: A Quantitative Viability Assay for Multicellular Spheroids

Joel Mario Vej-Nielsen and Adelina Rogowska-Wrzesinska

Abstract

Three-dimensional cell culture became an essential method in molecular and cell biology research. Accumulating results show that cells grown in 3D, display increased functionality and are capable of recapitulating physiological functions that are not observed in classical in vitro models. Spheroid-based cell culture allows the cells to establish their own extracellular matrix and intricate intercellular connections promoting a tissue-like growth environment.

In this paper we present the 3D-ViaFlow method that combines an optimised dual live-dead cell staining with flow cytometry to deliver a quantitative estimation of viability of cells in multicellular spheroids. The method is optimised for monolayer cultures and multicellular spheroids created from HepG2/C3A human hepatocytes or coculture of HepG2/C3A and endothelial cell line HMEC-1. It includes protocol for spheroids disassembling, labeling of cells with fluorescein diacetate and propidium iodide and instructions for flow cytometry gating optimized for analysis of heterogeneous cell populations form spheroids.

Key words 3D-ViaFlow, Viability, Spheroid, 3D cell culture, Morphology, FACS, Flow cytometry, Fluorescein diacetate, Propidium iodide, High throughput

1 Introduction

Growing mammalian cells outside organisms in systems that support formation of three dimensional (3D) structures have gained a lot of scientific attention in the last decade. Such systems allow cells to recover several physiological functions that are normally lost outside the microenvironment of tissues and organisms. Among these physiologically relevant changes observed in 3D cell cultures are an increase in cell polarization, cell–cell and cell–matrix interactions, reviewed by Pampaloni et al. [1]. These properties of 3D cell culture systems have been shown to influence essential cellular mechanisms that are commonly studied using in vitro models (e.g., differentiation, drug response, and synthesis of physiologically relevant molecules) [2–4].

Tiziana A.L. Brevini et al. (eds.), *Next Generation Culture Platforms for Reliable In Vitro Models: Methods and Protocols*,
Methods in Molecular Biology, vol. 2273, https://doi.org/10.1007/978-1-0716-1246-0_11,
© The Author(s), under exclusive license to Springer Science+Business Media, LLC, part of Springer Nature 2021

A current bottleneck for utilizing the true potential of 3D spheroid cultures is a lack of robust downstream analysis protocols. Many of the existing methods have been developed using mono-layer or suspension cultures and are not easily adaptable for cells cultured as spheroids [1, 5]. An example of such a method is the estimation of cell viability. Cell viability is an essential indicator of the general health status within a population of cultured cells and it should be monitored prior to and during an experiment. Various approaches are being used to estimate the viability of cells grown in 3D cell cultures: ATP levels, lactate dehydrogenase activity, acid phosphatase activity, trypan blue staining and fluorescent staining in combination with imaging of intact and disassembled spheroids [6–10]. Due to the lack of a universal and robust method it is not uncommon that different assays are used to evaluate viability of cells in monolayer and 3D cultures in the same study [11, 12] compli-cating comparison and evaluation of culture systems.

In this chapter, we present the 3D-ViaFlow method that employs a dual viability stain using fluorescein diacetate (FDA) and propidium iodide (PI) in combination with flow cytometry. Through strict size and shape discrimination this method allows for robust viability evaluation of cells from multicellular spheroids and monolayer cultures. An example of results obtained with this method is presented in Fig. 1.

3D-ViaFlow relies on specific characteristics of FDA and PI. As an exclusion dye, PI is unable to pass through the intact cell membrane of live cells. Dead cells, however, have impaired mem-brane integrity which allows PI to enter the cell, bind to DNA and become fluorescent (excitation maximum 493 nm, emission maxi-mum 636 nm). FDA is a cell-permeant esterase substrate, that upon cleavage yields the fluorescent product fluorescein (excitation max-imum 495 nm, emission maximum 512 nm). FDA, unlike fluores-cein, penetrates intact membranes, which results in intracellular retention of its fluorescent product in live cells. Using the two fluorescent stains in combination greatly increases the confidence in the assignment of each cell's state [13].

In contrast to monolayer cultures, multicellular spheroids are complex structures that can contain several thousands of cells assembled in hundreds of layers. The distribution of live and dead cells within a spheroid is heterogeneous and a larger number of cells must be evaluated in order to obtain an accurate measurement. Furthermore, other objects such as dead cells, vesicles and mem-brane encased cell remnants are observed within the spheroid. This poses a technical challenge for manual, microscopy-based live/dead cell counting methods such as trypan blue staining, which is also limited in the number of cells that can be evaluated at a time and is prone to operator bias, ultimately reducing confidence, reproduc-ibility and comparability of results attained in different laboratories.

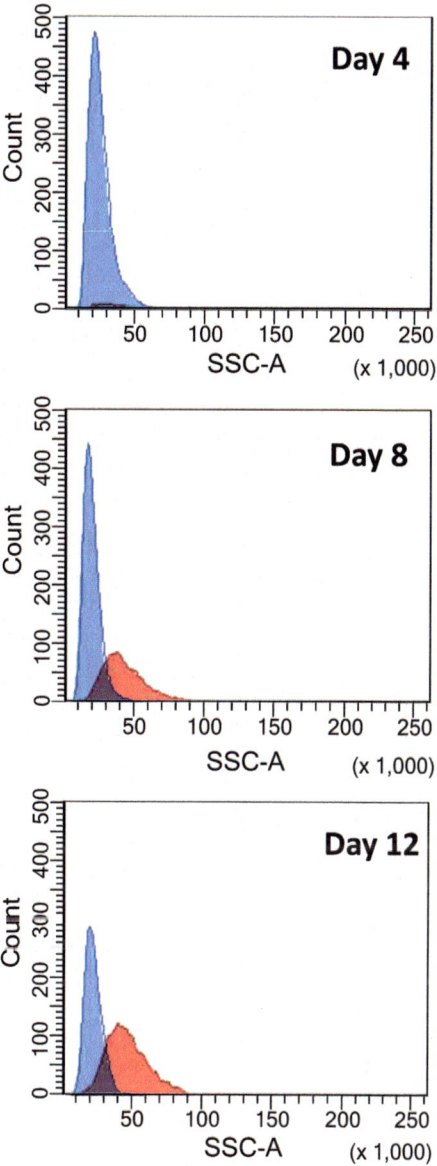

Fig. 1 Frequency histogram of live (red) and dead (blue) cells observed in HepG2/C3A spheroids at different ages as measured by 3D-ViaFlow method—the PI/FITC live dead staining combined with flow cytometry and spheroid disassembling method

Combination of the FDA and PI dual stain with flow cytometry overcomes those challenges and provides additional information that can be used to evaluate the heath state of the spheroids. Each flow cytometry measurement is based on 10,000 objects. Non–single-cell events and membrane remnants can be excluded by strict size and shape discrimination. Ambiguous objects that are either positive or negative for both stains can be excluded increasing the accuracy and comparability of the results between experiments and laboratories. Unlike viability estimations based on enzymatic activity and level of metabolites such as adenosine triphosphate (ATP), 3D-ViaFlow offers insight into the number of dead cells and provides morphological information that can be used to monitor cell health and behavior. Finally, the procedure for evaluating viability in monolayer and 3D cell culture, is almost identical making it ideal for comparative studies that utilize both models.

The 3D-ViaFlow protocol was tested on monoculture and coculture spheroids over 12 days in biological triplicates. The average viability observed, was in the range of 44.3% to 95.5% and 40.8% to 96.6% live cells, respectively. In our hands, the coefficient of variation of the presented method (and the viability of the 3D spheroids used in this study) was 2.87% and 2.86%. For reference, an average coefficient of variation of 9.11% and 6.65% was reported for the trypan blue method by Piccinini F. et al. [7]. This data was collected by two independent observers from 9 spheroid cultures in 5 technical replicates. This speaks to the high reproducibility and level of accuracy of 3D-ViaFlow as well as the low variation in viability within the spheroid population.

The additional strength of 3D-ViaFlow is that it allows for isolation of populations of live and dead cells while providing control of the amount of cell debris, cell aggregates and doublet cells included in the analysis, which improves the reliability of the estimation of cells state. This is especially important while evaluating viability of spheroids because they are characterized by a high level of heterogeneity of cell states (live, dead, proliferating, or in growth arrest). Evaluation of many cells is required to obtain results that are representative for an entire spheroid. Flow cytometry allows for the analysis of thousands of cells within a sample in less than 20 s. The automatic exclusion of noncell events by gating, further strengthens the reproducibility and accuracy of the method as all events are evaluated by identical criteria. Finally, 3D-ViaFlow employs a dual stain and evaluation of light scatter for increased confidence in the results. While ambiguous events are excluded from the viability estimation, the number of ambiguous events is used for quality control.

3D-ViaFlow was only tested on spheroids created without use of scaffolds or matrix materials and it is not established if such spheroids could be disassembled into single cells and analyzed with the presented protocol. Furthermore, it requires access to a flow cytometer, and is slow for measurements on less than three samples at a time.

The 3D-ViaFlow protocol provides reliable results only if the spheroid disassembling and PI and FITC staining procedures are optimized for the cells and spheroids in use. This can be monitored by following the number of cells that are either positive or negative for both viability stains. The doubly positive cells are likely cells that were alive at the beginning of the experiment, but had their membranes damaged during handling and spheroids disassembly. If a large number of doubly positive cells is observed, the incubation time for both staining and trypsinization should be decreased, use of other digestion enzymes (e.g., papain) and gentler handling should be considered. The doubly negative cells are likely noncell events that were not excluded in the initial gating (Subheading 3.1, steps 4–6) or cells that have not stained properly. If many doubly negative events are observed, increase the incubation time with PI and FITC (Subheading 3.1, step 2) and adjust gating for stricter exclusion of noncell events (Subheading 3.1, steps 4–6). In our hands, on average less than 2% of cells where either dual negative or dual positive.

2   Materials

2.1   Multicellular Spheroids and Monolayer Cultures

The monolayer cultures and spheroids used in the presented protocol were generated using AggreWell plates and cultured in rotary bioreactors as described by Fey and Wrzesinski [3] and described in detail by Wrzesinski et al. in this issue. Two types of multicellular spheroids were analyzed: monoculture spheroids from hepatocellular carcinoma cell line HepG2/C3A (ATCC® CRL-10741), henceforth, referred to as C3A and cocultured spheroids combining C3A and the endothelial cell line HMEC-1 (ATCC® CRL-3243). The cocultured spheroids were generated with 80% C3A cells and 20% HMEC-1 cells. All spheroids were cultured in growth media (Dulbecco's modified eagle medium (DMEM) supplemented with 10% fetal bovine serum, 1% nonessential amino acids, 0.5% penicillin–streptomycin, 1% GlutaMAX), media change was performed every other day on weekdays. All incubators in this protocol, were humidified with 5% CO_2 and 37 °C. Spheroids were washed in Hanks' Balanced Salt Solution (HBSS) when indicated in the protocols.

2.2   Prepare Flow Cytometer (See Note 1)

1. Fill ethanol tank with 96% ethanol and sheath tank with sterile phosphate-buffered saline (PBS).
2. Carefully wipe down deflection plates with dust free wipes.
3. Insert rubber tube in pinch valve.
4. Ensure that the neutral density filter matches the one used for cytometer setup and tracking (CS&T) baseline.

2.3   Start Flow Cytometer

1. Turn on flow cytometer (BD Bioscience—FACSAria III) and the connected computer.
2. Open the BD FACSDiva software (version 8.0.1, CST version 3.0.1, PLA version 2.0).
3. Run fluidics start-up (see Note 2).

4. Click "cytometer" then "view configurations" and select the installed nozzle.

5. Turn flow rate down to 1.

6. Turn on the stream and adjust frequency and amplitude to get drop 1 and gap of 150 and 10, respectively (*see* **Note 3**).

7. Wait until the drop pattern stabilizes (approx. 20 min), before readjusting the settings.

2.4 Calibrate Flow Cytometer

1. Mix 1 drop of CS&T beads with 350 µl of sterile PBS in a polystyrene round-bottom tube.

2. Load into the flow cytometer, click "use CST settings" to use the new calibration (*see* **Note 4**).

3. After CS&T calibration, insert the 2.0 neutral density filter (*see* **Note 5**).

2.5 Set up Flow Cytometer for a New Experiment (See Note 6)

1. Create a new experiment, select a blank experiment and click "OK."

2. Create new specimen in the new experiment and new tubes in the specimen (*see* **Note 7**).

3. Select tube of interest by clicking icon to the left of the tube name.

4. Open the "Parameter" tab in the cytometer window.

5. Remove all parameters except side scatter (SSC) and forward scatter (FSC), by selecting them and clicking delete.

6. Add PE (585/45) and FITC (530/30) to measure intensity of PI and FDA, respectively.

7. Generate three plots: FSC-A against SSC-A, FSC-A against FSC-H, and PE against FITC.

8. Load sample containing cells labeled with PI and FDA (Subheading 3.1, **steps 4–6**) and proceed to acquire data.

9. Open the parameter tab.

10. Adjust voltage of FSC and SSC while observing the FSC-A against SSC-A plot.

11. Adjust voltage of PE and FITC while observing the PE against FITC plot.

12. The events should be spread across the entire X- and Y-axis without loss of nonoutliers.

13. Stop acquiring data and remove the sample containing cells.

14. FACS is now ready for a new experiment.

2.6 Viability Stain (See Note 8)

1. Prepare stock solution of 1 mg of FDA in 1 ml DMEM and store at $-20\ ^{\circ}$C.

2. Mix 8 µl of 1 mg/ml FDA, 20 µl 1 mg/ml PI in 1 ml DMEM.

2.7 Trypsin Solution	1. Prepare stock solution of 0.5% trypsin in EDTA stored at −20 °C.
	2. Dilute trypsin stock to 0.05% in HBSS just prior to use (*see* **Notes 9** and **10**).

3 Methods

3.1 Disassemble Spheroids into Single Cells	1. Harvest 5 to 10 spheroids using a cut pipette tip (*see* **Notes 11** and **12**).
	2. Transfer them to a fresh 1.5 ml Eppendorf tube containing 1 ml 37 °C HBSS.
	3. Let the spheroids sediment to the bottom of the tube and gently collect them with a cut pipette tip.
	4. Transfer the spheroids to a fresh 1.5 ml Eppendorf tube containing 200 µl of 0.05% trypsin (*see* **Note 13**).
	5. Transfer the Eppendorf tube to an incubator for 5 min.
	6. Aspirate the spheroids into a p200 pipette tip (uncut) (*see* **Notes 14–16**) repeatedly.
	7. Continue until no visible spheroid fragments remain (*see* **Notes 17** and **18**).
	8. Pellet the cells by centrifugation at 37 °C and remove the supernatant (*see* **Note 19**).

3.2 Collect Cells from Monolayer Cultures	1. Grow monolayercultures to the desired density (*see* **Note 20**).
	2. Remove the growth media from the cells and wash twice with 5 ml HBSS.
	3. Add 3 ml of 0.05% trypsin and incubate for 3 min at 37 °C (*see* **Note 21**).
	4. Add 1 ml of fetal bovine serum to quench trypsin.
	5. Collect the cells and transfer to a sterile 15 ml falcon tube.
	6. Wash the flask with 5 ml growth media and transfer the media to falcon tube to collect the remaining cells.
	7. Pellet the cells by centrifugation at 37 °C and remove the supernatant (*see* **Note 19**).
	8. Use at least 15,000 cells for dual viability staining.

3.3 Apply Dual Viability Stain	1. Prepare viability stain (Subheading 2.3, **item 1**).
	2. Resuspend cell pellet prepared in Subheading 3.1, **steps 1** or **2** in 200 µl staining solution.
	3. Incubate the sample in the dark for 3 min at room temperature (*see* **Note 22**).
	4. Analyze sample on flow cytometer.

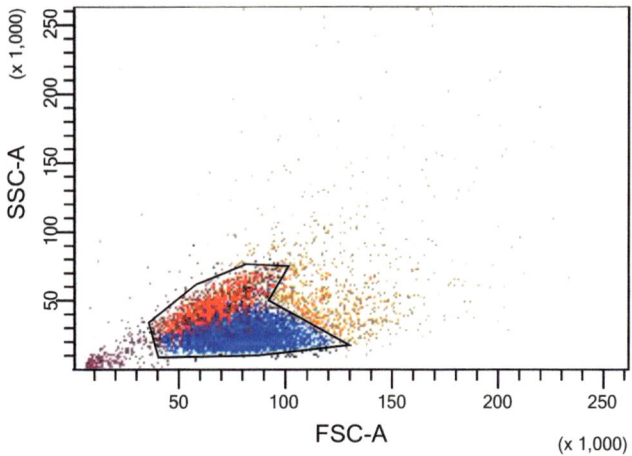

Fig. 2 Flow cytometry plot of forward scatter area (FSC-A) against side scatter area (SSC-A) from cells derived from 7-day-old HepG2/C3A spheroid obtained using the 3D-ViaFlow method. The purple dots are excluded from the analysis as suspected cell debris. The orange dots are excluded from the analysis as suspected cellular aggregates. The blue (live) and red (dead) cells within the gate are included in further analysis

5. Transfer the sample to a polystyrene round-bottom tube.

6. Briefly vortex the sample and load it onto the flow cytometer.

7. Acquire data and adjust flow rate to attain approximately 1500 events/s for most accurate results with a 100 μm nozzle.

8. Record 10.000 events and discard the sample.

3.4 Exclude Cellular Debris and Cellular Aggregates (See Notes 23 and 24)

1. Click the "Dot Plot" option in the global worksheet to generate a new plot.

2. Generate a dot plot with forward scatter (FSC-A) against side scatter (SSC-A).

3. Click "Polygon Gate" to set a new gate.

4. Place the gate over the densely packed area of the plot (Fig. 2) (*see* **Note 25**).

5. Adjust the gate to exclude the dense area off events with lower FSC-A and SSC-A (*see* **Note 26**).

6. Next, adjust the gate to exclude events with large values for FSC-A and SSC-A (*see* **Note 27**).

3.5 Exclude Doublet Cells

1. Click the "Dot Plot" option in the global worksheet to generate a new plot.

2. Generate a plot with forward scatter area (FSC-A) against forward scatter height (FSC-H).

3. Right click on the plot, click "Show population" and select P1 (*see* **Note 28**).

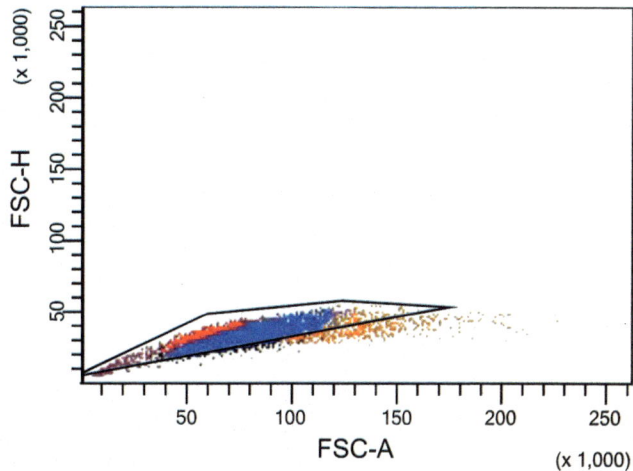

Fig. 3 Flow cytometry plot of forward scatter area (FSC-A) against forward scatter height (FSC-H) from cells derived from 7-day-old HepG2/C3A spheroid obtained using the 3D-ViaFlow method. The purple (cell remnants) and orange (cell aggregates) dots are previously excluded events. The black events outside the gate are excluded from further analysis (suspected doublet cells). The blue (live cells) and red (dead cells) dots are single cell events that are included in further analysis

4. Click "Polygon Gate" to set new gate.

5. Place the gate over the cone of events in the plot (Fig. 3).

6. Adjust the bottom of the gate to exclude single events with high FSC-A and low FSC-H (*see* **Note 29**).

3.6 Define Live and Dead Cell Populations

1. Click the "Dot Plot" option in the global worksheet to generate a new plot.

2. Generate a plot with red fluorescence (PI-A) against green fluorescence (FITC-A).

3. Right click on the plot, click "Show population" and select P2.

4. Click "Rectangle Gate" to generate two square gates.

5. Fit the gates to the population with high FITC-A values (top left) and high PI-A values (bottom right), respectively (Fig. 4) (*see* **Notes 30–32**).

3.7 Export High Resolution Plots (See Note 33)

1. Open the sample of interest.

2. Click "File" tab in the menu and then select the "Print" option.

3. Choose "Microsoft XPS Document Writer" option (*see* **Note 34**).

4. Open the newly generated PDF file with InkScape (Version 1.0).

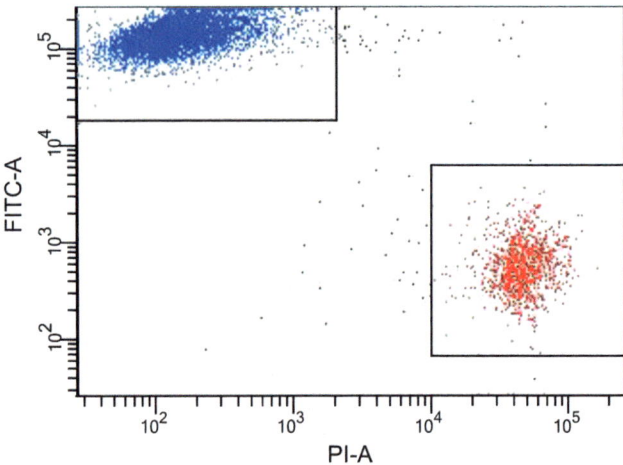

Fig. 4 Flow cytometry plot of red fluorescence (PI-A) against green fluorescence (FITC-A) from cells derived from 7-day old HepG2/C3A spheroid obtained using the 3D-ViaFlow method. Discrimination between live and dead cells is based on emitted fluorescence signal. Cells positive for FDA emit strong fluoresce at the wavelength corresponding to the FITC channel, cells positive for PI emit strong fluoresce at the wavelength corresponding to the PI channel. Cells within the gates are considered live (blue) and dead (red), respectively. The black dots are not included in the viability estimation but are used for assessing the quality of the staining process

5. Choose internal import, have all check boxes ticked and click OK.

6. Select a plot of interest, make sure the axis values and titles are not part of the selection.

7. Click "Filters," then "Fill and Transparency" and then "Light eraser."

8. Set Expansion at 750, Erosion at 950, Global opacity at 1, and click apply.

9. You can alter colors, the size and font of the text, and remove/move labels if necessary.

10. Finally export the image in the desired resolution.

4   Notes

1. Flow cytometer used in the experiments presented here was FACS Aria version III. Any flow cytometer with the PI and FITC channels can be used for this protocol.

2. C3A, HMEC-1 and most other eukaryotic cells use 100 μm nozzle.

3. These values are specific for 100 μm nozzle.

4. Ensure DeltaPMTV and %difference target value are smaller than 10. If not, clean the flow cytometer and repeat the calibration.

5. This filter is ideal for C3A, HMEC-1 and most eukaryotic cells, but might need to be changed for other cell types.

6. Data that you want to compare should be acquired with identical voltage settings.

7. The specimen and tube names will be displayed in the plot.

8. Use a light safe Eppendorf tube to protect the dye from photobleaching. FDA (Cat# F7378) and PI (Cat# P4170) was acquired from sigma.

9. Aliquot 0.5% trypsin in 1 ml for quick dilution by addition of 9 ml HBSS.

10. The diluted trypsin can be refrozen a single time before it should be discarded.

11. Samples should contain around 15,000 cells, adjust the number of spheroids used for each sample accordingly.

12. Cut p1000 pipette tip creating a tip diameter of minimum 2 mm or use commercially available pipette tips with larger opening.

13. In case many samples need to be processed simultaneously trypsin aliquots can be prepared and stored at 4 °C.

14. The trypsin can be inactivated by adding 50 µl FBS to avoid further digesting and damaging the cells.

15. To fully disassemble fractions of smaller spheroid, narrow the pipette opening by pressing the pipette tip against the bottom of the Eppendorf tube at an angle.

16. Set the pipette to 150 µl to avoid sucking liquid into the pipette during disassembly.

17. Check that the spheroids have been successfully reduced to single cell suspensions under light microscope.

18. If the spheroids are "intact" after 5–10 repetitions, extend the incubation time.

19. C3A and HMEC-1 cells are effectively pelleted at $140 \times g$ for 5 min.

20. Typically, 70% confluency is used. Note that cell viability generally is higher in monolayer culture as dead cells are not retained as they are within spheroids.

21. Check that the cells detach from the surface of the flask under light microscope.

22. It is crucial that incubation time is optimized for each cell and spheroid type, to avoid a large increase in false positive or false negative results (Subheading 1).

23. Flow cytometry data analysis is based on assigning events to cellular populations. Assignment is carried out by placing gates around events with common characteristics.

24. These exclusions are especially important when working with 3D cell cultures as spheroids contain cells at different life cycle stages and are difficult to reduce to a single cell suspension due to their extensive extracellular matrix.

25. Larger events will generate higher forward scatter while events with a high level of internal complexity will generate higher side scatter.

26. Objects with low forward and side scatter are most likely cellular debris that will mostly be confined to a small highly dense area in the bottom left corner of the plot.

27. Objects with large forward or side scatter are very often cell aggregates, these objects are typically scarcely distributed above and to the right of the dense main population, due to large variation in both their size and internal complexity.

28. This step removes the previously excluded events from the plot. "P1" is the automatically generated name for the first assigned population.

29. Singlet cells will generally have a spherical shape and as such the relationship between their height and area is linear. Doublet cells, however, will be elongated, resulting in a higher area-to-height ratio on the plot. This will result in them falling below the cone of singlet cells on the plot.

30. Ensure that there is no overlap between the two gates on either axis.

31. The gates will be saved and can be used for each subsequent measurement in the experiment. It is, however, important to evaluate whether the gates need to be adjusted when making new measurements.

32. For studies on spheroids at around 14 days, a decrease in PI intensity will be observed. As long as early measurements have a low number of dual negative cells, assume all dual negative events are dead cells for quantification of viability.

33. Many journals require high resolution figures (>300 dpi), which are not typically provided by the software associated with flow cytometers, in such cases high resolution plots can be obtained using other freely available tools. For extensive instructions *see* Weber et al. [14].

34. Install the free software PDFcreator as an alternative if your PDF reader does not support this option.

Acknowledgments

The authors acknowledge the financial support from the COST Action CA16119 (In vitro 3-D total cell guidance and fitness) and the Sino Danish Center for Education and Research for PhD grant to JMVN.

References

1. Pampaloni F, Reynaud EG, Stelzer EHK (2007) The third dimension bridges the gap between cell culture and live tissue. Nat Rev Mol Cell Biol 8:839–845

2. Wang W, Itaka K, Ohba S, Nishiyama N, Chung U-i, Yamasaki Y et al (2009) 3D spheroid culture system on micropatterned substrates for improved differentiation efficiency of multipotent mesenchymal stem cells. Biomaterials 30(14):2705–2715

3. Fey SJ, Wrzesinski K (2012) Determination of drug toxicity using 3D spheroids constructed from an immortal human hepatocyte cell line. Toxicol Sci 127(2):403–411

4. Wrzesinski K, Fey SJ (2013) After trypsinisation, 3D spheroids of C3A hepatocytes need 18 days to re-establish similar levels of key physiological functions to those seen in the liver. Toxicol Res (Camb) 2(2):123–135

5. Mehta G, Hsiao AY, Ingram M, Luker GD, Takayama S (2012) Opportunities and challenges for use of tumor spheroids as models to test drug delivery and efficacy. J Control Release 164(2):192–204

6. Sirenko O, Mitlo T, Hesley J, Luke S, Owens W, Cromwell EF (2015) High-content assays for characterizing the viability and morphology of 3D cancer spheroid cultures. Assay Drug Dev Technol 13(7):402–414

7. Piccinini F, Tesei A, Arienti C, Bevilacqua A (2017) Cell counting and viability assessment of 2D and 3D cell cultures: expected reliability of the trypan blue assay. Biol Proced Online 19(1):8

8. Madoux F, Tanner A, Vessels M, Willetts L, Hou S, Scampavia L et al (2017) A 1536-well 3D viability assay to assess the cytotoxic effect of drugs on spheroids. SLAS Discov Adv Sci Drug Discov 22(5):516–524

9. De Witt Hamer PC, Jonker A, Leenstra S, Ruijter JM, Van CJF N (2005) Quantification of viability in organotypic multicellular spheroids of human malignant glioma using lactate dehydrogenase activity: a rapid and reliable automated assay. J Histochem Cytochem 53(1):23–34

10. Friedrich J, Eder W, Castaneda J, Doss M, Huber E, Ebner R et al (2007) A reliable tool to determine cell viability in complex 3-D culture: the acid phosphatase assay. J Biomol Screen 12(7):925–937

11. Perard M, Tricot-Doleux S, Pellen-Mussi P, Meary F, Pérez F (2011) Evaluation of the cytotoxicity of pulp floor perforation filling materials by using in parallel 2d and 3d culture models. Bull Group Int Rech Sci Stomatol Odontol 50(2):42–43

12. Kim YE, Jeon HJ, Kim D, Lee SY, Kim KY, Hong J et al (2018) Quantitative proteomic analysis of 2D and 3D cultured colorectal cancer cells: profiling of Tankyrase inhibitor XAV939-induced proteome. Sci Rep 8(1):1–12

13. Jones KH, Senft JA (1985) An improved method to determine cell viability by simultaneous staining with fluorescein diacetate-propidium iodide. J Histochem Cytochem 33(1):77–79

14. Weber K, Fehse B (2010) Diva-fit: a step-by-step manual for generating high-resolution graphs and histogram overlays of flow cytometry data obtained with FACSDiva software. Cell Ther Transplant 1(4)

Chapter 12

Method to Disassemble Spheroids into Core and Rim for Downstream Applications Such as Flow Cytometry, Comet Assay, Transcriptomics, Proteomics, and Lipidomics

Helle Sedighi Frandsen, Martina Štampar, Joel Mario Vej-Nielsen, Bojana Žegura, and Adelina Rogowska-Wrzesinska

Abstract

Cells cultured in a monolayer have been a central tool in molecular and cell biology, toxicology, biochemistry, and so on. Therefore, most methods for adherent cells in cell biology are tailored to this format of cell culturing. Limitations and disadvantages of monolayer cultures, however, have resulted in the ongoing development of advanced cell culturing techniques. One such technique is culturing cells as multicellular spheroids, that had been shown to mimic the physiological conditions found in vivo more accurately. This chapter presents a novel method for separation of the spheroid rim and core in mature spheroids (>21 days) for further analysis using advanced molecular biology techniques such as flow cytometry, viability estimations, comet assay, transcriptomics, proteomics and lipidomic. This fast and gentle disassembly of intact spheroids into rim and core fractions, and further into viable single-cell suspension provides an opportunity to bridge the gap from 3D cell culture to current state-of-the-art analysis methods.

Key words 3D cell culture, HepG2/C3A, Spheroids, Rim and core, Flow cytometry, Comet assay, Transcriptomics, Proteomics, Lipidomic

1 Introduction

For more than a century monolayer (2D) cultures have been used as a valuable model for cell-based in vitro studies. Recently several limitations, such as the loss of tissue-specific architecture, mechanical and biochemical identity and cell-to-cell communication, have been increasingly recognized as a significant drawback of such systems [1]. At the same time, three-dimensional (3D) cell cultures have emerged as a new approach providing environment physiologically more relevant and data with more predictive power for in vivo conditions [2]. Cells in the 3D culture differ morphologically and physiologically from cells in monolayer cultures and

Tiziana A.L. Brevini et al. (eds.), *Next Generation Culture Platforms for Reliable In Vitro Models: Methods and Protocols*, Methods in Molecular Biology, vol. 2273, https://doi.org/10.1007/978-1-0716-1246-0_12,
© The Author(s), under exclusive license to Springer Science+Business Media, LLC, part of Springer Nature 2021

provide a more accurate insight into the responses to stimuli, drug metabolism, and general cell function [3]. Creation of multicellular spheroids or organoids is one of the ways for providing three-dimensional microenvironment for cells grown in cell cultures. Three main zones can be observed in those multilayer cell aggregates: an outer proliferating rime, viable nonproliferating zone and an inner core that contains cell remnants. Cells from different zones within the spheroid have different morphology [4]. It is currently understood that these zones are formed in response to the exposure of cells to different microenvironment factors such as access to nutrients, CO_2, and oxygen levels [5, 6]. In cancer research, the zones of spheroids are thought to represent the behavior of cells in larger tumors, where a necrotic core appears at the center of tissue mass. The different physiological states of cells mimic the cellular heterogeneity of solid in vivo tumors [7].

The availability of accurate, robust, and cost-effective in vitro assays for 3D cell cultures is an increasingly important challenge for applications in basic research, toxicity testing, and safety assessment [3]. The existing 3D cell models vary widely due to the diverse requirements of different cell lines and applications, and each model has its own advantages and limitations [8, 9]. The methods described in this chapter are applicable to spheroids obtained with a scaffold independent system and had been tested on HepG2/C3A spheroids cultured for 21 days.

The protocol provides an efficient method for disassembly of spheroids into the core and rim fractions and to obtain a single-cell suspension. This new approach allows for the application of a wide variety of existing techniques, enabling us to study the biological processes taking place in the inner and outer fraction of the spheroid. Within this protocol we provide guidelines for how to prepare samples from spheroids' core and rim for downstream applications techniques such as comet assay, flow cytometry, transcriptomics, proteomics, and lipidomics. Figure 1 presents the summary of methods described.

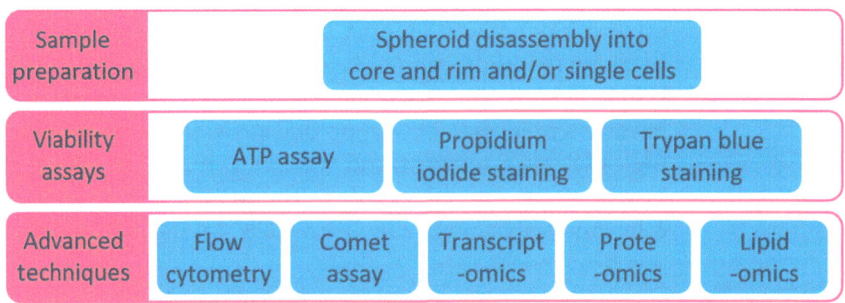

Fig. 1 Overview of the methods presented in the chapter. (**step 1**) spheroid processing and separation into core and rim or single-cell suspension; (**step 2**) evaluation of cell population by measuring cell viability; (**step 3**) analysis by comet assay, flow cytometry, transcriptomics, proteomics, and lipidomics

2   Materials

A sterile laminar flow hood, humidified incubator with 5% CO_2 at 37 °C and a centrifuge are required in several steps of the protocol.

2.1   Separation of Spheroids into Core and Rim Fractions and Single-Cell Suspension in Sections

Materials from this section are used in protocols described in Subheading 3.1.

1. Eppendorf tubes 1.5 mL.

2. 0.5% Trypsin–EDTA.

3. Hanks Balanced Salt Solution without calcium and magnesium (HBSS).

4. Cut P1000 tip (*see* **Note 1**) and cut P200 pipette tip (*see* **Note 2**).

5. Fetal bovine serum (FBS).

3   Methods

This chapter utilizes spheroids created using AggreWell plates (STEMCELL Technologies UK), rotary bioreactors and the CelVivo bioarray matrix (BAM) system (CelVivo Denmark) as described before by Fey and Wrzesinski [10] and by Wrzesinski et al. in this issue. It can also be applied to multicellular spheroids and organoids created with other systems. All handling is performed at room temperature; all liquids are preheated to 37 °C unless otherwise stated.

3.1   Spheroid Disassembly

This protocol presents a technique to separate the inner core and outer rim fractions of intact spheroids, and subsequently single-cell suspensions either from intact spheroid or core/rim fraction. Figure 2 shows the overview of the workflow of spheroid formation and disassembly.

3.1.1   Single-Cell Suspension from Spheroid

Preparing a single-cell suspension is a critical step, when spheroids had been grown for several weeks and tight connections between cells have been formed. Spheroid disassembly, enzymatic digestion, and mechanical dissociation are the major steps leading to the degradation of the intracellular connections and to the isolation of viable single cells.

1. Harvest desired number of spheroids at specific age (suggested at 21 days) and transfer them to an Eppendorf tube with 200 μL HBSS (*see* **Note 3**).

2. Create diluted trypsin–EDTA solution (0.05%) by mixing 9 mL cool HBSS into a 15 mL falcon tube with 1 mL trypsin/EDTA 0.5% (*see* **Note 4**).

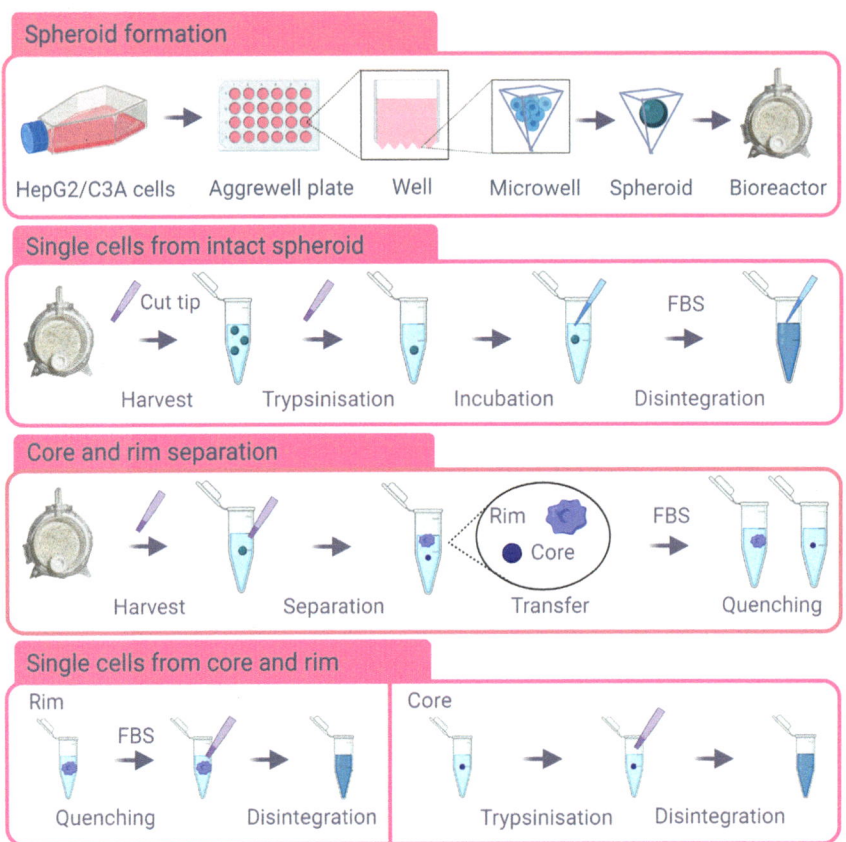

Fig. 2 Overview of the spheroid disassembly workflow steps, with key processes indicated

3. Transfer an appropriate number of spheroids to an Eppendorf tube containing 200 μL of 0.05% trypsin–EDTA in HBSS.

4. Incubate spheroids for 8 min at 37 °C (this refers to the spheroids of the age of 21 days) (*see* **Note 5**).

5. Gently aspirate the spheroids using a P200 pipette (uncut tip). Repeat approximately 50 times or until the spheroid breaks into smaller fragments.

6. Add 200 μL of FBS to deactivate/quench the trypsin.

7. Aspirate the solution into the pipette tip for additional five times to create a single-cell suspension.

8. After obtaining a single cell-suspension examine viability of cells using trypan blue staining (Subheading 3.2.1.1).

3.1.2 Separation of Spheroids into Core and Rim Fractions

This protocol can only be performed on spheroids with a mature core, for system described in this article the core will be mature after approximately 21 days of culturing.

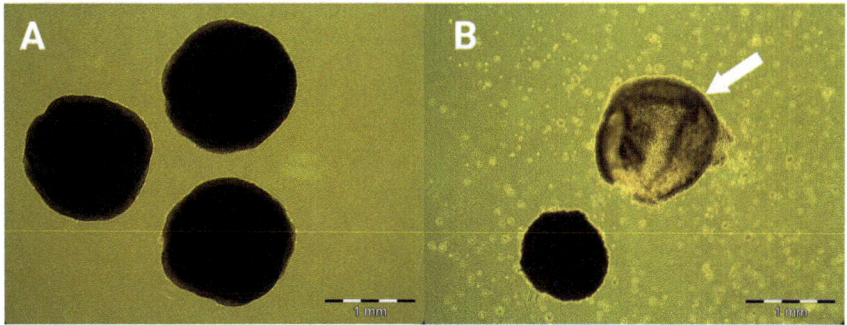

Fig. 3 Separation of 32-day-old HEPG2/C3A spheroids into core and rim structures. (**a**) Three individual intact spheroids aged 32 days; (**b**) one spheroid disassembled into rim (arrow) and core fractions. Scale bar equals 1 mm

1. Prepare diluted trypsin solution (0.05%) (Subheading 3.1.1, **step 2**) (*see* **Note 4**).

2. Collect the spheroids from the bioreactor and transfer them to a petri dish (35 mm) containing 3 mL growth media, use a cut pipette tip in all steps unless indicated otherwise (*see* **Notes 2, 6,** and **7**).

3. Select one spheroid at a time and transfer to an Eppendorf tube containing 200 μL HBSS.

4. After few seconds transfer the spheroid to a new Eppendorf tube containing 200 μL 0.05% trypsin–EDTA.

5. Incubate at 37 °C for 4 min (the time depends on spheroid size and compactness).

6. Aspirate the spheroid while still in trypsin–EDTA solution, aspirate and drain 30 times until rim peels off (Fig. 3) (*see* **Note 8**).

7. Immediately transfer the core to an Eppendorf tube containing 200 μL FBS.

8. Add 200 μL of FBS to the Eppendorf tube with the spheroid rim.

3.1.3 Single-Cell Suspension from Core and Rim

To generate single-cell suspensions from each spheroid part, first perform the separation of spheroids into core and rim fractions (Subheading 3.1.2, **steps 1–6**), then follow this procedure.

1. Transfer the core to a new Eppendorf tube containing 200 μL 0.05% trypsin–EDTA and incubate it for additional 2 min at room temperature.

2. After incubation, resuspend spheroid core ten times with a cut P200 pipette tip (*see* **Note 9**) until single-cell suspension in obtained. Add 100 μL of FBS.

3. Keep the spheroid rim in the original tube.

4. Resuspend it with a cut P200 pipette tip five times to obtain a single-cell solution (*see* **Note 10**). Add 100 μL FBS.

5. Centrifuge both spheroid parts (core/rim) for 5 min at $140 \times g$ at 37 °C.

6. Remove supernatant and resuspend cells in a solution of choice or snap-freeze cells for later analysis.

3.2 Applications

This section presents how samples created with the above protocols can be used to analyze cells for viability and by comet assay, flow cytometry, transcriptomics, proteomics and lipidomic.

3.2.1 Cell Viability

Estimating cell viability is an essential tool to evaluate the cell culture population. All viability estimations highlighted in this chapter, require disassembly of the spheroid. As such, sterile conditions are not necessary as the cells are subsequently not cultured. Table 1 summarizes commonly used viability methods with their strengths and weaknesses highlighted.

Trypan Blue Staining

This protocol was adopted from [11]. The single-cell suspension can also be used in automated cell counters.

1. Prepare a 0.2% solution of trypan blue, in HBSS with calcium and magnesium in an Eppendorf tube.

2. Prepare single-cell suspension from spheroid(s) of interest as described in Subheadings 3.1.1 or 3.1.3 (*see* **Note 11**).

Table 1
Benefits and shortcomings of three frequently used viability assays

	Advantages	Disadvantages
Trypan blue staining	Quantify number of cells Fast (automated)	Quickly overstains Chance of induced cell death Hard to discriminate live/dead
ATP assay	Works directly on intact spheroids, rim/core fraction, and single cells (low probability of induced cell death) Evaluates entire spheroid	No exact estimation of the number of dead cells Slow (40 min, incubation) Can over- or underestimate ATP since amount is not the same in core and rim Requires sophisticated equipment (luminometer)
Flow cytometry and propidium iodide staining	Quantify number of cells Fast Semiautomated High number of cells evaluated in short time, giving more reliable results	Possibility of induced cell death Possibility to overstain Requires sophisticated equipment (flow cytometer)

3. Transfer and mix single-cell suspension with staining solution to attain a final concentration of 0.2% trypan blue (*see* **Note 12**).

4. Use a hemocytometer and light microscopy to quantify the amount of dead and live cells in sample (*see* **Note 13**).

ATP Assay

ATP is present in all living cells and the amount of ATP within the spheroid can be used as an indicator of the relative number of viable cells. For more detailed protocol for conducting ATP assay for measurement of viability in spheroids follow the procedure as described before by Fey and Wrzesinski [12] and Wrzesinski et al., in this issue. As the assay utilizes cell lysis to release ATP from the cells, it can be applied directly on intact spheroids as well as spheroid fractions and single-cell suspension.

1. Mix CellTiter-Glo reagents and leave on a shaker for 5 min at room temperature at 300 rpm.

2. Harvest spheroids from a bioreactor and transfer them to a petri dish containing growth media (*see* **Note 14**).

3. Prepare an appropriate ATP standard for the size of the spheroids-of-interest. As spheroids grow during the experiment the standard may need to be adapted throughout the experiment (*see* **Note 15**).

4. Add samples, ATP standards and negative control (blank) to a 96-well plate, in triplicates.

5. Add CellTiter-Glo reagents to each well with content and mix gently by pipetting.

6. Facilitate thorough cell lysis by additional pipetting in wells containing spheroids and confirm complete spheroid destruction (only single cells visible) prior to incubation period by light microscopy.

7. Wrap the plate in aluminum foil and incubate for 40 min on shaker at 300 rpm at room temperature.

8. Determine luminescence signal of samples using plate reader.

Flow Cytometry Using Propidium Iodide

Viability can be estimated by combining spheroid disassembly with propidium iodide staining [13].

1. To analyze spheroids by flow cytometer, the single-cell suspension must be prepared from the spheroids (whole or core/rim) as described in Subheadings 3.1.1 or 3.1.3.

2. Mix cells with a solution of 10 μg/mL propidium iodide (PI) at a ratio of 1:1.

3. Load cells on flow cytometer.

4. Estimate viability based on 10,000 events.

3.2.2 *Comet Assay*   The comet assay (single cell gel electrophoresis) is a simple and sensitive technique for detecting DNA damage, caused by compounds of interest. With this method, different types of strand brakes can be observed such as DNA double-strand breaks (DSB) and single-strand breaks (SSB), alkali labile sites (ALS) like apurinic/apyrimidinic (AP) sites, DNA–DNA and DNA–protein cross-links, and SSB associated with incomplete excision repair. The protocol is presented as described previously by Zegura and Filipic [14].

1. Prepare a viable single-cell suspension by following the procedure described in Subheadings 3.1.1 or 3.1.3.

2. Evaluate viability of the cell suspension (Subheading 3.2.1). The comet assay can be performed up to 25% decrease of viability.

3. Mix 30 μL single-cell suspension with 70 μL 1% low melting point (LMP) agarose.

4. Transfer 70 μL of the mixture of LMP and cell suspension to fully frosted slides (or slides covered with 0.5% NMP) covered with a layer of 1% NMP (normal melting point) agarose.

5. Add lysis buffer (0.1 MEDTA, 2.5 M NaOH, pH 10, 0.01 M Tris and 1% Triton X-100) to the slides and incubate for 1 h at 4 °C in darkness (*see* **Note 16**).

6. Perform DNA denaturation for 20 min in alkaline solution (300 mM NaOH, 1 m M EDTA, pH 13) at 4 °C to allow DNA unwinding and subsequently conduct the electrophoresis for 20 min at 0.7–1 V/cm (*see* **Note 17**). Neutralize the nuclei in neutralization buffer (0.4 M Tris buffer; pH 7.5) for 15 min in darkness.

7. Stain the gels with GelRed or similar stain (e.g., acridine orange).

8. Analyze the comets using a fluorescence microscope and the image analysis software (e.g., Comet IV from Instem, UK).

9. The experiment should be carried out on at least five spheroids each considered as a separate unit and at least 50 nuclei should be analyzed from each spheroid.

3.2.3 *Flow Cytometry*   Flow cytometry is a powerful technology in cell biology as it allows for high-throughput analysis of large number of cells. It is a technique that utilizes fluorescent dyes to quantify and identify cellular components and cell types. In flow cytometry, cells treated with one or more fluorochrome(s) undergo monochromatic excitations by one or more laser(s) and the resulting fluorescence is collected by detectors in the machine [15]. The technique can effectively be applied to the 3D cell culture model as it can be disassembled into the highly homogenous viable single-cell suspension. This makes the tool ideal to keep track of cellular populations within the spheroid.

Table 2
Examples of flow cytometry applications used in cell culturing, adapted from [16]

Flow cytometry application	Source of fluorescence
Quantifying cells in specific cell cycle phases	Anti-Ki67, anti-PCNA, Hoechst 33342, 7AAD, and Chromomycin A3
Measure level of active proliferation	Bromodeoxy uridine (BrdU) and conjugated anti-BrdU dye Carboxyfluorescein succinimidyl ester (CFSE)
Quantifying cell viability	Fluorescent diacetate (FDA), propidium iodide (PI), and Hoechst 33342
Identifying differentiation states and quantifying amount of differentiation	Conjugated antibodies for differentiation markers
Quantifying cell types in cocultures	Endogenous fluorescence in reporter cell lines
Identifying or isolating cell types in immunology	Conjugated antibodies for lineage markers
Measuring cell activation and signaling	Calcium indicators: Indo-1 and fluo-3
Measuring specific antigen response	Biotinylated MHC multimers, in combination with fluorescent streptavidin
Identifying mechanism of cell death	JC-1, anti-APO 2.7, fluorogenic caspase substrates, and anti-annexin-V
Measuring phagocytosis	pH sensitive fluorescently tagged biomolecules or bacteria
Perform generational tracking	Carboxyfluorescein succinimidyl ester (CFSE)
Measure level of protein expression	Inducible expression of fluorescent protein

1. In order to analyze spheroids by flow cytometer, the viable single-cell suspension must be prepared first as described in Subheadings 3.1.1–3.1.3.

2. Prepare a staining solution by mixing fluorochromes of interest (Table 2) (*see* **Note 18**).

3. Mix single-cell suspension with staining solution and incubate in the dark for appropriate time (*see* **Notes 19** and **20**).

4. If you use multicolor staining each dye can contribute to the signal on several detectors. Therefore, it is essential to perform the calibration of the lasers prior to starting an experiment especially when applying a new cell line.

5. Run cell suspension through a strainer in order to remove larger cellular aggregates and to prevent clogging of the tubes in the flow cytometer.

6. Load cells on flow cytometer in an approximate concentration of 10^6 cells/mL.

3.2.4 Omics

Omics analysis have developed immensely in the last 10 years and are highly used technologies aiming to collectively characterize and quantify pools of biological molecules, subsequently translating into function and dynamics of an organism. Due to the heterogeneity of cells within spheroids, the rim and core separation protocol could be performed prior to omics analysis to increase information output.

Washing and Freezing of Samples

It is important to perform extensive washing after collecting the spheroids to avoid contaminating the sample with growth media components such as fetal calf serum which can interfere with the final results.

1. Transfer intact spheroids or spheroid core and rim (prepared as in Subheading 3.1.2) into separate Eppendorf tubes.

2. Remove growth media or any other solutions and gently wash with 500 μL HBSS.

3. For proteomics and lipidomic experiments repeat the washing step five times to remove contaminating FBS proteins and lipids.

4. Remove all liquid after the final wash step using for an example a gel loading tip.

5. Add RNA stabilization and storage solution such as RNAlater for samples destined for transcriptomics or store samples without any liquid for later proteomics and lipidomics analysis.

6. Snap-freeze samples in liquid nitrogen and store at −80 °C (*see* **Note 21**).

Sample Preparation for Transcriptomic

High-throughput quantitative, real-time, reverse-transcription PCR (QPCR) is the method of choice for measuring the relative level of expression of selected gene transcripts in a given tissue or cell type, and after pharmacologic or genotypic manipulation [17]. The presented protocol had been described in detail by Buh et al. and Stampar et al. [18, 19] previously. A short summary of the protocol follows.

1. Prepare spheroids or spheroid core and rim according to the protocols from Subheadings 3.2.4.1 and 3.1.2, respectively (*see* **Note 22**).

2. Each sample for transcriptomic analysis should contain at least 1 μg of RNA.

3. Isolate total mRNA using TRIzol Gibco BRL (Paisley, Scotland) or similar.

4. Measure the concentration and the purity of the isolated RNA (*see* **Note 23**).

5. Transfer 1 μg of total RNA from each sample to a fresh Eppendorf tube.

6. Apply the cDNA High Capacity Archive Kit (Biosystems, New Jersey, USA) or similar, to generate cDNA from each sample (reverse transcription).

7. Select TaqMan probes for genes of interest and carry out the preamplification of genes with the PreAmp GrandMasterMix (TATAA Biocenter AB, Göteborg, Sweden) or similar.

8. Prepare TaqMan Universal PCR Master Mix and add the mixture to each sample.

9. To evaluate the performance of a primer set and to eliminate the effect of the inhibition a serial of tenfold dilutions of each target gene should be performed.

10. Perform qPCR on 48.48 Dynamic Array™ IFC method (Bio-Mark HD machine system, Fluidigm, UK) or on a classic Q-PCR on 384 plate (VIA Real-Time PCR System machine, Applied Biosystems™) or similar equipment.

11. Analyze data using the relative quantification according to solvent control (*see* **Note 24**).

Sample Processing for Proteomics

Proteomics using mass spectrometry enables to quantification and identification of thousands of proteins in one experiment. This protocol presents a method of how to prepare samples from spheroids for bottom-up label free LC-MS using a one-pot buffer [21] and filter aided sample preparation [22]. This protocol is selected due to its speed and applicability to handling a high number of samples [23].

1. Prepare spheroids or spheroid core and rim according to the protocols from Subheadings 3.2.4.1 and 3.1.2, respectively.

2. Thaw the frozen samples on ice.

3. Add lysis buffer (50 mM TEAB, 1% SDC, 10 mM TCEP 40 mM chloroacetamide, protease and phosphatase inhibitors) so that the final protein concentration is within 1–5 μg/μL.

4. Disintegrate cells by pipetting using a regular P200 pipette tip (*see* **Note 25**).

5. Heat samples for 10 min (80 °C), vortex for 1 min and sonicate to lyse cells, fragment nucleic acids, and inactivate enzymes (*see* **Note 26**).

6. Determine protein concentration in the sample using for example amino acid analysis [24] or ProStain™ Protein Quantification Kit (ActiveMotif) or any other method compatible with the lysis buffer.

7. Transfer 100 μg protein to spin filter (Vivacon 500, Sartorius), dilute samples to equal total volume of buffer and mix gently by pipetting.

8. Centrifuge at 14,000 × g for 15 min, wash with 100 μL 50 mM ammonium bicarbonate (ABC), centrifuge at 14,000 × g for 15 min and repeat wash for a total of three washes (*see* **Note 27**).

9. Transfer the filters to a fresh low-binding Eppendorf tubes, add 1 μg trypsin and dilute sample in ABC to a volume of 50 μL. Incubate samples at 37 °C overnight (*see* **Note 28**).

10. Centrifuge at 14,000 × g for 10 min, add 1 μg trypsin diluted in 50 μL ABC and incubate for 4 h at 37 °C.

11. Centrifuge filter at 14,000 × g for 10 min, add 100 μL ABC to filters, centrifuge filters at 14,000 × g for 10 min, lyophilize peptides, and store at −20 °C.

12. Resuspend samples in 2% acetonitrile–0.1% trifluoroacetic acid, quantify the amount of peptides with for example amino acid analysis [24] and analyze 1 μg peptide using LC-MS/MS.

Sample Preparation for Lipidomics

Quantification and identification of several hundred of lipids in biological samples is achieve using mass spectrometry combined with liquid chromatography LC-MS [25]. We present a simple method for lipid extraction from intact spheroids and core and rim spheroid fractions. Here we present a brief summary of the protocol published by Matyash et al. [26]. All steps of the protocol should be performed in a laminar flow hood due to harmful vapors from the chemicals used.

1. Collect spheroids or spheroids' rim and core as described in Subheadings 3.2.4.1 and 3.1.2 respectively.

2. Add 100 μL PBS and thaw the samples on ice (*see* **Note 29**).

3. Disintegrate spheroids by repeated pipetting with a normal pipette tip.

4. Sonicate spheroid homogenates (*see* **Note 30**), determine protein content (*see* **Note 31**) and transfer equivalent of 200 μg protein to an Eppendorf tube.

5. Add internal standards (*see* **Note 32**) and vortex samples for 10 min.

6. Add 300 μL methanol, vortex for 10 min, add 500 μL methyl-*tert*-butyl ether, vortex for 10 min, add 250 μL H_2O and centrifuge at 12,000 × g for 10 min.

7. Collect upper organic layer in glass vial, repeat **step 5** and collect organic layer in same glass vial.

8. Lyophilize lipids and store at −20 °C with argon gas to void oxidation of lipids.

9. Resuspend sample in 30 μL chloroform/methanol (1:1, 10 mM ammonium acetate) and analyze using LC-MS/MS.

4   Notes

1. Cut the bottom end of pipette tip P1000 to create a larger opening of approximately 2 mm in diameter, keep the tip sterile.

2. Cut the bottom end of pipette tip P200 so the diameter of the opening is larger than the spheroid (approximately 1.5 mm in diameter for 21-day-old spheroids), keep the tip sterile.

3. The number of harvested spheroids depends on which method you intend to perform subsequently. Single-cell suspension from an intact spheroid can be obtained at any spheroid age.

4. Thaw trypsin at 4 °C to limit autodigestion.

5. Incubation time depends on the size and compactness of the spheroids. Optimize incubation time of the protocol for each spheroid type. Too long exposure to trypsin may damage the cell membrane and cause cell death.

6. Leave the petri dish with spheroids in a sterile incubator at 37 °C with humidified 0.5% CO_2 atmosphere, whenever it is not being used. Spheroids should not be kept in the petri dish for prolonged periods of time. Collection should be carried out in sterile conditions.

7. Cut P200 pipette tip is needed as the diameter of spheroids is too large to make it through the original opening.

8. Start with gentle resuspensions and gradually get more vigorous. The rim will look like a light and transparent sheet and the inner core will be a small dark compact lump of cells.

9. After 10 resuspensions, no visible pieces of core should be present. Keep in mind that core is more compact, so longer time or more resuspension is needed.

10. In this step, you will get the single-cell suspension of the rim, while the core is still incubating in trypsin.

11. Be aware that the cells can lose membrane integrity during disassembly to single-cell suspension if incubation in trypsin is not properly adjusted.

12. Single-cell suspension should be prepared immediately before trypan blue staining, as cells are being treated harshly and can die outside of the bioreactor.

13. Evaluate viability immediately after mixing staining solution and cell suspension as trypan blue will eventually penetrate the cell membrane of even live cells.

14. Ensure that after collection the spheroids are kept in growth media in a humidified incubator until analyzed.

15. It is preferable to analyze a single spheroid per well as the signal can be too high for the plate reader for larger spheroids.

16. The method should be adapted to optimal conditions for every cell type. Perform all the following steps of the protocol at 4 °C in the dark to prevent additional DNA damage occurring during the assay.

17. Always use identical settings to obtain comparable and reproducible results.

18. If needed exclude unspecific binding with isotopic controls.

19. Incubation is carried out in the dark as to avoid photo bleaching; the incubation time depends on characteristics of antibodies.

20. If you use multicolor staining each dye can contribute to the signal on several detectors. Therefore, it is essential to perform the calibration of the lasers prior to starting an experiment especially when applying a new cell line.

21. For experiments where several time points are analyzed, all samples can be stored frozen and subsequent sample preparation can be performed in parallel.

22. A pool of at least 25 spheroids or at least 35 parts of each fraction—core and rim at the age of 21 days is sufficient to obtain a desired amount of mRNA.

23. For reliable results obtained with qPCR, it is essential to check the purity and degradation (gel-electrophoresis) as well as the concentration of isolated mRNA (NanoDrop 1000 Spectrophotometer) before starting the experiment.

24. Freely available program such as QuantGenious [20] can be used.

25. Add as low volume of buffer as possible to keep a high protein concentration; however enough buffer to stop processes if necessary. Five spheroids aged 21 days can be properly dissolved in 100 μL buffer.

26. Cycle between 15 s of sonication and 30 s pause to avoid overheating of the sample.

27. If some samples do not fully pass through the filter, increase the number of washes.

28. If you intend to digest another amount of protein maintain a ratio of 1:100 of trypsin to protein.

29. Use as low volume of liquid as possible; however enough to dissolve it. For this protocol 100 μL is sufficient to dissolve 5 × 21-day-old spheroids.

30. Cycle between 15 s of sonication and 30 s pause to avoid overheating of the sample.

31. Lipid content is not measured directly but is assumed to be relative to the protein concentration. Each 21-day-old spheroid should contain at least 200 µg protein.

32. The internal standard should contain lipid species that are to be identified in the samples. Use as low concentration of the standard as possible to avoid losing peaks of interest. Include a lipid standard for each class of interest as the internal standard.

Acknowledgments

The authors acknowledge the financial support from the Sino Danish Research and Education Center for PhD project for HSF and JMVN, Slovenian Research Agency [research core funding J1-2465, and grant to young researchers MR-MStampar P1-0245], and COST Actions CA16119 (In vitro 3-D total cell guidance and fitness).

References

1. Pampaloni F, Reynaud EG, Stelzer EH (2007) The third dimension bridges the gap between cell culture and live tissue. Nat Rev Mol Cell Biol 8(10):839–845. https://doi.org/10.1038/nrm2236

2. Edmondson R, Broglie JJ, Adcock AF, Yang L (2014) Three-dimensional cell culture systems and their applications in drug discovery and cell-based biosensors. Assay Drug Dev Technol 12(4):207–218. https://doi.org/10.1089/adt.2014.573

3. Antoni D, Burckel H, Josset E, Noel G (2015) Three-dimensional cell culture: a breakthrough in vivo. Int J Mol Sci 16(3):5517–5527. https://doi.org/10.3390/ijms16035517

4. Wrzesinski K, Rogowska-Wrzesinska A, Kanlaya R et al (2014) The cultural divide: exponential growth in classical 2D and metabolic equilibrium in 3D environments. PLoS One 9(9):e106973. https://doi.org/10.1371/journal.pone.0106973

5. Mehta G, Hsiao AY, Ingram M et al (2012) Opportunities and challenges for use of tumor spheroids as models to test drug delivery and efficacy. J Control Release 164(2):192–204. https://doi.org/10.1016/j.jconrel.2012.04.045

6. Asthana A, Kisaalita WS (2012) Microtissue size and hypoxia in HTS with 3D cultures. Drug Discov Today 17(15-16):810–817. https://doi.org/10.1016/j.drudis.2012.03.004

7. Zanoni M, Piccinini F, Arienti C et al (2016) 3D tumor spheroid models for in vitro therapeutic screening: a systematic approach to enhance the biological relevance of data obtained. Sci Rep 6:19103. https://doi.org/10.1038/srep19103

8. Breslin S, O'Driscoll L (2013) Three-dimensional cell culture: the missing link in drug discovery. Drug Discov Today 18(5-6):240–249. https://doi.org/10.1016/j.drudis.2012.10.003

9. Godoy P, Hewitt NJ, Albrecht U et al (2013) Recent advances in 2D and 3D in vitro systems using primary hepatocytes, alternative hepatocyte sources and non-parenchymal liver cells and their use in investigating mechanisms of hepatotoxicity, cell signaling and ADME. Arch Toxicol 87(8):1315–1530. https://doi.org/10.1007/s00204-013-1078-5

10. Fey SJ, Wrzesinski K (2012) Determination of drug toxicity using 3D spheroids constructed from an immortal human hepatocyte cell line. Toxicol Sci 127(2):403–411. https://doi.org/10.1093/toxsci/kfs122

11. Strober W (2001) Trypan blue exclusion test of cell viability. Curr Protoc Immunol. https://doi.org/10.1002/0471142735.ima03bs21

12. Fey SJ, Wrzesinski K (2013) Determination of acute lethal and chronic lethal dose thresholds of Valproic acid using 3D spheroids constructed from the immortal human hepatocyte cell line HepG2/C3A. Valproic acid:

pharmacology, mechanisms of action and clinical implications (pp 141–165). Nova Science Publishers, Inc. ISBN: 978-162417952-5

13. Wigg AJ, Phillips JW, Wheatland L, Berry MN (2003) Assessment of cell concentration and viability of isolated hepatocytes using flow cytometry. Anal Biochem 317(1):19–25. https://doi.org/10.1016/s0003-2697(03)00057-5

14. Zegura B, Filipic M (2004) Application of in vitro comet assay for genotoxicity testing. In: Methods in pharmacology and toxicology. Humana Press, Totowa, New Jersey

15. Tung JW, Heydari K, Tirouvanziam R et al (2007) Modern flow cytometry: a practical approach. Clin Lab Med 27(3):453–468. https://doi.org/10.1016/j.cll.2007.05.001

16. McKinnon KM (2018) Flow cytometry: an overview. Curr Protoc Immunol 120:5.1.1–5.1.11. https://doi.org/10.1002/cpim.40

17. Bookout AL, Mangelsdorf DJ (2003) Quantitative real-time PCR protocol for analysis of nuclear receptor signaling pathways. Nucl Recept Signal 1:e012. https://doi.org/10.1621/nrs.01012

18. Buh Gasparic M, Cankar K, Zel J, Gruden K (2008) Comparison of different real-time PCR chemistries and their suitability for detection and quantification of genetically modified organisms. BMC Biotechnol 8:26. https://doi.org/10.1186/1472-6750-8-26

19. Štampar M, Tomc J, Filipič M, Žegura B (2019) Development of in vitro 3D cell model from hepatocellular carcinoma (HepG2) cell line and its application for genotoxicity testing. Arch Toxicol 93 (11):3321–3333. https://doi.org/10.1007/s00204-019-02576-6

20. Baebler Š, Svalina M, Petek M et al (2017) quantGenius: implementation of a decision support system for qPCR-based gene quantification. BMC Bioinformatics 18(1):276. https://doi.org/10.1186/s12859-017-1688-7

21. Kovalchuk SI, Jensen ON, Rogowska-Wrzesinska A (2019) FlashPack: fast and simple preparation of ultrahigh-performance capillary columns for LC-MS. Mol Cell Proteomics 18 (2):383–390. https://doi.org/10.1074/mcp.TIR118.000953

22. Wiśniewski JR, Zougman A, Nagaraj N, Mann M (2009) Universal sample preparation method for proteome analysis. Nat Methods 6 (5):359–362. https://doi.org/10.1038/nmeth.1322

23. Wang WQ, Jensen ON, Møller IM, Hebelstrup KH, Rogowska-Wrzesinska A (2015) Evaluation of sample preparation methods for mass spectrometry-based proteomic analysis of barley leaves. Plant Methods 14:72. https://doi.org/10.1186/s13007-018-0341-4

24. Højrup P (2015) Analysis of peptides and conjugates by amino acid analysis. Methods Mol Biol 1348:65–76. https://doi.org/10.1007/978-1-4939-2999-3_8

25. Lydic TA, Goo YH (2018) Lipidomics unveils the complexity of the lipidome in metabolic diseases. Clin Transl Med 7(1):4. https://doi.org/10.1186/s40169-018-0182-9

26. Matyash V, Liebisch G, Kurzchalia TV et al (2008) Lipid extraction by methyl-tert-butyl ether for high-throughput lipidomics. J Lipid Res 49(5):1137–1146. https://doi.org/10.1194/jlr.D700041-JLR200

Chapter 13

Isolation of Extracellular Vesicles (EVs) Using Size-Exclusion High-Performance Liquid Chromatography (SE-HPLC)

Keerthie Dissanayake, Kasun Godakumara, and Alireza Fazeli

Abstract

Extracellular vesicles (EVs), are membrane-bound nanoparticles of biological origin. These signature molecules of health and disease have raised remarkable attention of the biomedical arena due to its potential diagnostic and therapeutic applicability. Among the many different techniques available for EV isolation, size-exclusion chromatography (SEC) is widely accepted.

In this chapter, we present a protocol of size-exclusion high-performance liquid chromatography (SE-HPLC) as a method of EV isolation. This method can be adapted as a low cost but a reliable and scalable method of EV isolation in those laboratories having access to the HPLC systems.

Key words Extracellular vesicles, Size-exclusion high-performance liquid chromatography (SE-HPLC)

1   Introduction

One of the biggest challenges in EVs research is the lack of a gold standard method of EV isolation from biological fluids or cell culture media. Different methods have their own merits and limitations [1]. Of the available methods, size exclusion chromatography is widely used. Size exclusion chromatography (SEC) separates EVs from the other constituents, including proteins, in the media or biological fluids based on the principle of size exclusion. It is claimed that the functionality of EVs is more preserved with SEC compared to ultracentrifugation [2].

Most of the commercially available SEC columns are single-use, bench-top SEC columns. Considering the single-use, such methods are relatively expensive. Several studies have previously taken approaches to isolate EVs from high-performance liquid chromatography based SEC methods (SE-HPLC) [3, 4]. Here, we outline a method we have optimized for SEC based EV isolation

Tiziana A.L. Brevini et al. (eds.), *Next Generation Culture Platforms for Reliable In Vitro Models: Methods and Protocols*, Methods in Molecular Biology, vol. 2273, https://doi.org/10.1007/978-1-0716-1246-0_13,

supported by a high-performance liquid chromatography system (SE-HPLC) [5]. As many of the biotechnological laboratories have HPLC systems, this method could easily be adapted for EV isolation as this method proved to be low cost, efficient, and reliable.

2 Materials

Reagents are stored at room temperature unless indicated otherwise.

1. Phosphate buffered saline- PBS (pH 7.4) (500 ml). Commercial $1\times$ PBS or 10 times diluted $10\times$ PBS may be used (*see* **Note 1**).

2. Centrifuge.

3. 50 ml centrifuge tubes.

4. Serological pipettes (25 ml).

5. Pipette controller.

6. Amicon® Ultra-15 Centrifugal Filter devices (10 kDa cut-off).

7. Benchtop microcentrifuge.

8. Microcentrifuge tubes (1.5 ml).

9. Pipettes and pipette tips (100 μl and 1000 μl).

10. GE Healthcare Tricorn™ Gel Filtration Column (10/300) packed with Sepharose 4 Fast Flow® size exclusion chromatography resin (*see* **Note 2**).

11. Liquid Chromatography System (ÄKTAprime plus) and its accessories including round-tipped metal needle, sample loop and Luer union connectors (*see* **Note 3**).

12. Syringes (1 ml).

13. 20% ethanol (Filter with 0.22 μm filters).

14. Ultra-pure water (500 ml).

15. Wet ice.

3 Methods

All procedures are carried out at room temperature unless specified. The following protocol describes the EV isolation from in vitro cell culture conditioned media.

3.1 Preparing the Samples

1. Collect the conditioned media from the cell culture dishes/flasks to 50 ml centrifuge tubes using a 25 ml serological pipette.

2. Sequentially centrifuge the sample as follows to get rid of cells, cell debris, and other bigger particles from the media.

3. First, centrifuge the conditioned media at $400 \times g$ for 10 min at 4 °C. Transfer the supernatant to a new 50 ml centrifuge tube.

4. Secondly, centrifuge the supernatant at $2000 \times g$ for 10 min at 4 °C. Transfer the supernatant to a new 50 ml centrifuge tube.

5. Thirdly, centrifuge the supernatant at $5000 \times g$ for 10 min at 4 °C. Collect the supernatant.

6. Concentrate the supernatant by centrifuging at $3200 \times g$ at 4 °C using Amicon Ultra-15 Centrifugal Filters till the final volume reach 250 μl.

7. Collect the sample to a microcentrifuge tube and keep on wet ice.

8. Unless used for EV isolation on the same day, label and freeze the sample for later use. Never store the conditioned cell culture media intended for EV isolation before the sequential centrifugation.

3.2 Preparing the Liquid Chromatography System for EV Purification

The following steps outline the use of SEC in an ÄKTAprime plus chromatography system (Fig. 1) for EV isolation. ÄKTAprime plus is originally designed for purification of proteins at the laboratory scale. However, we use the same system to isolate EVs from culture media.

The following steps outline the operation of the ÄKTAprime plus for EV isolation (*see* **Note 4**).

1. Check the connectivity of the tubes (sample loop, A1 tube, and W1–3 waste capillaries).

2. Waste capillaries W1, W2, and W3 should be directed to the waste bottle. Discard the residue, if there is any, in the waste bottle.

3. Check if the inlet tube A1 is immersed in 20% ethanol (the flow path of the chromatography system is filled with ethanol during storage and in between uses).

4. Sample loop (500 μl) should be connected to ports 2 and 6 of the injection valve (Fig. 2).

5. Connect the "injection fill port" into the valve port 3 by threading it loosely. Insert the round-tipped injection needle into the injection fill port and tighten the injection fill port further till it makes a seal around the tip of the needle. However, this should not be too tight and should enable the insertion and removal of the needle easy (*see* **Note 5**). Make sure that position 1 on the injection valve is connected to the UV flow cell through a tubing.

6. Open PrimeView: the control software package that supervises the operations of the ÄKTAprime plus.

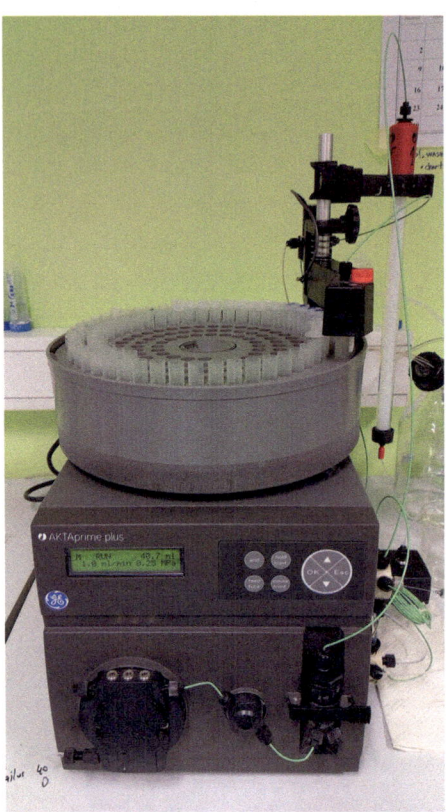

Fig. 1 ÄKTAprime plus Chromatography system

7. Switch on the Chromatography instrument. At first, the system will perform a set of self-tests automatically which will set all the parameters to factory default. Then the front LCD panel shows the Templates menu.

8. Select the 'manual run'.

9. Flush the flow path of the system with 20% ethanol at a higher flow rate for 1 min. For this, make the following flow parameter settings; flow rate:30 ml/min, injection valve position: WASTE (*see* **Note 6**). Injection valve position determines the flow path (Fig. 3). Leave all other changeable parameter settings at default and select "Run." This will also remove any air bubbles in the tubes. If further air bubbles are suspected, you may have to use the purge kit (*see* **Note 4**).

10. After the flush, lower the flow rate (1 ml/min) and pause the flow (press "pause" on the front panel).

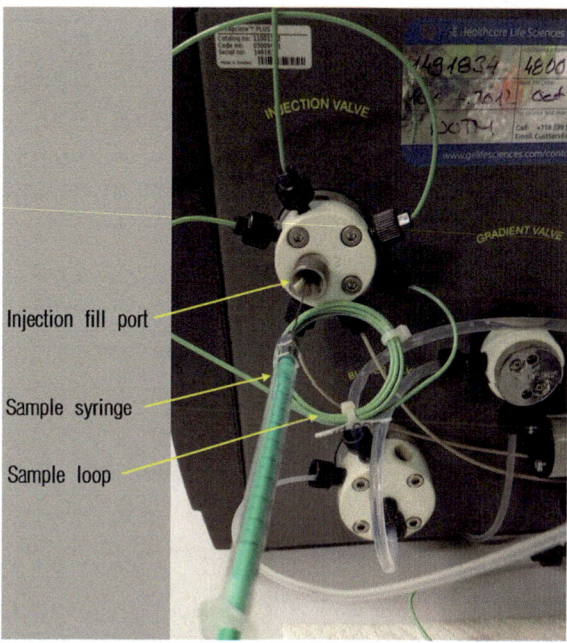

Fig. 2 Injection valve of the ÄKTAprime plus. The sample loop is mounted between port 2 and 6 of the injection valve. The sample is loaded to the sample loop via the injection fill port attached to port 3. Round tipped metal needle is used when injecting the sample

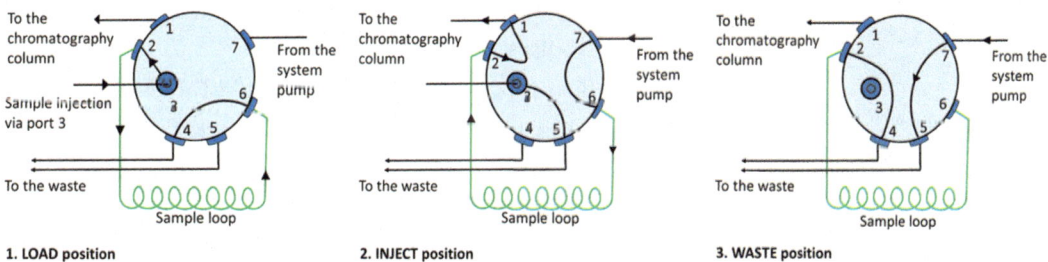

Fig. 3 Injection valve positions. The flow path varies depending on the selected injection valve position. The sample is injected when the injection valve is in LOAD position, and when it is changed to INJECT position, the sample is pumped to the chromatography column

3.3 Connecting the Size Exclusion Chromatography Column to the Chromatography System

1. Take out the GE Healthcare Tricorn™ Gel Filtration Column from the storage (generally kept at room temperature).

2. Check the column and its matrix for cracks, leaks or air bubbles by projecting a flashing light to the column.

3. Lower the flow rate of the chromatography instrument further to 0.2–0.4 ml/min.

4. Hold the column upright and remove the stop plug at the upper end of the column.

5. Loosen the tubing that is connected to the upper port of the "UV flow cell." Recommence the flow (Ethanol would be seen dripping from the free end of the tubing).

6. Fill the upper end of the column with ethanol dripping from the free end of tubing (This will prevent air getting into the column when connecting the tubing to the column).

7. Pause the flow.

8. Connect the free end of the tubing to the upper end of the column.

9. Remove the stop plug at the lower end of the column.

10. Continue the flow (Now the fluid (ethanol) would drip from the lower end of the column).

11. Using an additional tubing, connect the lower end of the column to the upper port of the UV flow cell.

12. Now, place the chromatography column in the column holder.

3.4 Flushing the Flow Path and the SEC Column with Distilled Water

1. Pause the flow and transfer the A1 tubing to distilled water.

2. Flush and fill the flow path with distilled water for 1 min. For this, make the following parameter changes: flow rate—30 ml/min, injection valve position—WASTE, fraction size—0 (*see* **Note 7**).

3. Continue the flow for 1 min. (This step will save time which is otherwise taken for filling the flow path with distilled water if run at a lower flow rate).

4. Lower the flow rate to 1 ml/min and pause the flow.

5. Change the flow parameter settings as follows: flow rate:1 ml/min, injection valve position: LOAD, and continue the flow (this will flush the column with water).

6. Wash the column with water for at least 1 column volume (CV) (24 ml). This will get rid of all the ethanol in the SEC column.

7. Flush the sample loop carefully with 5 sample loop volumes of water (2.5 ml). For this, inject the distilled water to the injection port of the injection valve. Now repeat the same using the buffer.

3.5 Equilibration of the SEC Column with the Buffer (PBS)

The SEC column should be equilibrated with the buffer as EV isolation takes place in the buffer.

1. Pause the flow and transfer the end of A1 tubing to PBS.

2. Flush and fill the flow path of the instrument with the buffer for 1 min. For this, set the following flow parameter settings; flow rate to 30 ml/min, injection valve position: WASTE.

3. Continue the flow for 1 min and then pause. This step will save time which is otherwise taken for filling the flow path with the buffer if run at a lower flow rate.

4. After 1 min, lower the flow rate to 1 ml/min and pause the flow.

5. Change the flow settings (flow rate—1 ml/min, injection valve position—LOAD, and fraction size—0) and recommence the flow (this will equilibrate the column with the buffer).

6. Observe the gradual rise in the conductivity as shown in the PrimeView® chromatogram (The conductivity of PBS is higher than water. Therefore, when the distilled water in the column is replaced with PBS, the conductivity gradually increases and reaches a plateau, and then stabilizes). Once the column is equilibrated with PBS, pause the flow.

3.6 Sample Loading and EV Isolation

Partial sample filling technique is used.

1. Check the fraction collector. Place at least 15 collection tubes (1.5 ml microcentrifuge tubes) in the bowl, starting from the first position. Rotate the fraction collector counterclockwise and position the delivery arm in a way that the tube sensor touches the first tube and the eluent tubing is over the centre of the first collection tube.

2. Use the partial filling technique as follows. As the sample loop is 500 µl, the maximum sample volume is 250 µl (max. 50% of the sample loop volume is recommended to load).

3. Set the following flow parameter settings: injection valve position: LOAD, flow rate: 0.5 ml/min, fraction size: 1 ml (*see* **Note 8**).

4. Fix a Luer union connector to a 5 ml syringe and fill with the buffer (Fig. 5a). Remove if there are any air bubbles in the syringe by gentle tapping.

5. Take 1 ml syringe and fix the round-tipped metal needle (Fig. 5b). Draw the sample (250 µl) to the syringe. Remove air bubbles in the syringe or needle, if there are any, by tapping gently.

6. With the buffer and the sample ready, first, inject 2.5 ml of buffer carefully followed by the sample through the injection fill port attached to the port 3 of the injection valve. Following injection, do not remove the syringe from the injection fill port (*see* **Note 9**).

7. After injecting the sample, change the injection valve position from LOAD to INJECT. This will direct the sample in the loop to the top of the column.

8. Continue the flow. This will commence the separation of the EVs from the sample. Observe the fluid dripping from the tip of the eluent tubing and collect it into the first fraction collection tube.

9. With each fraction collected into the respective tube, the fraction collector bowl rotates automatically clockwise to collect the next fraction.

10. With the aid of PrimeView®, observe the chromatogram that shows the separation of EVs from the rest of the non-EV particles based on the size. Usually, the first peak of the chromatogram, which appears corresponding to fractions 7–9, indicates the EVs in the sample. Identify these fractions from the tubes in the bowl and transfer them to ice until further processing.

11. After the first peak, there is a taller peak on the chromatogram. This corresponds to the proteins and other similar size biomolecules in the sample, which are not of our interest (Fig. 4).

12. After the 15th fraction was collected, make the following changes to the flow settings to stop the collection of fractions; fraction size:0 ml, flow rate:0.5 ml/min.

13. Raise and rotate the delivery arm out of the fraction collector.

14. Continue the flow at 0.5 ml/min to wash the column with PBS until the UV absorbance reaches the baseline.

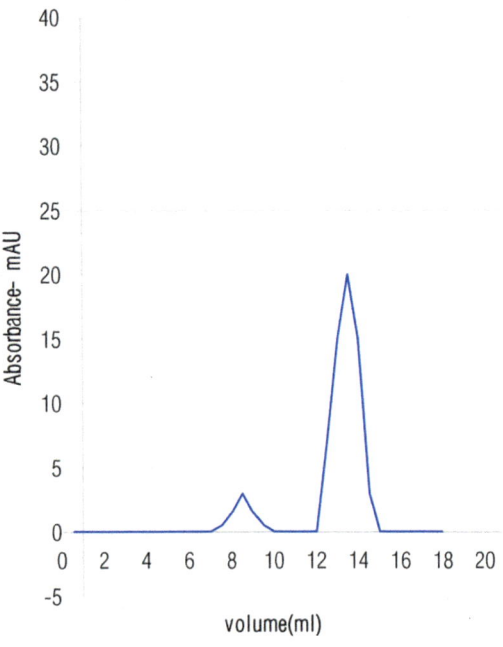

Fig. 4 Schematic view of a chromatogram illustrating the absorbance of EVs and other particles at 280 nm. The first peak corresponds to EVs whereas the second pear corresponds to non-EV constituents of the sample including proteins

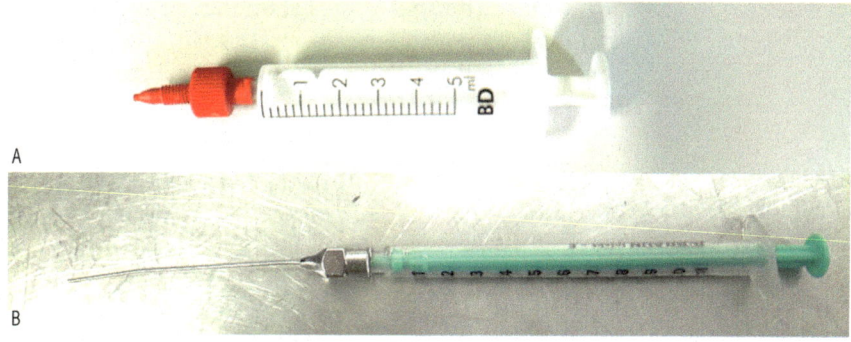

Fig. 5 Luer union connector (**a**) and the round-tipped metal needle (**b**)

3.7 Washing the Chromatography Column

1. After washing the column, the instrument is ready for another sample purification step. In such a case, place new tubes for the fraction collection in the fraction collector bowl.

2. Start from the sample loading and run the sample purification as outlined previously.

3. If there are no other samples, the run should end. For this, the buffer in the column and the flow path should be replaced with 20% ethanol.

4. Firstly, place the A1 tube in water.

5. First, flush and fill the flow path (except the column) with the water. For this set the injection valve position to WASTE and flow rate to 30 ml/min.

6. Continue the flow for 1 min and, then pause the flow. This step will save time which is otherwise used for filling the flow path with the water if run at a lower flow rate.

7. Then, make the following parameter settings: flow rate—1 ml/min., injection valve position—LOAD and fraction size—0, and commence the flow. This will flush out the buffer in the SEC column and tubes.

8. Observe the drop in the conductivity as the buffer is replaced with distilled water.

9. Once the conductivity dropped to the baseline, the flow path and the column should be loaded with 20% ethanol.

10. Pause the flow and place the A1 tube in 20% ethanol.

11. First, flush and fill the flow path (except the column) with the 20% ethanol. For this make the following parameter settings: injection valve position—WASTE and flow rate to 30 ml/min.

12. Continue the flow for 1 min, and then pause the flow. This step will save time which is otherwise used for filling the flow path with the ethanol at a lower flow rate.

13. Now, make the following parameter settings; flow rate:1 ml/min, injection valve position: LOAD, and recommence the flow. This will load the column with 20% ethanol. Continue the flow until 1 CV(24 ml) is passed through the column.

3.8 Disconnecting the Chromatography Column from the System and Ending the Run

1. Lower the flow rate to 0.2 ml/min.

2. Disconnect the tubing from the lower end of the SEC column. Now, ethanol would drip from the open lower end of the column.

3. Pause the flow and cap the lower end with a stop plug.

4. Disconnect the tubing from the upper end of the column. Continue the flow and fill the open, upper end of SEC column with the ethanol dribbling from the tubing. Cap the upper end with a stop plug.

5. Connect the free end of the tubing to the upper port of the UV flow cell.

6. Tighten the stop plugs of the SEC column and safely store the column.

7. Flush the sample loop by injecting water (2.5 ml) into the injection valve and then load the sample loop with 20% ethanol.

8. End the run by switching off the system and the software.

3.9 Post-isolation Handling of EVs

1. Pool the collected EV fractions (decided by the UV absorbance, first peak of the chromatogram, usually fraction 7–10).

2. Prime the Amicon Ultra-15 kDa spin filters (Millipore) by wetting the membrane with PBS.

3. Transfer the collected EV fraction to the Amicon filter and centrifuge at $3200 \times g$ until it is concentrated to the desired final volume.

4. Quantify the EVs using a nanoparticle tracking analyzer or other methods.

5. Store the EV sample at $-80\ °C$ until further use.

4   Notes

1. It is recommended to use commercial buffer (PBS) preparations as partially dissolved particles in PBS can interfere with the measurement of isolated EVs.

2. The bed volume (column volume) of the gel filtration column (Tricorn 10/300) is approximately 24 ml. Therefore, 1 CV refers to 24 ml.

3. ÄKTA prime plus is a compact liquid chromatography system designed for purification of proteins at laboratory scale. The method described here uses this chromatography system for EV isolation. Compared to conventional benchtop SEC methods where the flow is based on gravity, in this method, the flow takes place at a specific rate and a pressure gradient generated by a pump. The fractions corresponding to the separated EVs are indicated by the UV absorbance at 280 nm (viewed on a chromatogram).

4. More details regarding the operation of ÄKTAprime plus is available at https://www.manualslib.com/products/Ge-Healthcare-Aktaprime-Plus-8984704.html

5. When the injection valve position is LOAD position and the injection fill port is not occupied by the needle and the syringe, the liquid in the sample loop may empty due to gravity.

6. Certain parameters can be changed from the front LCD panel and the keys. This can be done for most of the parameters even during the run. This includes flow rate (ml/min, fraction size (ml), buffer valve position (pos 1 or 2) and the injection valve position (WASTE, LOAD, and INJECT) and the buffer B concentration(%). Generally, it is only required to change the flow rate, injection valve position and the fraction size.

7. Changing the injection valve position from LOAD to WASTE position disconnect the column from the main flow path. The column should not be exposed to such a higher flow rate. Therefore, the flow rate should not be higher when the injection valve is in the LOAD position.

8. In size exclusion chromatography, the separation of particles, which is based on size, is better at lower flow rates. Therefore, during the isolation of EVs, it is needed to keep the flow rate at a lower level such as 0.5 ml/min.

9. Minimize the time gap between the two injections. If there is a time gap between the injections and/or if the sample syringe is removed when in the LOAD position, self-drainage of the sample and entry of air into the sample loop will take place.

Acknowledgments

The authors would like to acknowledge, Dr. Reet Kurg, the director of the Institute of Technology, University of Tartu, Estonia for kind permission offered to make use of the ÄKTAprime plus chromatography system for EV isolation. Furthermore, the authors would like to acknowledge the European Union's Horizon 2020 research and innovation program (under the grant agreement No

668989—project TRANSGENO) for funding this work. The Authors are members of the COST Action CA16119 In vitro 3-D total cell guidance and fitness (CellFit).

References

1. Yamamoto T, Kosaka N, Ochiya T (2019) Latest advances in extracellular vesicles: from bench to bedside. Sci Technol Adv Mater 20(1):746–757

2. Mol EA, Goumans MJ, Doevendans PA et al (2017) Higher functionality of extracellular vesicles isolated using size-exclusion chromatography compared to ultracentrifugation. Nanomedicine 13(6):2061–2065

3. Huang T, He J (2017) Characterization of extracellular vesicles by size-exclusion high-performance liquid chromatography (HPLC). Methods Mol Biol 1660:191–199

4. Corso G, Mäger I, Lee Y et al (2017) Reproducible and scalable purification of extracellular vesicles using combined bind-elute and size exclusion chromatography. Sci Rep 7:11561

5. Es-Haghi M, Godakumara K et al (2019) Specific trophoblast transcripts transferred by extracellular vesicles affect gene expression in endometrial epithelial cells and may have a role in embryo-maternal crosstalk. Cell Commun Signal 19:146

Chapter 14

Isolation of Extracellular Vesicles (EVs) Using Benchtop Size Exclusion Chromatography (SEC) Columns

Qurat Ul Ain Reshi, Mohammad Mehedi Hasan, Keerthie Dissanayake, and Alireza Fazeli

Abstract

A diverse group of lipid bilayer enclosed nanoparticles, referred to as extracellular vesicles (EVs), are released by all eukaryotic and prokaryotic cells into the extracellular space. The population of EVs being heterogeneous poses a challenge in their efficient separation. Several methods have been employed for EV isolation. However, there is no single conventional method that could recover a high amount of EVs while retaining their purity and functionality. The merging of differential centrifugation with size exclusion chromatography (SEC) is one of the best practices for EV isolation/purification as it recovers a sufficient amount of EVs while retaining their functionality. Here, we describe a method of purification of EVs from bovine follicular fluid samples using benchtop SEC columns. In conclusion, the EV purification method should be chosen based on the downstream applications, as every method poses its own limitations.

Key words Extracellular vesicles, Benchtop size-exclusion chromatography, EV purification

1 Introduction

At the cellular level, biological information is exchanged between cells in the form of EVs, and this intercellular communication is one of the most critical processes in higher mammals [1]. There is a growing body of literature that recognizes the importance of intercellular communication mediated by EVs. EVs can be defined broadly as a heterogeneous population of nanoparticles released by different cell types in health and disease. This heterogeneous population encompasses three major subtypes: exosomes (typically 40–100 nm), microvesicles (typically 100–500 nm), and apoptotic bodies (typically 500 nm to 2 µm). This system of classification is based on their mode of biogenesis, where two kinds of secretory pathways are required to generate these EVs. Exosomes are produced by the fusion of multivesicular endosomes with the plasma

Tiziana A.L. Brevini et al. (eds.), *Next Generation Culture Platforms for Reliable In Vitro Models: Methods and Protocols*, Methods in Molecular Biology, vol. 2273, https://doi.org/10.1007/978-1-0716-1246-0_14,

membrane and then their release in the form of intraluminal vesicles. In contrast, microvesicles are shed by the outward vesiculation of the plasma membrane [2]. EVs shuttle lipids, RNA, DNA from a donor cell to the recipient cell, and the contents of cargo alter the physiological behavior of the recipient cell.

The major hindrance to the efficient separation of EVs is their heterogeneity and the lack of standardized methods to isolate them with higher efficacy. Several methods have been used for EV purification from different cell types, biological fluids, and conditioned cell culture media. Ultracentrifugation is one of the most widely used methods for EV isolation, which gives a better EV recovery rate. However, the limitation of this method is that it damages the functionality of EVs [3, 4]. The amalgamation of techniques like differential centrifugation and size-exclusion chromatography (SEC) has been considered as one of the best practices, which gives a better EV yield while keeping EVs functionality intact [3, 5, 6].

SEC, also known as molecular sieve chromatography, is a method in which molecules in the solution are separated based on their sizes, and in some cases, molecular weight. It is usually applied to large molecules or macromolecular complexes such as proteins and industrial polymers. Sepharose resin is widely used in the SEC columns for the separation of the diverse molecules with varying sizes. Sepharose is a commercial preparation of cross-linked, beaded-form of agarose, a polysaccharide polymer material extracted from seaweeds. It is a porous matrix of spherical particles that lacks reactivity and has adsorptive properties. After the sample has been applied in the column, molecules larger than the pores are unable to diffuse into the beads, so they elute first. The degree of penetration by the molecules into the pores depends on the varying sizes of the molecules present in the sample.

Benchtop columns (custom made), also known as the gravity-flow column, is one type of size-exclusion chromatography columns that are packed manually with Sepharose^{TM}.

Here, we outline a protocol we optimized and used for the isolation of extracellular vesicles (EVs) using benchtop size exclusion chromatography (SEC) columns [6].

2 Materials

1. Benchtop column (Econo-Pac® Disposable chromatography column).
2. Sepharose^{TM} (Sepharose 4 fast flow).
3. Amicon Ultra-15 Centrifugal filters 10 kDa (Millipore).
4. Phosphate buffered saline (PBS) (pH 7.4).

5. 0.1 M sodium hydroxide (NaOH).

6. 20% ethanol.

7. Milli-Q$^{\circledR}$.

8. Serological pipettes (25 ml).

9. Conical centrifuge tubes (50 ml).

10. Forceps.

11. Microcentrifuge tubes (1.5 ml).

12. Wet ice.

3 Methods

3.1 Sample Preparation

Despite the differences in sample types, all the samples need to be preprocessed before using the benchtop SEC columns. The larger particles and cell debris have to be removed from the samples using differential centrifugation. Centrifugation speed could vary, depending on the samples and must be optimized accordingly. The centrifugation speeds mentioned below are used for bovine follicular fluid samples.

1. First, centrifuge the sample at $300 \times g$ for 10 min at 4 °C to remove different and larger cell types.

2. Then, collect the supernatant and centrifuge again for 10 min at $2000 \times g$ at 4 °C in a new centrifuge tube.

3. Now, centrifuge the collected supernatant at $10,000 \times g$ for 30 min at 4 °C to remove the cell debris and apoptotic bodies. Transfer the supernatant to a new tube.

4. Now, using a centrifugal filter unit (10 kDa), concentrate the sample up to 500 μl by centrifuging at $3000 \times g$ for 20 min at 4 °C and keep it on ice. The samples are ready for isolation of EVs using benchtop SEC column (*see* **Note 1**).

3.2 Packing a Benchtop Column with Sepharose

1. Take out one benchtop column with a closed end.

2. Shake the Sepharose container well to mix ethanol with the beads (*see* **Note 2**).

3. Mount the column vertically.

4. Fill the column with 15ml of Sepharose, and let the column stand for 2 h or more until the Sepharose beads settle down and separate from the 20% ethanol, as shown in Fig. 1.

5. Put the filter carefully on the top of the Sepharose layer using forceps with blunt ends, without disrupting the layer of Sepharose and press the filter down with the forceps (*see* **Note 3**).

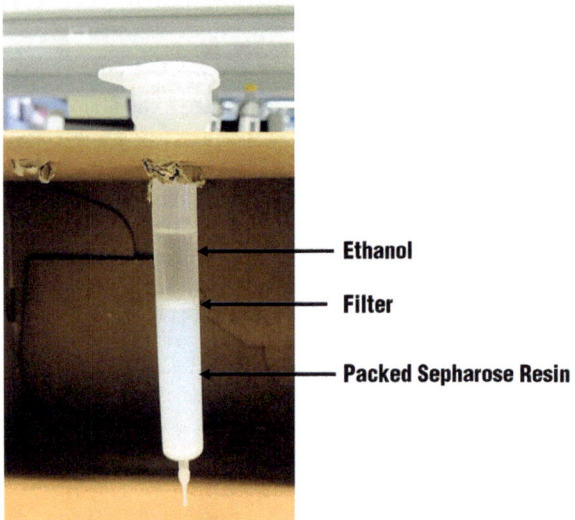

Fig. 1 A typical benchtop column (Econo-Pac®) packed with Sepharose resin. After ethanol separated from the beads, a filter is placed on top of the Sepharose

3.3 Prewash of the Column Before Use

1. Take out a benchtop column prepared by the protocol mentioned in Subheading 3.2.

2. Break the lower tip of the column and let the ethanol flow freely through the column.

3. When the level of ethanol is closer to the level of filter, prepare yourself to add 15 ml Milli-Q directly into the column. After the ethanol passes down the filter, immediately add 15 ml of Milli-Q, followed by 15 ml of PBS (*see* **Note 4**).

3.4 Purification of the EVs in the Sample

1. Start with labeling 1.5 ml microcentrifuge tubes and place them on ice until you start collecting the fractions.

2. Take the preprocessed sample (max 500 μl).

3. Add the sample on top of the filter inside the column.

4. Allow the sample to pass through the filter and add 10 ml of PBS directly into the column.

5. Immediately start the fraction collection (max 500 μl). In order to determine EV fractions from your sample, collect 20 fractions in 1.5 ml microcentrifuge tubes, each with 500 μl of the eluted sample.

6. Place the collected fractions on wet ice.

3.5 Analysis of the Samples

1. Quantify the nanoparticles (by nanoparticle tracking analyzer) and protein (by any protein assay) in all the collected fractions.

2. Plot a graph with particle concentration and protein concentration of each fraction collected (Fig. 2).

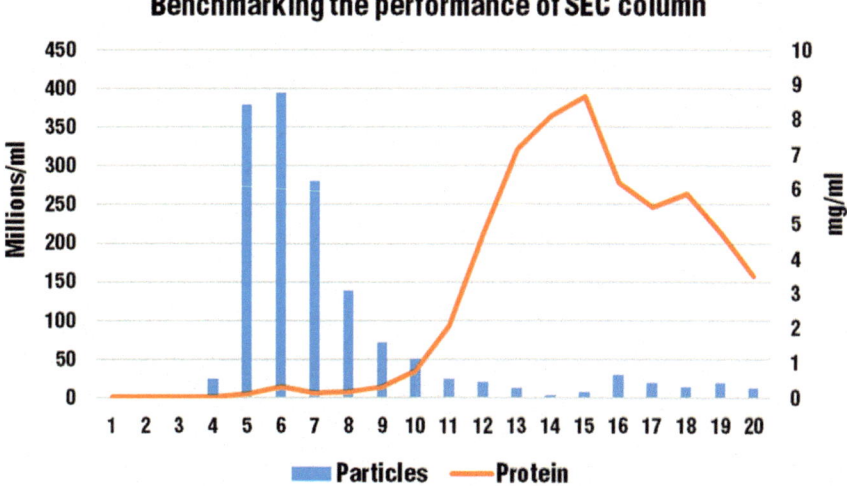

Fig. 2 Benchmarking the performance of the SEC column. The concentration of nanoparticles and proteins of each fraction is shown. Here, the first four fractions are considered the void volume as they have no particles. Fractions 5–8 have the highest EV concentrations with less or no protein contamination. Therefore, these four fractions (2 ml) will be considered as the EV fraction for further applications. Fractions 8–20 have no or fewer particles with higher proteins. The fraction numbers corresponding to EVs may vary depending on the type of sample; the sample used here is bovine follicular fluid

3.6 Washing the Benchtop Column for Reuse

1. After collecting all the fractions, let the PBS in the column drain.

2. Then add 15 ml of Milli-Q to wash the column.

3. Once Milli-Q ran through the column, add 15 ml 0.1 M NaOH.

4. Then wash the column twice with 15 ml of Milli-Q.

5. Rewash the column with 15 ml of PBS.

6. Now, load the new sample and isolate EVs, as mentioned previously (*see* **Note 5**).

4 Notes

1. Before concentrating, the samples could also be filtered using different filter sizes (0.2 μm) if any specific size of the EV population is aimed. Depending upon the type of sample, it might take more or less than 20 min to concentrate the sample up to 500 μl, and optimization with the duration of time must be done.

2. The Sepharose beads (Sepharose™ 4 fast flow) are supplied pre-swollen in 20% ethanol, and can be stored at room temperature. Over time, the beads settle down at the bottom and separate from the ethanol. So shake the Sepharose container to mix the beads and ethanol, before the preparation of the

column. It is important to shake the container gently, and not vigorously as vigorous shaking introduces bubbles which are difficult to get rid of.

3. It is recommended to keep the column overnight at 4 °C before use. Press the filter down if you notice any gap between the filter and the Sepharose resin.

4. Do not let the filter dry out. Prepare yourself to add the next fluid (PBS, Milli-Q, NaOH) in the column before the filter starts drying out. Add or top up the desired solution when the fluid level in the column reaches the level of the filter.

5. For reuse, please follow the steps mentioned in subheading 3.6. If there is no further use of the column, wash it with 20% ethanol, and discard it. To store the column, for reuse later, follow the **steps 1–5** in subheading 3.4, and then add 10 ml of 20% ethanol, let the ethanol to pass through the column and leave some of it on the top of the filter. Store the column at 4°C until the next use. The same column can be used several times but only for the same sample type.

Acknowledgments

The authors would like to thank Annika Häling for the technical assistance. Furthermore, the authors would like to acknowledge the European Union's Horizon 2020 research and innovation program (under the grant agreement No 668989—project TRANSGENO) for funding this work. The Authors are members of the COST Action CA16119 In vitro 3-D total cell guidance and fitness (CellFit).

References

1. Battistelli M, Falcieri E (2020) Apoptotic bodies: particular extracellular vesicles involved in intercellular communication. Biology 9(1):21

2. Théry C, Witwer KW, Aikawa E et al (2018) Minimal information for studies of extracellular vesicles 2018 (MISEV2018): a position statement of the International Society for Extracellular Vesicles and update of the MISEV2014 guidelines. J Extracell Vesicles 7(1):1535750

3. Benedikter BJ, Bouwman FG, Vajen T et al (2017) Ultrafiltration combined with size exclusion chromatography efficiently isolates extracellular vesicles from cell culture media for compositional and functional studies. Sci Rep 7(1):15297

4. Szatanek R, Baran J, Siedlar M, Baj-Krzyworzeka M (2015) Isolation of extracellular vesicles: determining the correct approach (Review). Int J Mol Med 36(1):11–17

5. Böing AN, van der Pol E, Grootemaat AE, Coumans FAW, Sturk A, Nieuwland R (2014) Single-step isolation of extracellular vesicles by size-exclusion chromatography. J Extracell Vesicles 3(1):23430

6. Hasan MM, Viil J, Lättekivi F, Ord J, Reshi QUA et al (2020) Bovine follicular fluid and extracellular vesicles derived from follicular fluid alter the bovine oviductal epithelial cells transcriptome. Int J Mol Sci 21(15):5365

Chapter 15

Measurement of the Size and Concentration and Zeta Potential of Extracellular Vesicles Using Nanoparticle Tracking Analyzer

Keerthie Dissanayake, Getnet Midekessa, Freddy Lättekivi, and Alireza Fazeli

Abstract

Extracellular vesicles (EVs) are membrane-bound nanoparticles that are secreted by most cell types with an emerging role in cellular communication and potential as biomarkers of disease. Nanoparticle tracking analysis (NTA) is a commonly used technique to measure the size and concentration of nanoparticles, such as EVs. Here, we present two protocols for the analysis of size profile concentration, and zeta potential (ZP) of well-characterized EVs derived from human choriocarcinoma JAr cells using NTA. These protocols describe how the size profile concentration, and ZP of JAr EVs are measured using optimized settings of NTA. With good experimental practices and consistent protocol, NTA measurements of EVs can provide reliable data that could potentially translate further uses of EVs for diagnostic and therapeutic applications.

Key words Extracellular vesicles, Nanoparticle tracking analysis, Size profile, Zeta potential

1 Introduction

Extracellular vesicles (EVs) are a heterogeneous group of membrane-confined biological nanoparticles that include endosome-derived exosomes as well as cell membrane-derived microvesicles [1]. The accurate quantification of biophysical properties of extracellular vesicles (EVs) such as size and concentration are vital for both their characterization and subsequent use [2]. Among the many technologies existing to measure the size and concentration of EVs (nanoparticles), light scattering technologies, such as nanoparticle tracking analysis (NTA) are widely used [3]. NTA facilitates the EV analysis on a single particle level not only in scatter mode but also in the fluorescent mode [4].

The principle behind the measurement of EVs based on NTA relies on the Brownian motion of these particles in a solution.

Tiziana A.L. Brevini et al. (eds.), *Next Generation Culture Platforms for Reliable In Vitro Models: Methods and Protocols*,
Methods in Molecular Biology, vol. 2273, https://doi.org/10.1007/978-1-0716-1246-0_15,
© The Author(s), under exclusive license to Springer Science+Business Media, LLC, part of Springer Nature 2021

Following the illumination from a laser, the light scattering from these particles is captured by a camera that records the movement path of the particles in a defined volume. Using the Stokes–Einstein equation, the size/hydrodynamic diameter of the tracked particles is determined based on the distance travelled. Moreover, tracking individual particles in the defined volume of the sample allows the calculation of the total particle concentration. In the scatter mode, the NTA tracks all the nanoparticles irrespective of whether they are EVs or not. Therefore, it is the responsibility of the researcher to isolate EVs from the biological samples or media with minimum contamination from other nanoparticles. Another physical property of interest regarding EVs that can be determined with NTA is the surface charge, which is described by zeta potential (ZP). This property is measured based on the particle movement within an applied electrical field. The interaction between particles in dispersed systems, such as suspensions, can be quite complex. To a greater extent this complexity may appear from the presence of electrically charged particles in dispersed systems. EVs as one group of nanoparticles carry a net negative surface charge due to the nature of membrane surface groups [5].

Our group has used nanoparticle tracking analyzer-ZetaView (by Particle Metrix, Germany) for the measurement of size, concentration, and Zeta potential of EVs in different sample types [5–7]. In a comparative study by Bachurski et al. the higher accuracy of ZetaView in determining the concentration of EVs has been demonstrated [8]. Here, we outline a protocol we optimized and used for the measurement of size, concentration, and zeta potential of JAr cell-derived EVs using nanoparticle tracking analyzer-ZetaView. Although this protocol is mainly based on our experiences gained with ZetaView, its principles can be applied to any other NTA machine currently used or being developed in the future and introduced to the field.

2   Materials

Analytical grade reagents are kept at room temperature, and the EV samples should be brought to room temperature before the measurements.

1. Falcon™ centrifuge tubes (15 mL and 50 mL).

2. Standard calibration 100 nm polystyrene beads (Applied Microspheres B. V., Netherlands).

3. Eppendorf safe-lock microcentrifuge tubes (1.5 mL).

4. Dulbecco's phosphate buffered saline—PBS (pH 7.4) (500 mL). Commercial $1\times$ PBS or 10 times diluted $10\times$ PBS may be used.

5. Ultrapure water (500 mL).

6. Pipettes and pipette tips (10 µL, 100 µL, and 1000 µL).

7. EV samples.

8. ZetaView PMX-110 V3.0 instrument (Particle Metrix GmbH, Germany) with proprietary ZetaView NTA software (8.04.02 SP2) for data analysis.

9. pH meter (Seven Compact™ pH/Ion S220, Mettler-Toledo AG, Schwerzenbach, Switzerland), 0.1 M NaOH and 0.1 M HCl for pH adjustments.

3   Methods

3.1   Preparation of the ZetaView Instrument for EV Measurements

The preparation of the ZetaView instrument is the same for measuring both the size and the ZP of EV samples. For the NTA measurement of EVs, we recommend using the PMX-110 monolaser instrument equipped with a CCD/CMOS camera, 488 nm laser, and 504 nm long-pass filter. Note that the instrument is precalibrated for scattering mode. Zetaview-NTA instrument performance check includes the following steps.

1. Fill the bottle A and B with distilled water and the buffer, respectively (*see* **Note 1**). Empty the waste bottle, if full. Check to make sure all of the tubings are properly attached to the instrument.

2. Switch on the Zetaview instrument and open the Zetaview software program (Fig. 1) in the computer/laptop connected to the instrument (*see* **Note 2**).

3. Run pump 1 (go to "Pump and Temp," set the flow rate at 5 mL/min, and select "Run"). This will wash the cell (500 µL chamber into which the samples are loaded for the measurement) with distilled water (*see* **Note 3**).

4. Freshly prepare the alignment beads by diluting 1:250,000 in distilled water (*see* **Note 4**).

5. Next three steps (Step 1, 2, and 3) evaluate the cell for its suitability for measuring the samples and verify the focus of the instrument by using 100 nm polystyrene alignment suspension of known size and concentration. The software guides the operator through these steps automatically.

   (a) *Step 1: Cell quality check*: The instrument requests to fill the cell with distilled water to check the quality of the cell. Inject 1–2 mL of distilled water to the cell through the injection port using a syringe (*see* **Note 5**). After injecting distilled water, click "OK." The instrument analyzes the cell quality and if the cell quality is reported as Good, move to step 2 in **step (b)** (*see* **Note 6**).

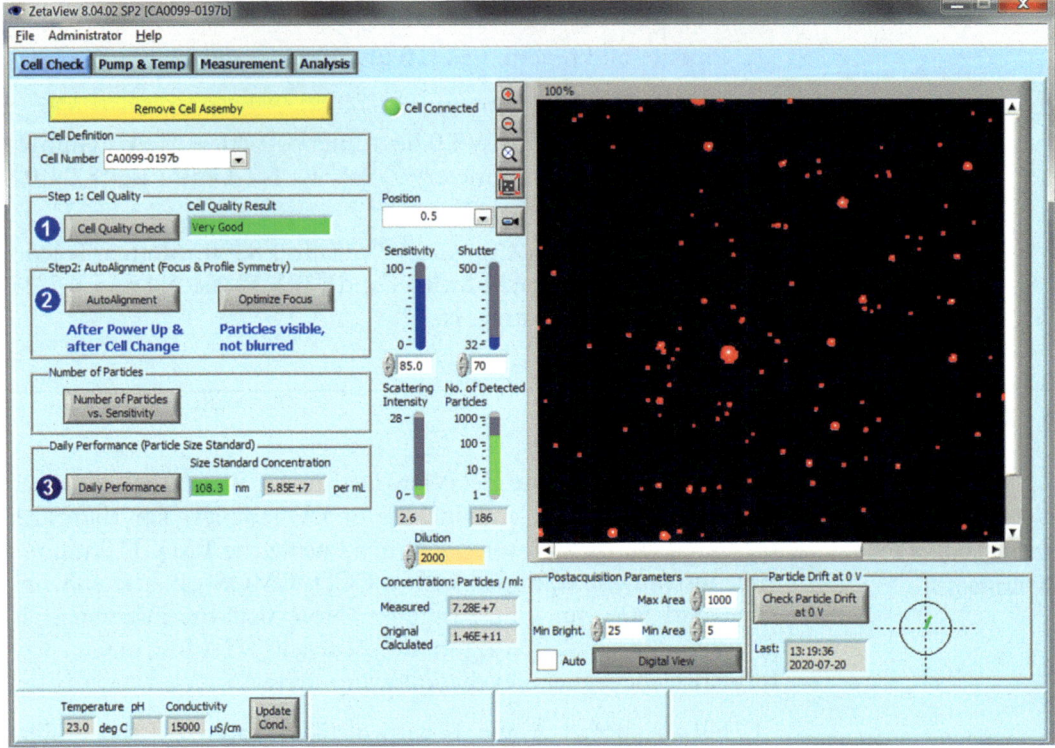

Fig. 1 The user interface of the ZetaView software. There are four areas of operation: Cell check, Pump & Temp, Measurement, and Analysis. All the operations of the instrument can be mediated through this software. The black square is the live camera view and EVs rather nanoparticles are visible as red dots

(b) *Step 2: Autoalignment*: If the cell quality is good, the instrument requests to fill the cell with 100 nm alignment suspension (1:250,000 dilution). Thus, inject the freshly prepared bead alignment suspension to the cell and click "OK." The instrument performs the alignment and auto-focus. If the autoalignment is good, move to step 3 in **step (c)**.

(c) *Step 3: Daily performance*: While the alignment bead suspension is still in the cell, click the "daily performance." When the daily performance is completed, the results are reported as the trueness and precision of the measurement of beads. The results should be "good" or "very good" to proceed (*see* **Note 7**).

6. Prime the cell with an ample volume of distilled water to remove the alignment bead suspension (use either pump 1 or directly inject water into the cell via the injection port, using a syringe). Using the "number of detected particles per frame" indicator on the user interface as a guide, monitor the drop in the particle concentration following the flushing. After the

flushing, the detected number of particles per frame should be less than five. Prime the fluidics of the ZetaView instrument with distilled water, to remove any air bubbles that are in the cell chamber and tubing.

7. Load the cell with the buffer (*see* **Note 8**) using pump B or by direct injection using a syringe. Now the instrument is ready for the measurement of samples.

8. Keep the temperature of the cell constant during the measurement by setting the temperature at room temperature (Go to "Pump and Temp," set the temperature at room temperature and press "OK").

3.2 Measurement of EV Samples for Size and Concentration

1. Prime the ZetaView instrument with PBS before the measurement of EVs in the scatter mode. EVs are diluted in PBS before performing an NTA scatter measurement under flow conditions of the EV stock in PBS to obtain an appropriate concentration. Note that the main EV stock is kept on ice. Samples may need to be diluted in the buffer if the original sample is highly concentrated with EVs. For example, if the original sample EV concentration is expected to be about 1×10^{10} particles/mL, the sample may need to be diluted 1:1000 times to a final volume of 700–1000 µL).

2. Draw the sample (800–1000 µL) into a 1 mL syringe. By gentle tapping, get rid of the air that has entered the syringe. Inject the sample into the cell without introducing air-bubbles (*see* **Note 9**).

3. For accurate sample measurements by ZetaView, it is recommended that, on average, 100–150 particles per frame is maintained. If higher particle density is observed, the sample needs to be further diluted (*see* **Note 10**).

4. Enter the dilution factor of the sample.

5. Click on the "check particle drift" button to control the allowed range of particles drifts in the field of view. In a couple of minutes, the red line indicating the particle drift gets shorter and turns green when the particles in the cell reach the normal Brownian motion. Then, click on the "check particle drift" again and proceed to the measurements.

6. Go to the "Measurement" and click "run video acquisition."

7. Select the desired folder where the measurement data should be stored and properly name the sample to be measured.

8. Select the experiment parameters. The following experiment and camera control settings are recommended (Fig. 2) for the measurement of the size and concentration of EVs (*see* **Note 11**). Make sure that other parameters remain unchanged.
   Experiment: size (Fig. 2).

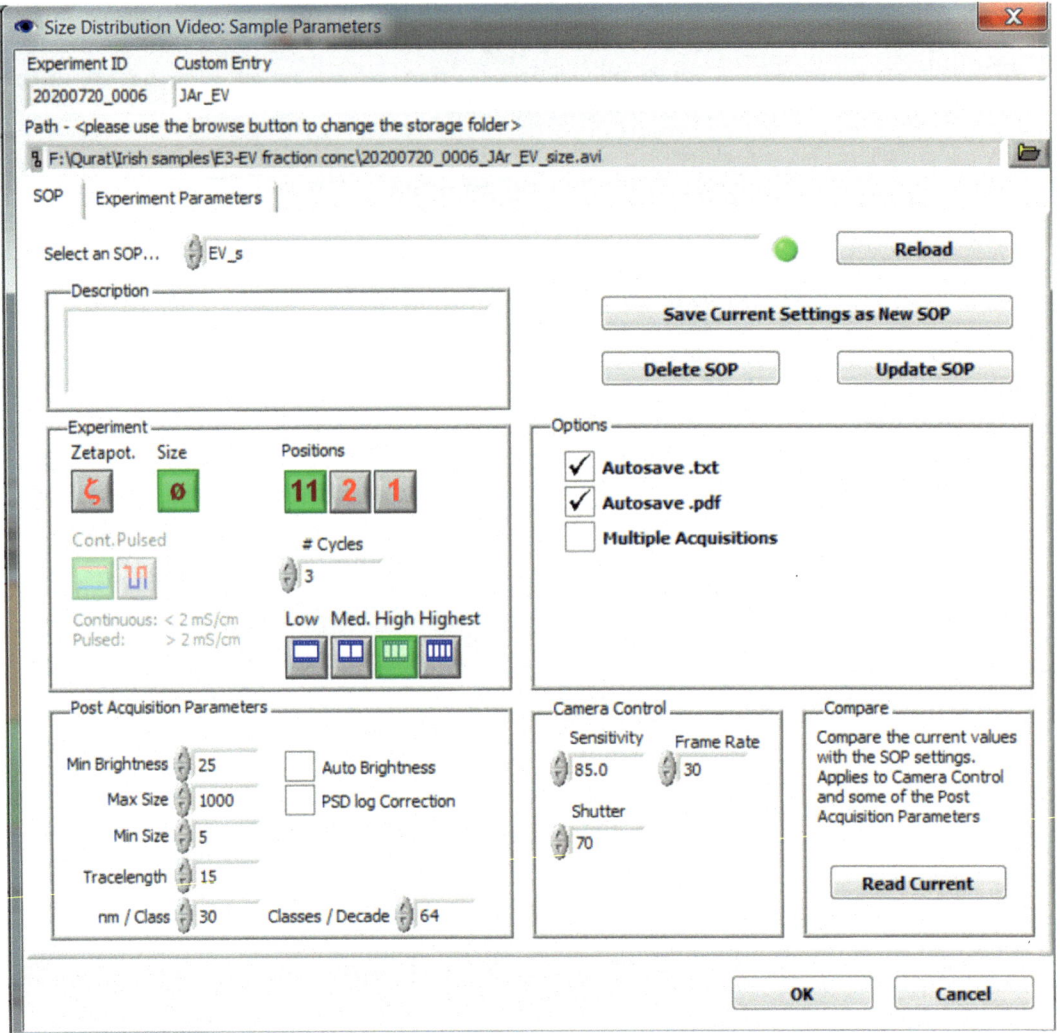

Fig. 2 Experiment and camera control settings for size and concentration of EVs. These settings can be saved as an SOP to ease future sample measurements. Note—Nanoparticle tracking analysis counts all the nanoparticles in the sample including EVs. Therefore, the purity of the measured sample is crucial

(a) Positions: 11.

(b) Cycles: 3.

(c) High/Medium.

(d) Sensitivity: 85.

(e) Frame rate: 30 fps.

(f) Shutter: 70.

These settings could be saved as an SOP (Standard Operating Procedure) for convenient measurement of subsequent samples.

9. Click "OK" and the instrument commences the measurements. After the measurement of the size profile of EVs, rather nanoparticles, in the cell, the instrument generates an initial report indicating the measurements of the 11 frames (*see* **Note 12**).

10. Click "Continue," and subsequently the software generates all the result reports (*see* **Note 13**).

3.3 Measurement of the Zeta Potential of EVs

1. Prepare the samples to be measured beforehand and bring them to room temperature. For accurate sample measurements by ZetaView, it is recommended that, on average, 100–150 particles per frame is maintained. If higher particle density is observed, the sample needs to be further diluted.

2. Measure the pH of the prepared sample and adjust it to 6.9 if different. Use either diluted acid (HCl) or base (NaOH) as needed for the adjustment of sample pH (*see* **Note 14**).

3. Draw the sample (800–1000 μL) into a 1 mL syringe. By gentle tapping, get rid of the air that has entered the syringe. Inject the sample into the cell without introducing air bubbles.

4. Enter the dilution factor of the sample.

5. Click on the "check particle drift." In a couple of minutes, the red line indicating the particle drift gets shorter and turns green when the particles in the cell reach the Brownian motion. Then, click on the "check particle drift" again and proceed to the measurements.

6. Go to the "Measurement" and click "run video acquisition."

7. Select the desired folder where the measurement data files should be stored, and properly name the sample to be measured.

8. Select the experiment parameters. The following experiment and camera control settings are recommended for the measurement of the size and concentration of EVs (*see* **Note 10**).

(a) Experiment: Zeta potential (Fig. 3).

(b) Positions: 2.

(c) Conductivity: Choose the continuous mode if the conductivity level of your suspension medium is below 2 mS/cm. Otherwise, a pulsed mode is chosen for suspensions with higher conductivity levels.

(d) Cycles: 5.

(e) High/Medium.

(f) Sensitivity: 85.

(g) Frame rate: 30 fps.

(h) Shutter: 70.

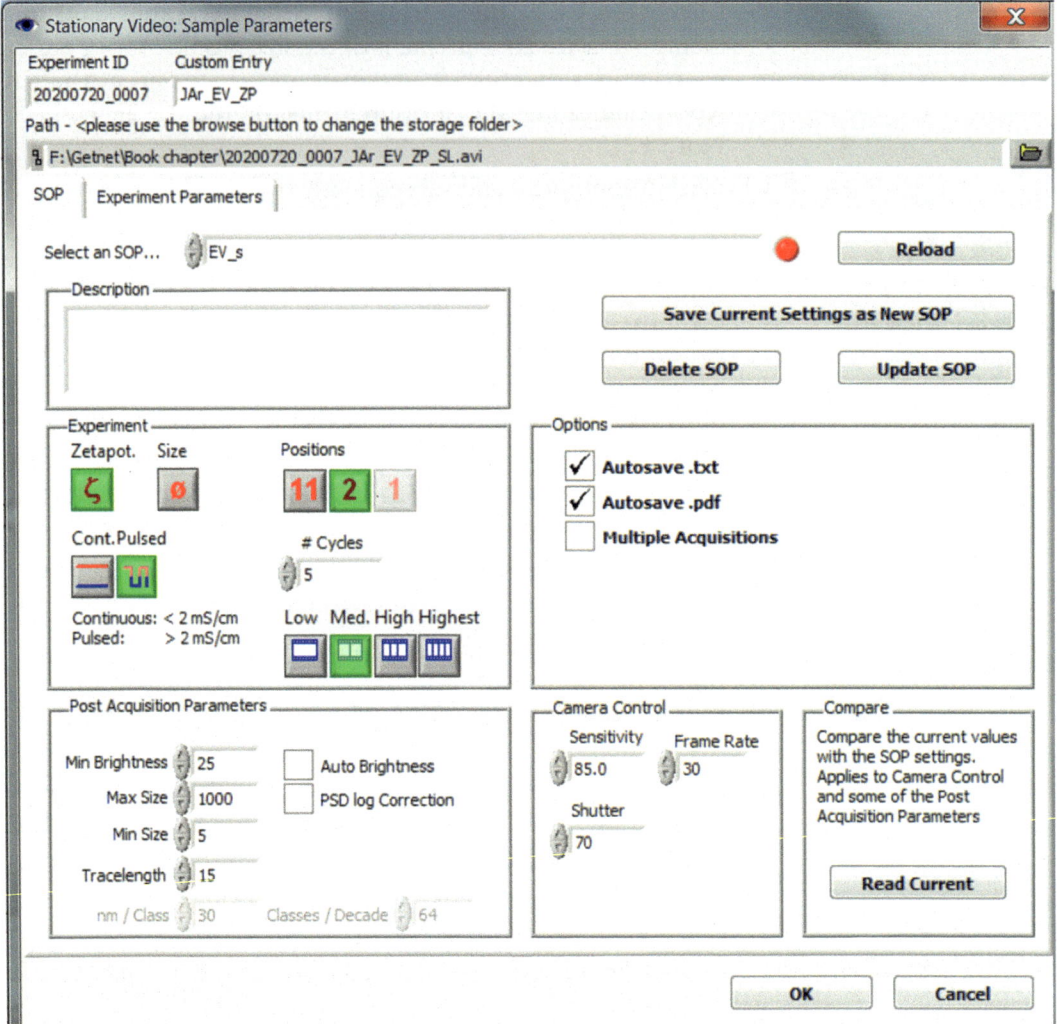

Fig. 3 Experiment and camera control settings for ZP measurement of EVs. These settings can be saved as an SOP (standard operating procedure) to ease future sample measurements. Select the continuous or pulse mode depending on the conductivity of the sample

9. Click "OK" and the instrument measures the zeta potential of the EVs of the sample.

10. The results of the measurement of the ZP of the EV sample will be reported (*see* **Note 15**).

3.4 Cleaning the Cell Chamber After the Sample Measurements

1. After measuring the samples, flush the cell with 5–10 mL of distilled water (use either pump 1 or a syringe).

2. Using an empty 5 mL syringe, inject air into the cell to push out all the liquid retaining inside the cell (*see* **Note 16**).

3. Switch off the ZetaView.

3.5 Analysis of the Results Obtained from the ZetaView

The ZetaView proprietary software (8.04.02 SP2) produces three types of output files.

1. PDF reports. These reports contain all of the information one needs for interpreting individual sample-levels results as a graphical overview of the acquired data is presented.

    In the case of size measurement reports, these graphs present:

    (a) nanoparticle abundance (as particles/mL) across different size ranges (nm);

    (b) the volume occupied by the nanoparticles across different size ranges.

    In the case of ZP measurement reports, these graphs present:

    (a) nanoparticle abundance (as number of observed particles) across different ZP ranges (mV);

    (b) the volume occupied by the nanoparticles across different ZP ranges (mV).

    In addition, the reports present descriptive statistics describing the measured sample such as nanoparticle concentration, mean particle size, or peak ZP value. These sample-level statistics can further be used for additional statistical analysis in case of experiments with multiple samples and/or groups of samples. For example, one could use Student's t-test (given normality) to test whether an observed difference in the mean size of particles between two groups of samples is statistically significant.

2. TXT files. These files contain a tab-separated table that is the basis of the graphs presented in the PDF reports. One can use this binned distribution of size or ZP values to conduct further calculations such as calculating the mean size/ZP values of the samples or comparing (groups of) samples in specific size/ZP ranges. When comparing the distribution of size or ZP values between (groups of) samples, it is advisable to normalize the sample-level binned data for the total nanoparticle concentration. The total nanoparticle concentration is often subjected to random loss of sample during complex purification procedures and can, therefore, introduced unwelcome bias into the analysis and biological interpretation of size/ZP distributions.

3. FCS files. These files contain nanoparticle-level measurements of variable optical and motion-based characteristics in addition to size and ZP values for all the nanoparticles tracked by the ZetaView instrument during a measurement. This binary output file is the rawest form of data accessible for the user of a ZetaView instrument. Several open-source libraries are available to extract this information from the FCS files, such as FlowCytometryTools for Python and flowCore in R.

4   Notes

1. Bottle A and B are connected to the pump 1 and pump 2 of the instrument respectively. When activated these pumps drain the relevant liquid and fill the cell (chamber used for loading the sample for the measurements). The capacity of the cell is about 500 μL. Excess liquid is directed from the cell to the waste bottle.

2. All the operations of the ZetaView instrument are carried out using the commands of the ZetaView software. The user interface of the ZetaView software has four tabs: Cell check, Pump & Temp, Measurement, and Analysis under which all these operations are done.

3. Alternatively, fill a 10 mL syringe with distilled water and inject it into the cell chamber.

4. Transfer 10 μL of 100 nm polystyrene (PS) beads to 10 mL of particle-free water and mix gently by flipping the tube up and down. This PS dilution yields to a 1:1000 dilution. Prepare 1:250,000 PS alignment suspension by pipetting 6 μL from 1:1000 PS dilution into 1494 μL of particle-free water.

5. By injecting a larger volume of liquid, the air trapped in the cell is sent out. During all the downstream measurements the cell should be free of air bubbles as they interfere with the measurements.

6. Sometimes the cell quality results appear as "Not acceptable." In such circumstances, the cell should be flushed with an ample amount of water and repeat the cell quality check. If still the results do not turn "good," the cell may have to be taken out of the instrument and washed manually.

7. If the results of the "daily performance" are not good, prepare the bead suspension again and start from the begininig.

8. The buffer used should be the same as the buffer in which the EVs are suspended. It is recommended to use commercial PBS preparations, as they contain low quantities of nanoparticles that are undissolved ingredients of PBS. Otherwise, these nanoparticles may interfere with the measurement of the EVs.

9. It is possible to view the different positions of the cell by manually changing the camera measuring position. This enables us to view some, but not all, air bubbles that have entered the cell chamber.

10. The concentrations of the EVs in samples vary based on many factors such as the type of cells and the number of cells releasing the EVs. Therefore, it is recommended to dilute the

samples with a higher dilution factor such as 1:1000 at the beginning and adjust accordingly.

11. The experimental parameters need to be reported when publishing the study results.

12. This report provides information about the concentration of the particles in all the 11 sections of the cell chamber at which the measurements were taken. Also, it provides information on sections of the cell from which the taken measurements were ignored along with the reasons.

13. The generated results reports are the two text documents (an intermediate report and the final, detailed report of the measurements), one AVI file (video recording of the particle tracking), one PDF file (summary of the measurements), and one FCS file. (An FCS file is generated if you want to display a scatter plot. In other ways, the "FCS export" should be ticked/switched if you want to display the scatter plot immediately before analyzing the data in detail. Also, this FCS file has both size and ZP files together.)

14. pH of the sample is a crucial factor when measuring and interpreting the ZP of EVs. Therefore, it is needed to know the pH of the EV samples in which ZP is measured. It must be uniform in all the samples when the ZP of multiple samples are measured.

15. The generated results reports are the two text documents (an intermediate report and the final, detailed report of the measurements), one AVI file (video recording of the particle tracking), one pdf file (summary of the measurements), one FCS file (it is the same files with two different measurement settings, i.e., size and ZP).

16. The cell must be dried after the use. This can be achieved by pumping air into the chamber and leaving the sample injection port open.

Acknowledgments

The authors would like to acknowledge Particle Metrix, Germany for the continuous training, troubleshooting, and resources provided for ZetaView. Annika Häling of the Univeristy of Tartu is also acknowledged for the technical assistance provided. Furthermore, the authors would like to acknowledge European Union's Horizon 2020 research and innovation program (under the grant agreement No 668989—project TRANSGENO) for funding this work. The Authors are members of the COST Action CA16119 In vitro 3-D total cell guidance and fitness (CellFit).

References

1. Raposo G, Stoorvogel W (2013) Extracellular vesicles: exosomes, microvesicles, and friends. J Cell Biol 200(4):373–383

2. Théry C, Witwer KW, Aikawa E et al (2019) Minimal information for studies of extracellular vesicles 2018 (MISEV2018): a position statement of the International Society for Extracellular Vesicles and update of the MISEV2014 guidelines. J Extracell Vesicles 7:1535750

3. Gardiner C, Ferreira YJ, Dragovic RA et al (2013) Extracellular vesicle sizing and enumeration by nanoparticle tracking analysis. J Extracell Vesicles 2:19671

4. Thane KE, Davis AM, Hoffman AM (2019) Improved methods for fluorescent labeling and detection of single extracellular vesicles using nanoparticle tracking analysis. Sci Rep 9:12295

5. Midekessa G et al (2020) Zeta potential of extracellular vesicles: toward understanding the attributes that determine colloidal stability. ACS Omega 5(27):16701–16710

6. Es-Haghi M, Godakumara K, Häling A et al (2019) Specific trophoblast transcripts transferred by extracellular vesicles affect gene expression in endometrial epithelial cells and may have a role in embryo-maternal crosstalk. Cell Commun Signal 19:146

7. Dissanayake K et al (2020) Individually cultured bovine embryos produce extracellular vesicles that have the potential to be used as non-invasive embryo quality markers. Theriogenology 149:104–116

8. Bachurski D et al (2019) Extracellular vesicle measurements with nanoparticle tracking analysis—an accuracy and repeatability comparison between NanoSight NS300 and ZetaView. J Extracell Vesicles 8:1596016

Isolation, Characterization, and MicroRNA Analysis of Extracellular Vesicles from Bovine Oviduct and Uterine Fluids

Karina Cañón-Beltrán, Meriem Hamdi, Rosane Mazzarella, Yulia N. Cajas, Claudia L. V. Leal, Alfonso Gutiérrez-Adán, Encina M. González, Juliano C. da Silveira, and Dimitrios Rizos

Abstract

Intercellular communication can be carried out by circulating systemic and/or locally released extracellular vesicles (EVs), produced by nearly every cell type and tissue, and are involved in physiological and pathological processes. In recent years, EVs have been identified in reproductive tissues, such as oviduct and uterus, and have been shown to be related to several events important for reproductive success. The understanding of their functions in reproduction has important implications for assisted reproductive technologies, for the treatment of infertility in humans and improvement of reproduction efficiency in animals. To study such EVs, it is necessary to isolate and concentrate them from fluid samples, which in the case of reproductive tissues, are usually of limited volume. Several methods for EV isolation are available such as chromatography, ultracentrifugation, polymer-based precipitation, and immunoaffinity.

Outcomes can be variable in terms of the amount and quality of isolated EVs, due to the type of isolation method. The choice of method, or a different combination of methods, may depend on the type of sample and scientific question to be addressed in a given study. In this chapter, we describe a method for isolation of EVs from bovine oviductal and uterine fluids for use in functional studies. The method combines size exclusion chromatography and ultracentrifugation. We also describe the different protocols for characterization of isolated EVs (transmission electron microscopy, nanoparticle tracking analysis, and western blot), as well as the isolation of RNA content in EVs, and their miRNAs profiling for functional studies.

Key words Exosome, miRNAs, Chromatography, Transmission electron microscopy, Western blot, Cattle

1 Introduction

Research in the field of intercellular communication in the reproductive tract has increased exponentially in the last decade, especially with the discovery of extracellular vesicles (EVs) in the reproductive fluids (oviductal and uterine), suggesting their

Tiziana A.L. Brevini et al. (eds.), *Next Generation Culture Platforms for Reliable In Vitro Models: Methods and Protocols*, Methods in Molecular Biology, vol. 2273, https://doi.org/10.1007/978-1-0716-1246-0_16,
© The Author(s), under exclusive license to Springer Science+Business Media, LLC, part of Springer Nature 2021

implication in important events such as: fertilization, early embryonic development and conceptus implantation [1]. Most recent papers mention two different EV types, which differ in size range, composition, and formation: exosomes and microvesicles (MVs) [2, 3]. Exosomes are nanoparticles of endosomal origin with a size range between 40 and 100 nm, whereas MVs have a diameter between 100 and 1000 nm and are directly shed from the plasma membrane [2–4]. In the following text, both exosomes and MVs are referred to as EVs, as commonly used separation methods do not allow precise discrimination between them [5]. EVs serve as vehicles for the transfer of functional molecules, for example, proteins, lipids, and RNAs, between cells locally (autocrine and paracrine) and remotely [6, 7]. Due to this ability of transferring these types of molecules, EVs have lately been proposed as new messengers in embryo–maternal interaction [8]. Indeed, EVs derived from the bovine oviductal fluid (OF), as well as from conditioned medium of bovine oviductal epithelial cells, demonstrated a positive influence on embryonic development in vitro [9–11]. Furthermore, during the period of maternal recognition of pregnancy in ewes, day 14 of the estrous cycle, EVs were isolated from the uterine fluid (UF), fluorescently labeled and infused back into the uterus of day 6 pregnant ewes, and then at day 14, the labeled EVs were observed inside the conceptus trophectoderm cells. Such observation demonstrates that EVs are involved in paracrine communication between the endometrium and conceptus during early pregnancy [12]. Therefore, there is a critical need for further characterization, functional and mechanistic studies on EVs from OF and UF, to provide conclusive evidence on their role as mediators of intercellular communication in the reproductive tract.

Unraveling the molecular content of OF and UF EVs seems to be an important requisite to understand their possible roles in reproductive success. Besides the DNA and mRNA molecules, miRNAs are considered to be specifically enriched in EVs [13]. MicroRNAs are ~21 nt RNA molecules that bind to the 3'UTR of target mRNA to affect their translation. They regulate the expression of as many as 30% of mRNA [14, 15] suggesting that EVs can serve as a pathway for the transfer of genetic information from one cell to another. Because of importance of miRNAs in a variety of tissues, to understand their role in the reproductive environment will provide insight into how we can better enhance reproductive efficiency [16].

Nevertheless, before shedding light on the EV content, it is important to consider the methodology used for EV isolation. Indeed, there is currently no consensus on a "gold standard" method to isolate and/or purify EVs. The scientists should be aware that the methods that are most efficient probably depend on (a) the specific scientific question asked and (b) on the

Table 1
Different methods used to isolate EVs

Method characteristics	Method description
High recovery, low specificity	Methods that recover the highest amount of extracellular material, whatever its vesicular or nonvesicular nature, that is, whole or near-whole concentrated secretome. Examples of protocols include, but are not limited to, precipitation/polymer-based kits, low molecular weight cutoff centrifugal filters with no further separation step, and lengthy or very high-speed ultracentrifugation without previous lower-speed steps.
Intermediate recovery, intermediate specificity	Methods that recover mixed EVs along with some amount of free proteins, ribonucleoproteins, and lipoproteins, depending on the matrix. Examples of protocols: size exclusion chromatography [21–24], high molecular weight centrifugal filters [25], differential ultracentrifugation using intermediate time/speed with or without wash, tangential flow filtration, and membrane affinity columns [24–26].
Low recovery, high specificity	Methods that recover a subtype (or a few subtypes) of EVs with as few nonvesicular components as possible. Subtypes of EVs can be separated by their size (e.g., by filtration, which must be combined with another method such as SEC to eliminate non EV components), their density upon either flotation or pelleting in a density gradient, their surface protein, sugar, or lipid composition (immunoaffinity or other affinity isolation including flow cytometry for large particles), or other biophysical properties such as surface charge. Note that the designation of "low recovery" is relative to total EVs, and that high recovery of specific subtypes may be possible using these techniques.

downstream applications used [17]. Purification of EVs has been successfully achieved by at least four different principles: chromatography [18], ultracentrifugation, polymer-based precipitation [19], and immunoaffinity [20]. The table below summarizes the different methods used to isolate EVs, according to their recovery and specificity (Table 1). Note that the degree of specificity of a method might vary depending on the type of biofluid from which EVs are separated. Ultracentrifugation is the most common and well-established method for EVs purification [19]. This method consists of a two-stage process. The first is a medium–high speed spin ($\sim 10,000 \times g$) to remove cell debris. A subsequent high-speed spin ($\sim 100,000 \times g$) for ≥ 1 h is used to pellet the EV fraction which can then be isolated and resuspended [27, 28]. However, isolating EVs by this method is time-consuming and usually results in a low yield of degraded products [29]. Size exclusion chromatography (SEC) is a method which isolates particles based on their size [30]. Large particles cannot enter the resin matrix and as a result, pass through the column in lower volumes; inversely, smaller particles are eluted after greater volumes [21]. SEC is a very quick method and aids in the retention of EVs from the circulating proteins and does not affect their original shape and functionality [22].

Working with EVs is technically challenging, due to the heterogeneity of published protocols and the interpretation of results, the different experimental protocols, sources of biological specimens, investigator's experience and instrumentation used to isolate and characterize EVs [17]. Therefore, the Executive Committee of the International Society for Extracellular Vesicles (ISEV) established a framework of criteria, providing the researchers with a minimal set of biochemical, biophysical, and functional standards to guide them in discriminating EVs from non-EV components. One of these criteria includes the analysis and the characterization of proteins expected to be present in the EVs of interest, especially transmembrane proteins, and cytosolic proteins with membrane binding capacity. Analytic approaches can include western blots, flow cytometry or global proteomic analysis using mass spectrometry techniques to identify, for example, transmembrane proteins. Also, a characterization of single vesicles within a mixture is recommended, to provide an indication of the heterogeneity of the EV preparation studied. Generally, at least two different technologies should be used to characterize individual EVs: (1) By transmission electron microscopy (TEM) or atomic force microscopy (AFM), images should show a wide field encompassing multiple vesicles in addition to close-up images of single vesicles. (2) By size distribution measurements of EVs, such as nanoparticle tracking analysis, dynamic light scattering, or resistive pulse sensing, the diameters of many vesicles should be provided. However, the values acquired with these techniques should be compared with microscopy techniques, since they do not distinguish membrane vesicles from coisolated nonmembranous particles of similar size [17].

Regarding the link between functional activity and isolated EVs, one approach may be based on the use of miRNA expression analysis that can be performed using different techniques, such as microarray, RNAseq, TaqMan assays, or primer-based real-time PCR [23].

Our group has isolated, purified, and quantified EVs from different sources including cell media and biological fluids, using ultracentrifugation, SEC, or combination of both techniques. These purified EVs have been successfully detected by NanoSight analysis, electron microscopy and western blot, while the content has been analyzed by real-time PCR. In this chapter, we will describe these methods in detail focusing on EV isolation from OF and UF, combining SEC with ultracentrifugation techniques. We will also describe the isolation of EVs RNAs, and miRNAs profiling using real-time PCR.

2 Materials

2.1 Oviduct and Uterine Flushing Collection

1. Ice.
2. Scalpel and scalpel handle, forceps, scissors.
3. Plastic culture dish (100 mm, Corning©).
4. Syringe 1 mL and 3 mL.
5. Needles 21G and 18G.
6. Cold phosphate buffered saline without calcium and magnesium (PBS).
7. 2 mL and 1.5 mL propylene centrifuge tubes (Eppendorf®).
8. Refrigerated centrifuge (20,000 × g).

2.2 EV Isolation from OF and UF by Size Exclusion Chromatography and Ultracentrifugation

1. PURE-EVs: size exclusion chromatography columns (HBM–PEV).
2. PBS.
3. Retort stand.
4. Ultracentrifuge with compatible fixed angle or swinging bucket rotor and adequate ultracentrifuge tubes:
   (a) Beckman Coulter Optima L-100 XP ultracentrifuge.
   (b) Swinging-bucket rotor "SW41 Ti".
   (c) Beckman Coulter thin-wall polypropylene tubes (Ref: 331372).

2.3 Nanoparticle Tracking Analysis (NTA)

1. Pipettes and pipette tips.
2. Syringes 1 mL and 20 mL.
3. Cold PBS.
4. 1.5 mL propylene centrifuge tubes (Eppendorf®).
5. NanoSight LM-10 system equipped with a CCD video camera and particle-tracking software NTA 3.1 (NanoSight, Malvern Panalytical, Worcestershire, United Kingdom).
6. Computer.

2.4 Transmission Electron Microscopy (TEM)

1. Pipettes and pipette tips.
2. Parafilm.
3. High precision antimagnetic tweezers.
4. Aqueous uranyl acetate solution 2% (w/v).
5. Grid box.
6. Formvar 200 Mesh Copper Grids (Agar Scientific, Essex, UK).

7. JEOL JEM1010 (100 kV) Transmission Electron Microscope (Jeol Ltd., Tokyo, Japan) equipped with a CCD megaviewG2 camera integrated into iTEM, Olympus Soft Imaging Solutions software (Olympus, Tokyo, Japan).

8. Computer.

2.5 Western Blot Analysis

1. Lysis buffer [(1 M Tris–HCl pH 6.8, 2% SDS supplemented with 1× protease inhibitor (cOmplete™, EDTA-free Protease Inhibitor Cocktail, Roche®))].

2. 4–10% SDS polyacrylamide gradient gel [1.5 M Tris–HCl pH 8.8, 1 M Tris–HCl pH 6.8, Acrylamide/Bis-acrylamide 30% solution (Ref: A3699 Sigma), 10% ammonium persulfate (APS), N,N,N′,N′-Tetramethylethylenediamine (TEMED), 10% Sodium dodecyl sulfate (SDS)].

3. 2× Loading buffer (200 mg SDS, 20% glycerol, 20% Tris–10% SDS pH 6.8, bromophenol blue solution 0.001% in H_2O).

4. 10× Electrophoresis buffer (30 g Tris, 144 g glycine, and 10 g SDS in 1 L Milli-Q H_2O).

5. 1× Transfer buffer (3.03 g Tris, 14.4 g glycine, 2.5 mL 10% SDS, 900 mL Milli-Q H_2O, and 100 mL Methanol).

6. PBS-T (0.1% Tween 20).

7. PageRuler™ Prestained Protein Ladder, 10–180 kDa (Thermo Scientific).

8. Wet transfer apparatus (Criterion™ blotter, Bio-Rad).

9. Nitrocellulose blotting membrane (Amersham™ Protran™ 0.45 NC) and filter papers.

10. Ponceau S staining solution (Sigma-Aldrich).

11. Blocking and primary antibody dilution solution (PBS-T + 3% BSA: dissolve 3 g of bovine serum albumin (BSA) in 100 mL PBS-T).

12. Primary antibody specific to bovine EVs: antigen specific (mouse anti-bovine CD63 (Bio-Rad, MCA2042GA), anti-heat shock protein 70 (anti-HSP70 mAb, C92F3A-5, Enzo), anti-tetraspanin cell surface protein CD9 antigen (anti-CD9 mAb D3H4P, #13403, Cell Signaling Technology).

13. Secondary Antibody Conjugated to HRP: anti-mouse (Santa Cruz Biotechnology, 2005), anti-rabbit (Santa Cruz Biotechnology, 2004).

14. Chemiluminescence reagent (RPN2109, ECL, Amersham).

15. Chemiluminescence Image Analyzer (ImageQuant LAS500, GE LifeSciences).

2.6 RNA Isolation	1. Pipettes and filtered pipette tips (Eppendorf®).
	2. Centrifuge.
	3. Tri reagent (Life Technologies®).
	4. Polyacryl reagent.
	5. miRNeasy Mini Kit (Thermo Fisher Scientific®).
	6. NanoDrop spectrophotometer (Thermo Fisher Scientific®).

2.7 miRNA Profiling

1. miRNA cDNA synthesis kit (miScript® II RT Kit).
2. Master mix reagents for miRNA analysis.
3. miRNA forward primers.
4. Real-time PCR machine.

2.8 miRNA Expression Analysis and Bioinformatics Analysis

1. Computer and internet access.
2. Preanalyzed miRNA spreadsheet data.

---

3 Methods

3.1 Oviduct and Uterine Flushing Collection

1. Collect the reproductive tracts from slaughtered heifers based on the corpus luteum (CL) morphology (*see* **Note 1**) and transport on ice to the laboratory within 3 h of slaughter.

2. Clean the reproductive tract from blood to avoid sample contamination. Cut the cervix and uterine body (until the fork of the uterine horns) and separate the tract between ipsilateral and contralateral side to the CL. After the removal of the ovary and connective tissue from the ipsilateral side, separate the uterine horn from the oviduct at the uterine–tubal junction and straighten the oviduct by "tearing" the connective tissue around the fimbria and ampulla to facilitate the flushing (Fig. 1).

3. Oviduct and uterine horn are flushed with 1 and 2.2 mL of cold PBS, respectively (Fig. 1). Use 1 mL syringe +21G needle and 3 mL syringe +18G needle to flush the oviduct and uterine horn respectively.

4. Collect flushed fluids into microcentrifuge tubes (2.5 mL) placed on ice and process as soon as possible.

5. To pellet cells, centrifuge the collected flushing during 7 min at 4 °C and $300 \times g$.

6. Collect the supernatant without disturbing the pellet and centrifuge at $10,000 \times g$ during 30 min at 4 °C to further remove cells and cell debris.

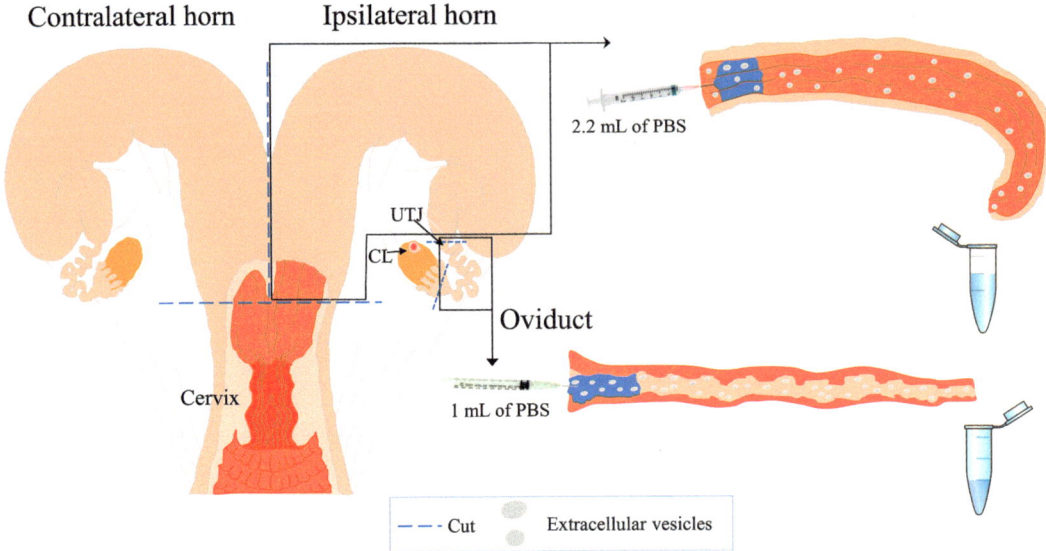

Fig. 1 Schematic illustration indicating the oviduct and uterine horn ipsilateral to the CL separation and flushing for EV isolation. *UTJ* uterine–tubal junction, *CL* corpus luteum

7. Filter the supernatant with 0.22 μm filter to remove particles larger than 220 nm [28].

8. Proceed for EV isolation with SEC column or keep the sample at 4 °C (24 h) or −80 °C until processing (avoid long EV storage for functional studies).

3.2 EV Isolation from OF and UF by Size Exclusion Chromatography and Ultracentrifugation

After assessing the efficiency of two SEC commercial kits (*see* **Note 2**), we selected the PURE-EVs SEC columns (HBM-PEV; Hansa-BioMed Life Sciences) as this one was the easiest and fastest for EV isolation from an amount of biological fluids up to 2 mL.

1. Secure the column in a vertical position by hanging it out in a retort stand. Ensure that there is sufficient space underneath the column to place the collection tubes.

2. PURE-EVs columns are provided with a layer of a preservative buffer. Open the upper and the lower cap of the PURE-EVs column and let flow almost all the buffer through the column, avoiding drying the surface of the gel.

3. Once all the buffer is inside the column gel add the sample obtained from the OF (≃1 mL) or UF (≃2 mL) collection.

4. Once the sample is inside the gel, keep adding PBS into the column to avoid the gel drying (*see* **Note 3**).

5. Collect the sample in 500 μL fractions (*see* **Note 4**): discard the first 6 fractions (3 mL) and save the following 5 fractions containing EVs (2.5 mL). The last 10 fractions correspond to proteins sample (5 mL).

6. After collecting all fractions wash the column from the residues of sample with 20 mL of PBS. Then add 1 mL of PBS, close the caps and store at 4 °C (*see* **Note 5**).

7. Pool the 5 EV fractions collected previously, and ultracentrifuge for 1 h at 100,000 × *g* and 4 °C to precipitate their content.

8. Discard the supernatant and resuspend the pelleted EVs in 100 μL of cold PBS. For maximum EV retrieval, pipet repeatedly the EV pellet.

9. Thirty microliters of the obtained pellet is used for EV characterization (maintain at 4 °C) and (70 μL is used for RNA extraction and analysis (stored at −80 °C until processing).

3.3 EV Characterization from OF and UF

3.3.1 NTA

1. Add 45 μL of cold PBS to the 5 μL of purified EVs (*see* **Note 6**).

2. Wash the NanoSight LM-10 chamber, repeatedly, with a 20 mL syringe filled with distilled water.

3. Standardize the measurement parameters of the NanoSight LM-10 with a 0.2 mL solution of 100 nm polystyrene beads (Malvern Panalytical, Worcestershire, UK) at a concentration of 1×10^8, in deionized water. The solution is loaded into the cuvette with a 1 mL syringe, with the laser power turned off (*see* **Notes 7** and **8**).

4. Turn on the laser power. The LM-10 instrument illuminates the particles from the bottom of the cuvette, made of glass, using a specially aligned and focused laser beam with $\lambda - 638$ nm. The light will be scattered by all nanoparticles present in the solution and viewed with a 20× microscope lens.

5. Adjust CCD video camera settings. Use a detection threshold of 3, temperature of 22 °C and viscosity similar to that of water (1 cP). The rest of the parameters such as blur size and maximum jump distance were set in automatic mode.

6. Fill a 1 mL syringe with 0.2 mL of the same PBS used to dilute their sample and loaded it into de chamber.

7. Clean the chamber between samples and between different dilutions of the same sample by introducing clean PBS in order to avoid the risk of presence of residual sample.

8. Each 50 μL sample of EVs is brought to a final volume of 0.2 mL with PBS (1/40 dilution) (*see* **Notes 9** and **10**). In order to keep the EVs in suspension and disaggregated as much as possible, pipet the sample repeatedly before placing it in a 1 mL syringe.

9. Capture three videos of 60 s per measurement at different focal planes within the cuvette using the script control model to examine 1800 frames per sample. These three videos are used

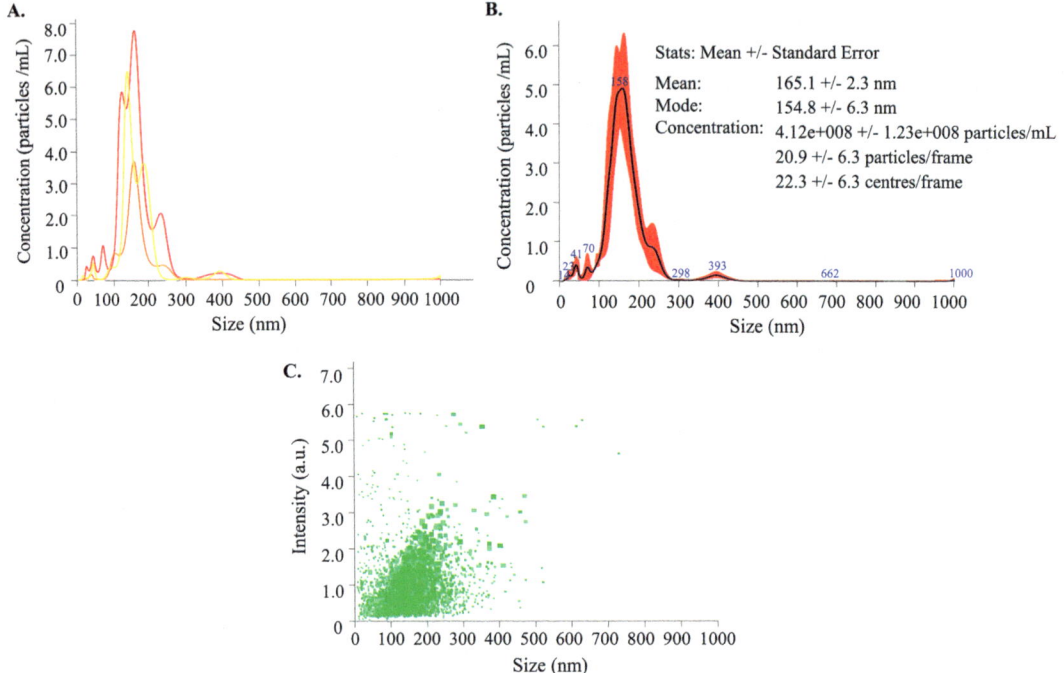

Fig. 2 Example of results obtained by NTA. (**a**) Finite track length adjustment (FTLA) Concentration/Size graphs for NTA analysis of particles isolated from oviductal fluid. The different colors of the curves express the measurements taken in triplicate for each sample, at different focal planes. (**b**) Averaged FTLA Concentration/ Size for sample (mean, mode, polydispersity, and concentration of particles/mL). (**c**) Diagram of the relative light scattering intensity plotted against particle size

by NTA 3.1 software to determine EVs size and concentration averages of the analyzed sample. Analyze each video to give the mean and mode vesicle size together with an estimate of the concentration (Fig. 2).

10. Repeat **steps 3–7** for each sample.

3.3.2   TEM

1. Add 45 µL of cold PBS to the 5 µL sample of purified EVs.

2. Prepare a rectangle of Parafilm and label it with a waterproof marker to specify the samples, keeping enough space between rows to correctly process them (*see* **Note 11**).

3. Pipet gently each sample to resuspend EVs.

4. Place two drops of 25 µL of each sample to have a duplicate and two drops of 50 µL of Milli-Q water per sample in labeled rows on the Parafilm.

5. Put a formvar/carbon-coated 200 mesh copper grids over each sample on the Parafilm using an antimagnetic precision tweezers and incubate the grid for 1 min (*see* **Note 12**).

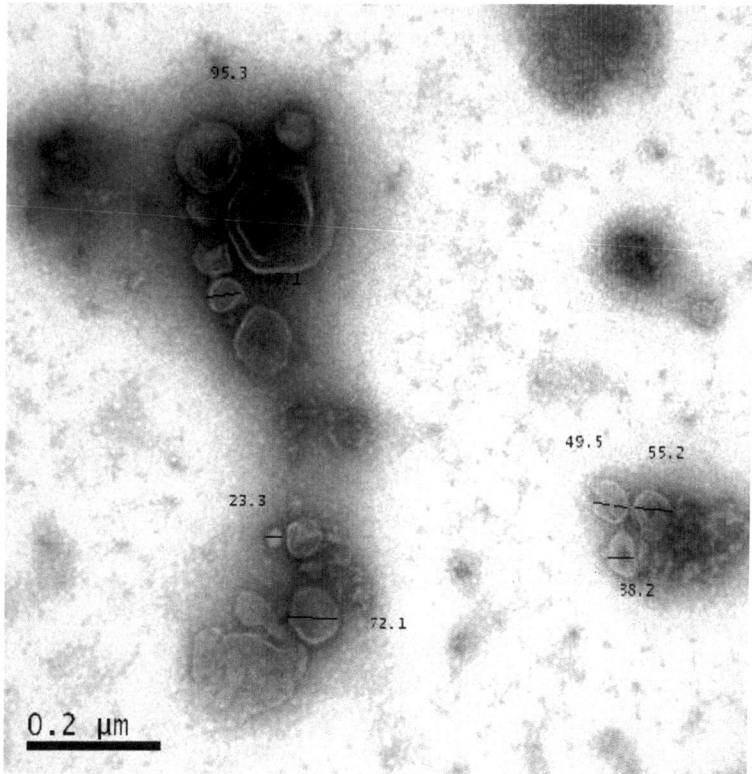

Fig. 3 Transmission electron microscopy image at 100 kW of EVs negatively stained with aqueous uranyl acetate 2%. Observation of the characteristic cup-shaped morphology of EVs. Measurement of approximate diameter of visualized EVs with the Olympus software platform

6. Wash the sample twice by brief contact with MilliQ water to remove the PBS from the grid and prevent visualization of salt precipitates, followed by blotting to remove excess liquid.

7. Next, the grid is placed on 50 µL of aqueous uranyl acetate 2% for 1 min, followed by blotting to remove excess staining and let it air-dry.

8. Store the stained grids in a box kept inside a desiccator to avoid humidity until we analyzed them.

9. Acquire the images using a JEOL JEM1010 Microscope equipped with a CCD megaview2 camera at 100K magnification, viewed with the iTEM Olympus software platform, and export images in .tiff format to measure the diameter of different particles with the Olympus software platform (Fig. 3).

3.3.3 Western Blot Analysis

1. To extract proteins from EVs isolated from OF and UF, add 180 µL of lysis buffer supplemented with 1× loading buffer to the pellet of EVs obtained by ultracentrifugation and keep at 4 °C overnight—to detect up to 4 EV proteins the obtention of item and UF from 12 reproductive tracts is required.

Table 2
Gel composition to make a 4% stacking and 10% running gel

		Stacking gel 4%	Running gel 10%
Milli-Q water	mL	1.8	3.2
1.5 M Tris–HCl pH 8.8	mL		2
1 M Tris–HCl pH 6.8	mL	0.76	
SDS 10%	μL		80
Acrylamide–bisacrylamide 30%	mL	0.40	2.7
TEMED	μL	3.0	4
APS 10%	μL	15.0	40

2. Prepare running and stacking gel solutions according to Table 2. Volumes given are for one gel in the Mini-PROTEAN Tetra system. Pour the running gel into the 1.5 mm thick glass chamber with a pipette until the height of the solution reaches three-fourths of that of the small glass plate (approximately 7 mL).

3. Overlay the solution with 3 mL of propanol; pour the propanol from the top of the gel (*see* **Note 13**). Let the solution polymerize for 30 min and then rinse the surface twice with distilled H$_2$O (*see* **Note 14**).

4. Add APS and TEMED to the stacking gel solution (Table 2) and pour immediately into the glass chamber until the solution reaches the top of the small glass plate. Insert a 1.5-mm comb into the chamber. The solution should polymerize within 10 min (*see* **Note 15**).

5. Transfer the gels to the cassette for electrophoresis (U-shaped) with the combs inward. Pour 1× electrophoresis buffer between the space of the two gels and carefully remove the comb, making sure not to damage wells. Load marker (5 μL) followed by samples (≃45 μL) into each well.

6. Attach the cassette to the electrophoresis system according to the instruction manual and fill with electrophoresis solution up to the indicated mark (it depends if you use 2 or 4 gels) (*see* **Note 16**). Connect the electrophoresis system to the power supply and set voltage (100 V) for stacking gel; use a higher voltage (150 V) for running gel. Stop the power supply when the dye front runs off the bottom of the gel.

7. Disassemble the glass sandwich of the electrophoresis system and take out the gels. Place the gel in a tray containing 1× transfer buffer for 15 min.

8. Cut the membrane and the filter paper so that their length and width are corresponding to the size of the gel. Soak the membrane, filter papers, and blotting pads in 1× transfer buffer.

9. Create a transfer sandwich starting at the black side, as follows: 1 pad, 1 filter paper, gel, membrane, 1 filter paper, 1 pad. Remove bubbles from between the membrane and the gel by rolling a plastic pipette on the sandwich from one end to the other. Repeat this in the other direction.

10. Place the sandwich to the transfer equipment, make sure the red part of the cassette is towards the positive pole, and add transfer buffer to the equipment up to the max indication mark. Connect the blotting equipment to a power supply, set at 300 mA for 1.5 h and place it on ice to maintain 4 °C during the process.

11. Remove the membrane from the sandwich and wash twice with distilled H_2O, then add Ponceau S solution just enough to cover the membrane for 5 min. Wash the membrane briefly with distilled H_2O until that bands are visible. Photograph the membrane using the colorimetry option.

12. Rinse the membrane in PBS-T until the staining is completely washed off.

13. Add blocking solution in a shallow container so that the membrane can be soaked and incubate at room temperature for 30 min with continuous gentle shaking.

14. Add primary antibody diluted to 1:1000 with the blocking solution in the container and maintain it overnight with continuous gentle shaking at 4 °C.

15. Wash the membrane with PBS-T at room temp for 5 min. Repeat this procedure three times.

16. Prepare secondary antibody diluted to 1:2500 in PBS-T + 1% BSA. Incubate at room temperature for 2 h with gentle shaking.

17. Wash the membrane three times with PBS-T for 5 min.

18. Prepare ECL mix (following the proportion of solution A and B provided by the manufacturer) and cover the membrane. Incubate the membrane for 1–2 min (see **Note 17**).

19. Record the signal by exposing the membrane in the imaging system device. Signals should be visible within 30 min.

3.4 MicroRNA Isolation and Analyses

3.4.1 RNA Isolation and Profile of microRNA

The profile of microRNA can be performed using different approaches such as RNA sequencing, microarrays, TaqMan microRNA assays, or real-time quantitative reverse transcription PCR (qRT-PCR) [23]. Here we will focus on qRT-PCR performed with the miScript PCR System for miRNA analysis of 383 mature miRNA sequences in extracellular vesicles isolated from bovine oviductal and uterine fluids. RNA isolation protocol was performed according to Da Silveira [31], with minor modifications and following the manufacturer's instructions of the miRNeasy Mini Kit.

1. Add 200 μL Tri–reagent to the 70 μL diluted EVs, add 8 μL PolyAcryl carrier (Cat#PC152) to the mixture. Homogenize and incubate the mixture at room temperature for 5 min.

2. Add 140 μL chloroform and vortex the sample tube for 15 s, then incubate at room temperature for 5 min and centrifuge at 12,000 × g for 15 min at 4 °C.

3. Move the aqueous phase to a new RNase/DNase-free tube and add 500 μL of 100% ethanol and transfer 700 μL sample into an RNeasy Mini column and continue with the commercial protocol.

   For reverse-transcription use miScript® II RT Kit with 100–200 ng of total RNA per sample, following the manufacturer's instructions, with minor modifications, as described below (*see* **Note 18**).

4. Incubate total RNA, containing the miRNAs, with 5× miScript Hispec Buffer (*see* **Note 19**), 10× miScript Nucleic mix, RNase-free water, and miScript reverse transcriptase at 37 °C for 60 min, followed by 5 min at 95 °C to inactivate miScript Reverse (Table 3) (*see* **Note 20**).

5. Prepare the real time PCR mix for a 6 μL reaction containing 2× QuantiTec SYBR Green, 10× miScript Universal Primer, RNase-free water, and cDNA (Table 4).

6. In a 384 well plate, add 1 μL of the forward primer (*see* **Note 21**) of each 383 miRNAs in each well. Add 1 μL RNase-free water to the last well of the plate (blank).

7. Add 5 μL of PCR Mix (**step 2**) in each 384 well for a 6 μL final reaction (*see* **Note 22**).

8. Place the real-time PCR plate in the PCR machine and set the PCR cycle conditions: 95 °C for 15 min; 45 cycles of 94 °C for 10 s, 55 °C for 30 s, and 70 °C for 30 s. Perform fluorescence data collection in the step at 70 °C.

3.4.2 miRNA Expression Analysis

To analyze the relative levels of miRNA isolated from EVs, data from miRNAs with suitable threshold cycle (Ct) and an appropriate melting curve are considered and normalized to the geometric mean of internal controls as previously described by Da Silveira

Table 3
Reverse–transcription reaction components and volume per reaction

Components	Volume/reaction
5× miScript HiSpec Buffer	2 μL
10× miScript Nucleics Mix	1 μL
miScript Reverse Transcriptase Mix	1 μL
Total RNA	100–200 ng
RNase-free water	Variable
Total volume	10 μL

Table 4
384-well real-time PCR reaction, volume per reaction, and PCR mix

Components	Volume/reaction	PCR mix (×400 reactions)
2× QuantiTect SYBR Green PCR Master Mix	3 μL	1200 μL
10× miScript Universal Primer	0.6 μL	240 μL
cDNA	–	10 μL
RNase-free water	1.37 μL	550 μL

[31]. The miRNA data obtained as described below allows the identification and analysis of miRNAs exclusively expressed in each experimental group and miRNAs differentially expressed among the groups (*see* **Note 23**).

1. Only raw Ct data values <37 are considered as detected per sample.

2. The melting curve of mature miRNA, performed with miScript HiSpec Buffer, from a specific amplification shows a single peak.

3. Internal controls have to present a standard deviation <1 Ct across comparison groups. As examples of previously used internal controls, we suggest Hm/Ms./Rt U1 snRNA and bta-miR-99b as a reference for extracellular vesicles miRNAs isolated from bovine oviductal and uterine fluids.

3.4.3 Bioinformatics Analysis

Bioinformatics analysis allows the identification of biological pathways enriched by an individual miRNA or by a group of miRNAs (*see* **Note 24**). For the study of bovine miRNAs we recommend the use of the miRWalk database (http://mirwalk.umm.uni-heidelberg.de/search_mirnas/). However, the use of multiple databases to evaluate selected miRNAs supports the identification

Table 5
Summary of computational tools and databases used for prediction and validation of miRNA targets

Databases	Description	Website
miRWalk	Prediction of possible miRNA binding sites, target genes and biological pathway enriched by bovine sequences	http://mirwalk.umm.uni-heidelberg.de/
DIANA–miRPath	Interactions and the targeted genes of human miRNAs individually or in group, predicting the biological pathways enriched by them	http://snf-515788.vm.okeanos.grnet.gr/
miRBase	Provides microRNA sequences allowing the identification of similarities between different species	http://www.mirbase.org/
miRTarBase	Identification of validated microRNA-target interactions	http://mirtarbase.mbc.nctu.edu.tw/php/index.php
TargetScan	Provides prediction of microRNA targets and their gene-regulatory network in mammals	http://www.targetscan.org/vert_72/

of common pathways among the different databases, suggesting these pathways as strong candidates for functional analysis [32, 33]. Here, we will outline the approach using mirWalk and DIANA-miRPath (http://snf-515788.vm.okeanos.grnet.gr/), a miRNA pathway analysis web server, to identify pathways regulated by EVs miRNAs (Table 5).

1. Insert the full miRNAs names into mirWalk (bta-mir-xxx), after selecting "cow" as the specie of interest and proceed with the search.

2. Select the option GSEA (gene set enrichment analysis) to access their predicted pathways and target genes.

3. Before the bioinformatics analysis on DIANA-miRPath, it is important to compare the bovine miRNAs sequences with human miRNAs sequences in order to identify sequence similarities using miRBase, http://www.mirbase.org [34]. Only use miRNAs with 95–100% sequence similarity, otherwise DIANA-miRPath cannot be used for a particular bovine miRNA.

4. Once you confirm sequence similarity you can submit the full miRNAs names (hsa-mir-xxx) into DIANA–miRPath after selecting "human" as the species of interest in order to identify predicted pathways and target genes of all identified miRNAs.

5. The miRTarBase database can be used to confirm if your bioinformatics results are valid for the species of interest, based on luciferase, protein and qPCR assays [35].

6. The TargetScan can also be used to identify multiple target genes for individual miRNA, if necessary [36].

7. Also, based on 3'UTR conservation of the target genes and the miRNA sequence conservation it is possible further evaluated the mRNA–miRNA interaction by other approaches.

4   Notes

1. Whether the study is focused on one or four stages of the estrous cycle, it is important to distinguish between them based on the CL morphology, following the classification of [37] and [38]: Stage 1 [S1]: showing recently ovulated follicle characterized either by a bloody appearance that fills the remnant of the follicular lumen or formation of the apex of a new small CL (Day 1 to 4); Stage 2 [S2]: early luteal development with vasculature visible around its periphery, red in color with medium or large follicles or both present (Day 5 to 10); Stage 3 [S3]: fully functional CL with complete vasculature, orange or dark yellow in color (Day 11 to 17); Stage 4 [S4]: regressing CL with no vasculature in its surface and a large preovulatory follicle present (Day 18 to 20).

2. NanoSight and TEM analysis were performed to compare the efficiency on OF and UF EV concentration and size between both commercial kits.

3. An amount of 11 mL of PBS was necessary to keep the column flowing and collect all the fractions.

4. If the study is aiming to analyze each EV fraction individually, it is important to collect them separately, otherwise collect the 5 EV fractions together (2.5 mL) to increase concentration.

5. According to the manufacturer one column could be used up to five times; however, it is possible to start to observe dirtiness since the third use; it is important then to wash the column with NaOH and measure the pH before and after washing.

6. The EVs dilution could be kept at 4 °C for a maximum of 5 days, before NTA. The particles could also be measured after freezing at −20 °C although their analysis reveals some size variations and a tendency to form more aggregates.

7. It is very important to verify that there are no bubbles inside the syringe before injecting the distilled water or the sample into the NanoSight chamber.

8. Be careful to slowly introduce the solution/sample into the cuvette to avoid bypassing its seals that could cause damage to the laser.

9. Although the company recommends using at least 0.3 mL of sample, we have obtained good results loading only 0.2 mL.

10. The number of particles per frame to obtain a good signal to noise ratio, while at the same time avoiding multiple scattering interferences, should be between 10 and 200, so if necessary, more dilutions should be made in the suspended EVs.

11. If you are not very familiar with TEM techniques, we recommend processing a maximum of six samples per time.

12. Two grids are prepared per sample.

13. The propanol layer ensures that the top of the gel is flat after polymerization.

14. Remaining H_2O may cause the stacking gel to shrink.

15. Pay attention not to introduce bubbles during the process. Because acrylamide polymerizes quickly, the whole procedure should be done as quickly as possible.

16. The top and bottom of the sandwich should be submerged in buffer for electric current to flow through the gel matrix.

17. Ensure that ECL covers all the membrane surface for optimal capture signaling.

18. This protocol is for use with a final concentration of cDNA per well between 50 pg to 3 ng.

19. Two buffers are provided with the miScript II RT Kit of the miScript PCR System. Use the miScript HiSpec Buffer for quantification of mature miRNA only or the miScript Hiflex Buffer for real-time PCR quantification of mature and precursor miRNA.

20. Place the final cDNA reaction on ice for a short period until use or store at −20 °C.

21. Mature miRNA sequences can be downloaded from miRBase database [34].

22. After placing the optical adhesive and before placing the plate in the PCR machine, centrifuge the plate at $2000 \times g$ for 3 min to move the contents to the bottom of the wells.

23. Differentially expressed miRNAs are found after appropriated statistical tests analysis, according to the characteristics of the experiment.

24. Although bioinformatics analysis can be performed with miRNAs individually, for identification of pathways it is recommended the approach in "group," since it is how miRNAs are present in extracellular vesicles (multiple miRNAs in one vesicle or multiples vesicles carrying different individual miRNAs [39]).

Acknowledgments

This work was supported by the Spanish Ministry of Science and Innovation (AGL-2015-70140-R, PID2019-111641RB-I00, and RTI2018-093548-B-I00); Y.N.C. was supported by Secretaría de Educación Superior, Ciencia y Tecnología e Innovación (SENESCYT-Ecuador); São Paulo Research Foundation, Brazil (FAPESP; #2017/20339-3, #2014/22887-0, and #2019/04981-2); and National Council for Scientific and Technological Development—CNPq, Brazil (grant number #420152/2018-0). The Authors are members of the COST Action CA16119 In vitro 3-D total cell guidance and fitness (CellFit).

References

1. Machtinger R, Laurent LC, Baccarelli AA (2016) Extracellular vesicles: roles in gamete maturation, fertilization and embryo implantation. Hum Reprod Update 22:182–193

2. Edgar JR (2016) Q&A: what are exosomes, exactly? BMC Biol 14:46

3. Van Niel G, D'Angelo G, Raposo G (2018) Shedding light on the cell biology of extracellular vesicles. Nat Rev Mol Cell Biol 19 (4):213–228

4. Lee Y, EL Andaloussi S, Wood MJA (2012) Exosomes and microvesicles: extracellular vesicles for genetic information transfer and gene therapy. Hum Mol Genet 21:R125–R134

5. Théry C, Witwer KW, Aikawa E et al (2018) Minimal information for studies of extracellular vesicles 2018 (MISEV2018): a position statement of the International Society for Extracellular Vesicles and update of the MISEV2014 guidelines. J Extracell Vesicles 8:1535750

6. Zhang M, Ouyang H, Xia G (2009) The signal pathway of gonadotrophins-induced mammalian oocyte meiotic resumption. Mol Hum Reprod 15:399–409

7. Raposo G, Stoorvogel W (2013) Extracellular vesicles: exosomes, microvesicles, and friends. J Cell Biol 200:373–383

8. Pavani KC, Alminana C, Wydooghe E et al (2017) Emerging role of extracellular vesicles in communication of preimplantation embryos in vitro. Reprod Fertil Dev 29:66–83

9. Lopera-Vásquez R, Hamdi M, Fernandez-Fuertes B et al (2016) Extracellular vesicles from BOEC in in vitro embryo development and quality. PLoS One 11:e0148083

10. Lopera-Vasquez R, Hamdi M, Maillo V et al (2017) Effect of bovine oviductal extracellular vesicles on embryo development and quality in vitro. Reproduction 153:461–470

11. Almiñana C, Corbin E, Tsikis G et al (2017) Oviduct extracellular vesicles protein content and their role during oviduct-embryo crosstalk. Reproduction 154:153–168

12. Burns GW, Brooks KE, Spencer TE (2016) Extracellular vesicles originate from the conceptus and uterus during early pregnancy in sheep. Biol Reprod 94:56

13. Kurian NK, Modi D (2019) Extracellular vesicle mediated embryo-endometrial cross talk during implantation and in pregnancy. J Assist Reprod Genet 36:189–198

14. Berezikov E, Guryev V, van de Belt J et al (2005) Extracellular vesicles originate from the conceptus and uterus during early pregnancy in sheep. Cell 120:21 24

15. Lewis BP, Burge CB, Bartel DP (2005) Conserved seed pairing, often flanked by adenosines, indicates that thousands of human genes are microRNA targets. Cell 120:15–20

16. Carletti MZ, Christenson LK (2009) MicroRNA in the ovary and female reproductive tract. J Anim Sci 87:E29–E38

17. Lötvall J, Hill AF, Hochberg F et al (2014) Minimal experimental requirements for definition of extracellular vesicles and their functions: a position statement from the International Society for Extracellular Vesicles. J Extracell Vesicles 3:26913

18. Taylor DD, Shah S (2015) Methods of isolating extracellular vesicles impact down-stream analyses of their cargoes. Methods 87:3–10

19. Lane RE, Korbie D, Anderson W et al (2015) Analysis of exosome purification methods using a model liposome system and tunable-resistive pulse sensing. Sci Rep 5:7639

20. Mathivanan S, Lim JWE, Tauro BJ et al (2010) Proteomics analysis of A33 immunoaffinity-purified exosomes released from the human

colon tumor cell line LIM1215 reveals a tissue-specific protein signature. Mol Cell Proteomics 9:197–208

21. Böing AN, van der PE, Grootemaat AE et al (2014) Single-step isolation of extracellular vesicles by size-exclusion chromatography. J Extracell Vesicles 3

22. Mol EA, Goumans M-J, Doevendans PA et al (2017) Higher functionality of extracellular vesicles isolated using size-exclusion chromatography compared to ultracentrifugation. Nanomedicine 13:2061–2065

23. Pritchard CC, Cheng HH, Tewari M (2012) MicroRNA profiling: approaches and considerations. Nat Rev Genet 13:358–369

24. Stranska R, Gysbrechts L, Wouters J et al (2018) Comparison of membrane affinity-based method with size-exclusion chromatography for isolation of exosome-like vesicles from human plasma. J Transl Med 16:1

25. Vergauwen G, Dhondt B, Van Deun J et al (2017) Confounding factors of ultrafiltration and protein analysis in extracellular vesicle research. Sci Rep 7:2704

26. Enderle D, Spiel A, Coticchia CM et al (2015) Characterization of RNA from exosomes and other extracellular vesicles isolated by a novel spin column-based method. PLoS One 10: e0136133

27. Théry C, Amigorena S, Raposo G et al (2006) Isolation and characterization of exosomes from cell culture supernatants and biological fluids. Curr Protoc Cell Biol 30:3.22.1–3.22.29

28. Lässer C, Eldh M, Lötvall J (2012) Isolation and characterization of RNA-containing exosomes. J Vis Exp:e3037

29. Jeppesen DK, Hvam ML, Primdahl-Bengtson B et al (2014) Comparative analysis of discrete exosome fractions obtained by differential centrifugation. J Extracell Vesicles 3:25011

30. Witwer KW, Buzás EI, Bemis LT et al (2013) Standardization of sample collection, isolation and analysis methods in extracellular vesicle research. J Extracell Vesicles 2:20360

31. Da Silveira J, Andrade GM, Perecin F et al (2018) Isolation and analysis of exosomal microRNAs from ovarian follicular fluid. In: Ying S-Y (ed) MicroRNA protocols. Springer, New York, NY, pp 53–63

32. Chen L, Heikkinen L, Wang C et al (2019) Trends in the development of miRNA bioinformatics tools. Brief Bioinform 20:1836–1852

33. Eulalio A, Mano M (2015) MicroRNA screening and the quest for biologically relevant targets. J Biomol Screen 20:1003–1017

34. Kozomara A, Griffiths-Jones S (2014) miR-Base: annotating high confidence microRNAs using deep sequencing data. Nucleic Acids Res 42:D68–D73

35. Huang H-Y, Lin Y-C-D, Li J et al (2020) miRTarBase 2020: updates to the experimentally validated microRNA–target interaction database. Nucleic Acids Res 48:D148–D154

36. Agarwal V, Bell GW, Nam J-W et al (2015) Predicting effective microRNA target sites in mammalian mRNAs. elife 4:e05005

37. Ireland JJ, Murphee RL, Coulson PB (1980) Accuracy of predicting stages of bovine estrous cycle by gross appearance of the corpus luteum. J Dairy Sci 63:155–160

38. Senger PL (2005) Pathways to pregnancy and parturition. Current Conceptions, Pullman, WA

39. Pegtel DM, Cosmopoulos K, Thorley-Lawson DA et al (2010) Functional delivery of viral miRNAs via exosomes. PNAS 107:6328–6333

Chapter 17

Biomimetic 3D-Bone Tissue Model

Mahmut Parmaksiz, Ayşe Eser Elçin, and Yaşar Murat Elçin

Abstract

Various approaches have been evaluated for developing three-dimensional (3D) scaffolds for modeling or engineering of the bone tissue. However, most of such attempts have come up short in mimicking the natural bone tissue extracellular matrix (ECM) microenvironment, especially its natural bioactive content. Here we describe the methodology for the preparation of a natural ECM-based multichannel construct as a biomimetic 3D bone tissue model. We elucidate the construction of the composite scaffold incorporating decellularized small intestinal submucosa ECM, synthetic hydroxyapatite and poly(ε-caprolactone), and the mechanical stimulation of the cell-seeded construct under bioreactor culture.

Key words 3D bone tissue model, Biomimetic bone scaffold, Decellularization, Extracellular matrix, Composite matrices, Mechanical cue, Mesenchymal stem cell, Regenerative medicine

1 Introduction

Natural bone tissue can be considered as a composite organic-based living structure reinforced with inorganic components [1]. Owing to its intricate composition, bone can regain its natural complex anatomical structure following minor fractures or damages caused by trauma or diseases [2]. However, bone tissue can fall short in properly regenerating or even repairing itself after occurrence of major defects or damages. In such cases, autografts and allografts are the primary treatment options in the clinical practice. Nevertheless, each of these options has some limitations, such as donor site morbidity and limited availability during autograft procurement, and the risk of disease transmission or immune rejection after allograft transplantations. Apart from auto/allografts, a great number of studies have evaluated the bone repair efficiency of various biomaterials including ceramics, metals, polymers, and composites. In particular, synthetic bioceramics and their polymeric composites have attracted attention. Investigations have started with the use of bioinert ceramics, continued with bioactive and

Tiziana A.L. Brevini et al. (eds.), *Next Generation Culture Platforms for Reliable In Vitro Models: Methods and Protocols*,
Methods in Molecular Biology, vol. 2273, https://doi.org/10.1007/978-1-0716-1246-0_17,

resorbable counterparts, and more recently with designed third-generation bioceramics. At the present time, studies to develop an effective treatment option for major bone conditions are still ongoing [2, 3].

Tissue engineering concept has emerged as a potential regenerative option for eliminating the previously described issues of current clinical treatments. Bone tissue engineering has been considered as a potential alternative to the traditional bone grafting technology, and has become a major focus of tissue engineering studies. The primary objective of bone tissue engineering approach has been to develop osteogenic biomimetic scaffolds capable of mimicking the architectural, mechanical, and bioactive elements of the natural bone tissue, interacting with seeded and surrounding cells, and responding to biochemical and mechanical cues [4]. Scaffolds designed for this purpose are expected to emulate the natural bone extracellular matrix (ECM) in vitro, in terms of bioactive components and morphological structure. As a matter of fact, advances in biomaterials fabrication technologies have led to the development of bone tissue scaffolds providing better mechanical and functional support; however, limitations still exist in mimicking the bioactive ECM components of the natural bone [5].

As known, the remarkable properties of the natural bone tissue are associated with its complex ECM content. Collagen (mainly type 1) is the predominant part (one third) of the organic components of bone ECM. Different than other tissues, bone ECM also contains a large proportion (two thirds) of inorganic components, primarily in the form of calcium hydroxyapatite (HAp). Typical bone tissue scaffolds usually provide these two major components [4, 5]. However, the inorganic part of bone ECM also incorporates smaller amounts of proteoglycans, matrix proteins (osteocalcin, osteonectin, and osteopontin), cytokines and growth factors, which all play critical roles in processes such as binding of growth factors, ECM remodeling, angiogenesis, and new bone formation [6, 7].

Thus far, bioceramic-based (i.e., HAp and β-tricalcium phosphate) bone scaffolds with suitable inorganic and architectural form have been successfully fabricated. Besides, tissue templates by use of natural polymers such as collagen, silk fibroin, and chitosan have also been developed with suitable porosity and surface topography. However, bone tissue templates incorporating optimal bioactive content with appropriate mechanical properties have not been achieved by simultaneous use of bioceramics, biopolymers, and synthetic growth factors to date.

In the last decade, the use of decellularized (cell-free) natural tissues incorporating the complex bioactive ECM content has initiated a new era for regenerative medicine. Decellularization technology has enabled the use of various types of tissues, organs, or even cell cultures from allogenic or xenogenic sources for regenerative applications; thus the technology has been transferred to the clinics for different tissue repairs [8, 9]. Among many ECM sources, the small intestinal submucosa (SIS) layer rich in bioactive ECM content has succeeded to pass to the clinics, and the porcine SIS product is used in treatments such as venous ulcers, diabetic ulcers, skin injuries, vascular repair, and bladder wall repair [8, 10, 11]. Besides, preclinical studies have shown that decellularized bovine SIS (bSIS) is effective in regenerating critical-sized full-thickness skin defects [12].

Recently, a composite multilayered bone tissue scaffold incorporating HAp microparticles and the bioactive ECM component (bSIS) was developed by our group which also can serve as a biomimetic 3D bone tissue model for studying the healthy or disease processes of the bone tissue in vitro [13]. Here, the major fabrication and characterization steps of the biomimetic composite construct, as well as the applied dynamic culture conditions on mesenchymal stem cell-laden bone tissue model are described.

2 Materials

2.1 Equipment

1. Mechanical testing machine (Shimadzu AGS-X, Tokyo, Japan).
2. CO_2 incubator (Thermo Scientific Heracell™ 240i, Waltham, MA, USA).
3. Centrifuge (Beckman Coulter, Allegra X-15R, Brea, CA, USA).
4. Multiplate reader (Molecular Devices, SpectraMax M2, San Jose, CA, USA).
5. Laminar flow cabinet (Labconco, Class II, Type A2, Kansas City, MO, USA).
6. Water bath sonicator (Elma Elmasonic, Singen (Hohentwiel), Germany).
7. Inverted light microscope (Zeiss Axio Vert A1, Oberkochen, Germany).
8. Upright Microscope (Zeiss Axio Imager Z2, Oberkochen, Germany).
9. Automated cell counter (TC20™, Bio-Rad Labs., Hercules, CA, USA).
10. Micropipettes (Eppendorf, Hamburg, Germany).
11. Ultralow temperature freezer (Thermo Scientific, Revco, Waltham, MA, USA).
12. Hand drill (OmniDrill, WPI, Sarasota, FL, USA).
13. Sterile scissors and blades.

2.2 Reagents and Materials

1. Hydroxyapatite microparticles (~30 μm-size, Plasma Biotal, Buxton, UK).

2. Decellularized bovine SIS (Matrisis™, Biovalda Health Technologies, Ankara, Turkey).

3. Poly(ε-caprolactone) (Mw: 80,000; Sigma-Aldrich, St. Louis, MO, USA).

4. Cell culture medium and supplements (fetal bovine serum, penicillin, streptomycin) (Lonza, Basel, Switzerland).

5. Dichloromethane, methanol, ethanol (Sigma-Aldrich, St. Louis, MO, USA).

6. Tissue culture plastic: sterile pipets, culture flasks (Corning, Corning, NY, USA).

7. Sterile syringes (50 mL volume) (BD, New Jersey, USA).

8. Syringe filters (0.22 μm) (Corning Life Sciences, Corning, NY, USA).

9. QuantiChrom™ Alkaline Phosphatase (ALP) Assay Kit (Bio-Assay System, CA, USA).

3 Methods

3.1 Preparation of Biomimetic Bone Scaffold

The construction steps of the biomimetic bone scaffold are summarized in Fig. 1.

1. Cut decellularized bSIS (Matrisis™) membrane in the form of ~5 mm-diameter discs to prepare the layers. The bSIS discs will serve as the main support structure resembling the organic component (collagen and bioactive molecules) of the natural bone [13, 14] (*see* **Note 1**).

2. Prepare 10% PCL solution (prepared in DCM–methanol, 3:1, v:v) which will be used as the binder for holding together the multilayered discs. Then, drop 8–10 μL of PCL solution at four distinct points of each disc by using a micropipette for superposing the layers (*see* **Notes 2–4**).

3. Superpose around 50 bSIS discs on top of each other to obtain a multilayered construct of ~8–10 mm in height.

4. Perforate the construct from top to the bottom at five different equidistant points by using a surgical hand drill with a pin bit of ~500 μm. The holes will positively influence nutrient transfer through the constructs during both the static and dynamic culture experiments (*see* **Note 5**).

5. Sterilize the constructs by keeping in 70% ethanol solution (aq.) for 2 h. Then excessively wash with sterile PBS (5 × 50 mL) to remove the remnants of ethanol. Remove

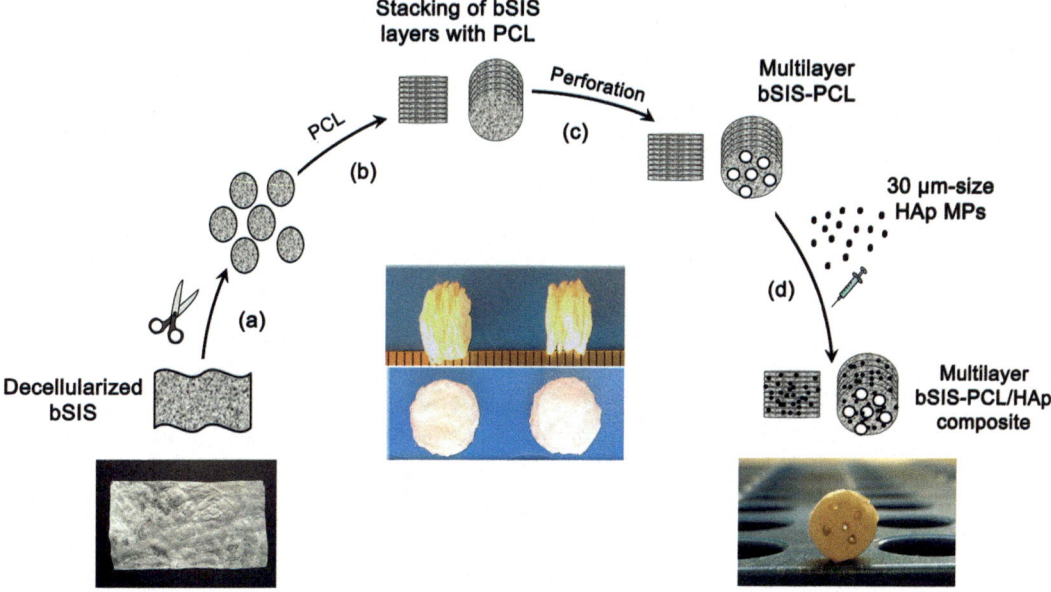

Fig. 1 Scheme demonstrating the construction steps of the biomimetic scaffold: (**a**) cutting of bSIS discs, (**b**) stacking of discs by the aid of PCL microdroplets, (**c**) perforation of the multilayered construct, and (**d**) incorporation of HAp microparticles (cited from [13] [*Parmaksiz M, Elçin AE, Elçin YM (2019). Mat Sci Eng C-Mater 94:788–797*])

excess PBS using a sterile paper towel and let the constructs to air dry inside the sterile laminar flow cabinet (*see* **Note 6**).

6. By aid of a sterile syringe, incorporate sterile HAp microparticles suspended in sterile PBS to the construct (*see* **Note 7**).

7. The HAp microparticle suspension (equivalent to ~0.1 mg dry HAp per construct) should be evenly transferred to different regions of the construct (*see* **Note 8**). Then, air dry the HAp suspension-soaked construct inside the sterile laminar flow cabinet. Ensure that the inorganic component is completely soaked up by the construct.

3.2 Cell Expansion and Seeding of Cells on the Construct

1. Isolate rat bone marrow-derived mesenchymal stem cells (rBM-MSCs) by standard methods [12] and cryopreserve the cells.

2. Thaw the stock frozen rBM-MSCs (at passage 2) inside a water bath adjusted to 37 °C, and transfer the cells to a 15 mL centrifuge tube containing 5 mL growth medium (i.e., Dulbecco's Modified Eagle Medium [DMEM] supplemented with 10% FCS, 100 U/ mL penicillin and 100 μg/mL streptomycin). Centrifuge the cell suspension at $250 \times g$ for 7 min, collect the cell pellet and plate them for routine cell culture (*see* **Note 9**).

3. When rBM-MSC cultures reach a confluence of ~80%, enzymatically collect the cells by using a solution of 0.05% trypsin– 0.02% EDTA. Then, centrifuge the cell suspension at $250 \times g$

for 7 min, discard the supernatant, and count the cells (*see* **Note 10**).

4. Place the sterile biomimetic constructs to 24-well culture dishes (one construct per well) for subsequent cell seeding and cultivation steps.

5. Seed the cells suspended in the growth medium onto multilayered composite constructs (at a density of ~4.0×10^6 cells per construct) from various regions by using a sterile injector (*see* **Note 11**).

6. Thirty minutes after cell seeding, add 2 mL of growth medium to each well containing the cell-laden construct. Maintain the culture at 37 °C under ambient conditions of 5% CO_2, 95% air, and 90% humidity. For the static culture, exchange the medium with fresh medium every other day for the duration of the experiment.

7. For dynamic culture of cell-laden constructs, use a modified version of a lab-made flow bioreactor system described elsewhere [15–17] (Fig. 2).

8. The flow bioreactor basically consists of a cassette system with a double o-ring that allows certain compression where the construct will be trapped. The bioreactor also includes an integrated flow-controlled pump system (*see* **Note 12**).

9. Considering the dimensions of the multilayered cell-laden construct, the cassette partition should be designed 5 mm wide and 7–8 mm high when put into place (*see* **Notes 13** and **14**).

10. Transfer the cell-laden constructs (~4.0×10^6 cells per scaffold) from static culture to the sterile cassette of the bioreactor system after 48 h (*see* **Notes 15** and **16**).

11. Connect the silicone tubing to the upper screw portion of the cassette chamber (*see* **Notes 17** and **18**).

12. Place the chambers containing cell-laden multilayered constructs into an incubator set to 37 °C, 5% CO_2, and >90% humidity. Maintain dynamic culture at a medium flow rate of 300–500 µL/min (*see* **Notes 19** and **20**).

13. Evaluate the proliferation status and the metabolic activity of the cells within the constructs at predetermined time points by using the MTT assay.

14. For this, wash the cell-laden construct with sterile PBS and add the MTT solution diluted in serum-free growth medium (1:10). Then, incubate the construct in MTT solution for 4 h at +37 °C, 5% CO_2 inside an incubator.

15. Later, add 1 mL of the dissociation solution on each construct to dissolve and then collect the formazan crystals by serial

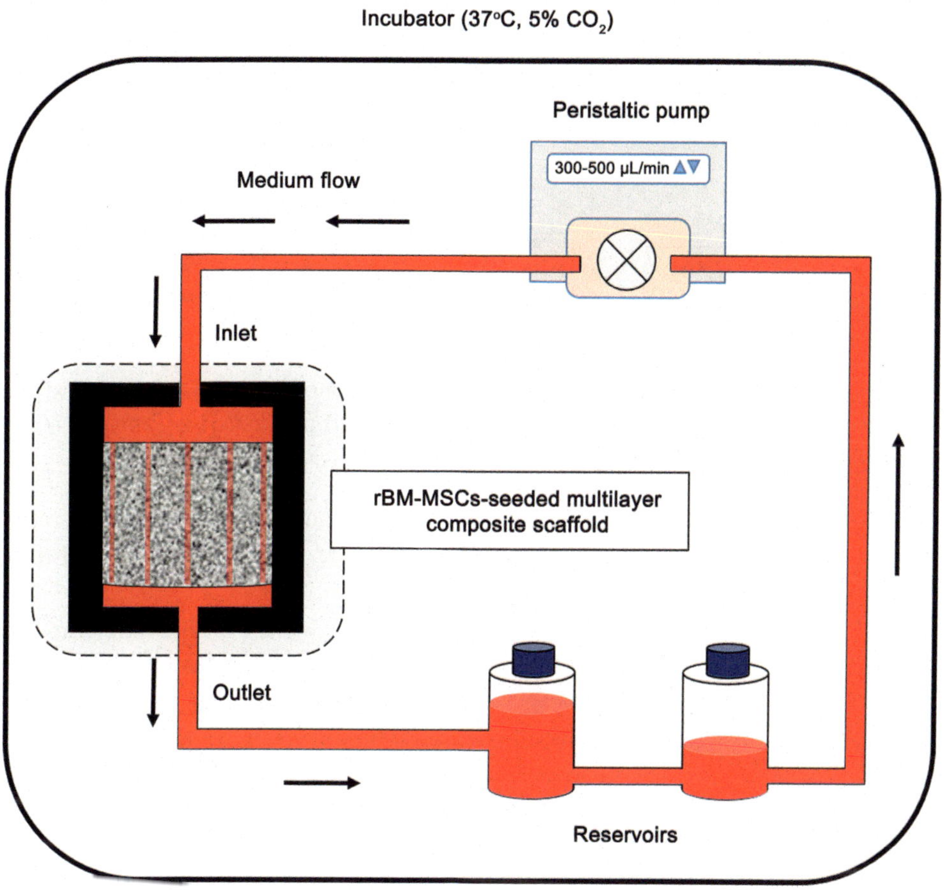

Fig. 2 Scheme demonstrating the flow bioreactor incorporating the cell-laden construct placed in the cassette, silicone tubing connecting the flow chamber to the pump and to culture medium reservoirs (Inspired by [15] [*Bancroft GN, Sikavitsas VI, Mikos AG (2003) Tissue Eng 9(3):549–554*])

pipetting. Measure the absorbance of the solution containing formazan at 570 nm by using a multiplate reader (*see* **Note 21**).

3.3 Mechanical, Morphological, and Biochemical Characterization

1. A compressive strength test can be performed on cell-laden multilayered constructs using a universal mechanical testing device (Shimadzu AGS-X) in order to evaluate the changes in mechanical properties of the bioartificial bone tissue constructs retrieved at different time points from different experimental groups.

2. Analyze the mechanical properties of the experimental groups. For example, the compressive strengths of the dry and wet cell-devoid scaffolds, as well as the rBM-MSCs-laden construct cultured for 21 days can be compared (Fig. 3).

3. Install 500 N load-cell to the device and perform the auto-calibration process before analyzing the specimens.

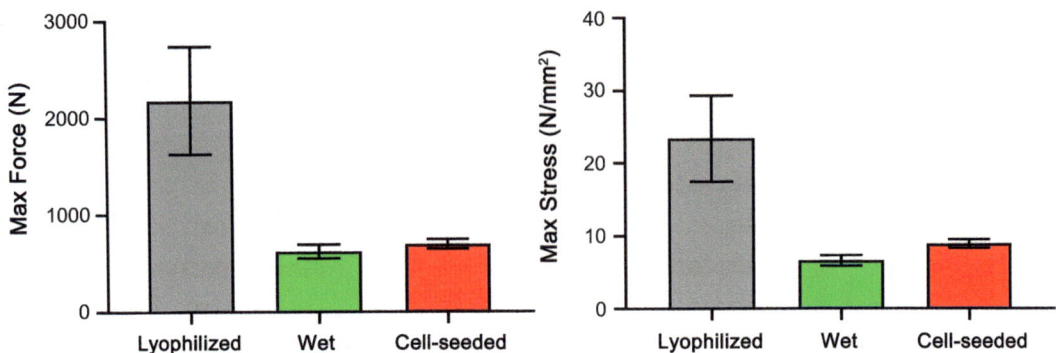

Fig. 3 Maximum force and maximum stress values of cell-devoid [in dry (lyophilized) and wet forms] and rBM-MSCs-laden constructs cultured for 21 days ($n = 3$ for each group) (adapted from [13] [*Parmaksiz M, Elçin AE, Elçin YM (2019). Mat Sci Eng C-Mater 94:788–797*])

4. Place the specimen at the center of the bottom plate. Then, set the gauge length according to construct height (~8 mm) (*see* **Note 22**).

5. After the preload test, perform the compression test at room temperature with the operation conditions of 1 mm/min crosshead speed (*see* **Note 23**).

6. For morphological characterization, transfer the cell-devoid (control) and cell-laden constructs into 2.5% glutaraldehyde solution prepared in PBS for fixation at predetermined time points.

7. Dehydrate the samples in 2 mL of ethanol series (from 60% to 95%) after fixation (*see* **Note 24**).

8. Sputter-coat the samples with a thin layer of gold, place them on SEM stubs, and analyze using a scanning electron microscope (JEOL JSM 5600, Tokyo, Japan) (*see* **Note 25**) (Fig. 4).

9. For the biochemical evaluation of osteogenic properties of cell-seeded constructs, use an alkaline phosphatase assay, thus ALP is one of the key enzymes of bone tissue ECM microenvironment. In this context, a colorimetric-based commercial kit is used which allows measurement of the ALP activity from the cell culture media without the need for any pretreatment [18]. ALP analysis should be carried out in accordance with the instructions of the commercial kit.

10. At predetermined time points transfer 50 μL of sample (cell culture medium) from the static/dynamic culture into a clear bottom 96-well plate. In addition, parallel tests should be performed using 200 μL of distilled water and 200 μL calibrator in different wells.

11. Then, carefully add 150 μL working solution to the samples and gently mix them (*see* **Note 26**).

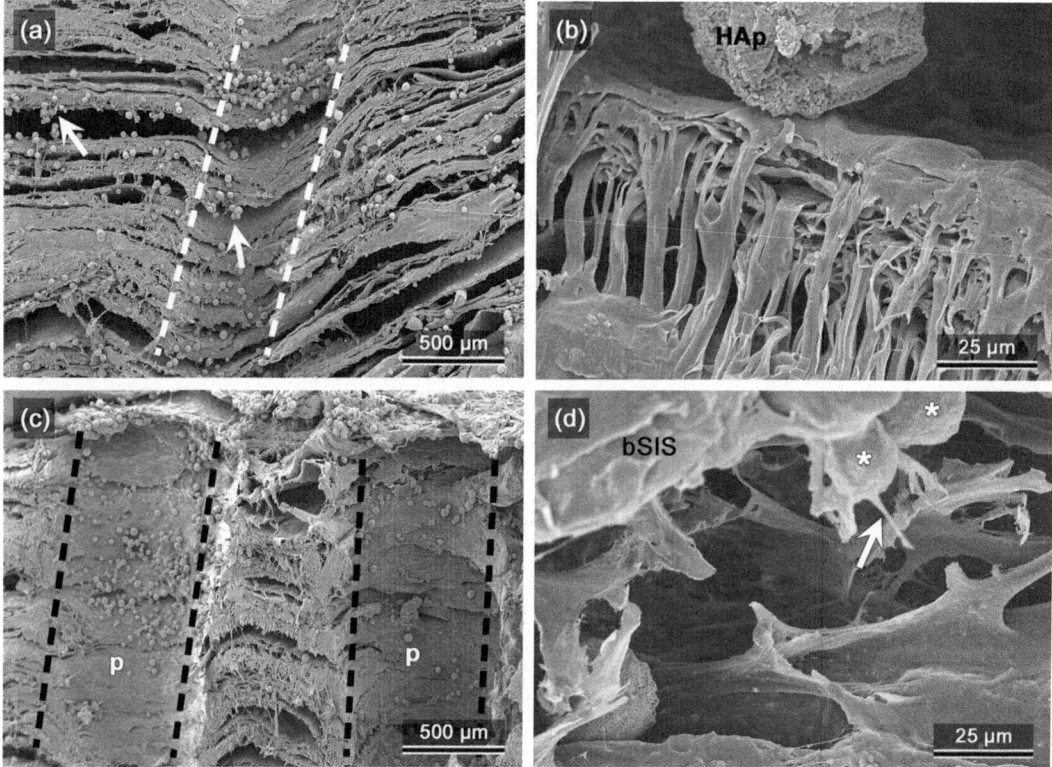

Fig. 4 Scanning electron micrographs of cell-devoid (**a**, **b**) and rBM-MSC-laden (**c**, **d**) biomimetic scaffold. While stacked layers are regularly spaced with gaps in the cell-devoid construct, they become integrated when cells are cultured on it (compare **a** with **c**). HAp microparticles are homogenously distributed even at construct perforations (p). Proliferating cells with cellular extensions are also visible (arrow in **d**) (cited from [13] [*Parmaksiz M, Elçin AE, Elçin YM (2019). Mat Sci Eng C-Mater 94:788–797*])

12. Finally perform the spectrophotometric measurements at a wavelength of 405 nm using a multiplate reader, immediately ($t = 0$) and after 4 min ($t = 4$), and calculate the ALP activity ($IU/L = \mu mol/(L \cdot min)$) of each sample using Eq. 1.

$$\text{ALP activity} = \frac{(\text{ODSAMPLE } t - \text{ODSAMPLE } o) \cdot \text{Reaction Volume}}{(\text{ODCALIBRATOR} - \text{ODH2O}) \cdot \text{Sample Volume} \cdot t} \times 35.3 \quad (1)$$

4 Notes

1. Decellularized bSIS membrane may have different thicknesses due to its natural structure. It should be selected from regions with equivalent thickness. After cutting, each piece should be weighed and the discs of equal weight and size should be used.

2. The polymer solution not exceeding 10 μL should be dropped on the first layer. Then, the subsequent layer should be placed on the preceding layer without pressing while the scaffolds are

superposed on top of each other. Otherwise, polymer micro-droplets may spread on the membrane surface and the layered structure could be lost.

3. During preparation of the PCL solution, the polymer should be gradually added to the DCM–methanol mixture and vor-texed. Care should be taken to prevent foam formation inside the polymer solution.

4. PCL solution should be stored at +4 °C. However, it should be used at room temperature (otherwise, the placement of PCL microdroplets will be difficult related to the decrease in solu-tion flowability).

5. Constructs should be prewetted with PBS before perforations. If they remain dry, the bioactive ECM content and the struc-tural integrity could be damaged due to heating.

6. The constructs should not be agitated during ethanol steriliza-tion. Otherwise this can lead to impairment of construct integrity.

7. If sterile HAp powder is not available, it can be sterilized by spreading evenly in a sterile petri dish and exposing to UV light (at 254 nm) for at least 30 min.

8. While preparing the HAp solution, a sonicator should be used to prevent any aggregate formation. The solution should be vortexed at each stage of use and injected gradually onto the scaffolds.

9. During thawing, the cells should not be completely defrosted and a small amount of ice particle should be kept.

10. Add serum-containing sterile PBS solution to halt excessive enzyme activity which could harm the dissociated cells.

11. Cell seeding through pipetting should be performed gradually and slowly, otherwise cell death may occur depending on pressure.

12. O-rings should be securely attached to the cassette with a screwed top.

13. The bioreactor provides a semiperfusive flow effect, since the constructs are perforated. Still, there should be no gaps on sides of the hopper to prevent flow around the construct.

14. The bioreactor culture unit must be made from a sterilizable and preferably transparent material (e.g., plexiglass) that allows monitoring of the construct during culture.

15. All operations must be carried out under a Class II Type A2 biosafety cabinet.

16. Constructs should be transferred with care using a sterile spat-ula, and should be press-fit into the cassette.

17. Use of silicone tubing is recommended, considering the protein binding capacity and gas permeability.

18. Care should be taken that there are no air bubbles kept in the chamber of the system.

19. It will be appropriate to use a parallel multichannel system with at least six chambers for simultaneous experimentation with different groups and time points.

20. The continuity of the medium flow should be monitored from the waste chamber. The perforated constructs are suitable for dynamic culture, and are mechanically stable for several weeks under the stated flow conditions. However, it should be noted that, these properties may change in longer cultures.

21. Cell-laden construct retrieved from static culture should be repetitively washed in another flask for dissociation. Otherwise, cells that might have attached to the culture flask could lead to errors in the results.

22. The center of the bottom plate should be measured and marked. The emplacement of different samples at the same plate position is important. Otherwise, there may be significant errors in the accuracy of the measurements.

23. Test completion parameters should be carefully set, considering the construct dimensions. It is recommended to apply a preload test and work with multiple repetitions.

24. Samples must be fixed for at least 12 h in 2.5% GA solution.

25. Cross-sections of the constructs should also be taken to visualize the interlayer structure and perforated channels.

26. The working solution should contain 200 μL of buffer, 5 μL of magnesium acetate and 2 μL of pNPP. This solution can be stored for a maximum of 24 h at room temperature. A multichannel pipettor should be used and the assay should be performed in triplicate.

Competing Interests

Y.M.E. is the founder and director of Biovalda, Inc. (Ankara, Turkey). The authors have intellectual property rights related to decellularized tissues.

Acknowledgments

The Authors are members of the COST Action CA16119 In vitro 3-D total cell guidance and fitness (CellFit).

References

1. Peng Z, Zhao T, Zhou Y, Li S, Li J, Leblanc RM (2020) Bone tissue engineering via carbon-based nanomaterials. Adv Healthc Mater 9(5):1901495

2. Zhang M, Lin R, Wang X, Xue J, Deng C, Feng C, Zhuang H, Ma J, Qin C, Wan L, Chang J (2020) 3D printing of Haversian bone–mimicking scaffolds for multicellular delivery in bone regeneration. Sci Adv 6(12): eaaz6725

3. Vurat MT, Elçin AE, Elçin YM (2018) Osteogenic composite nanocoating based on nano-hydroxyapatite, strontium ranelate and polycaprolactone for titanium implants. T Nonferr Metal Soc 28(9):1763–1773

4. Koç A, Emin N, Elçin AE, Elçin YM (2008) In vitro osteogenic differentiation of rat mesenchymal stem cells in a microgravity bioreactor. J Bioact Compat Pol 23(3):244–261

5. Carvalho MS, Poundarik AA, Cabral JM, da Silva CL, Vashishth D (2018) Biomimetic matrices for rapidly forming mineralized bone tissue based on stem cell-mediated osteogenesis. Sci Rep 8(1):1–6

6. Lee J-Y, Choi B, Wu B, Lee M (2013) Customized biomimetic scaffolds created by indirect three-dimensional printing for tissue engineering. Biofabrication 5:045003–045012

7. Thibault RA, Mikos AG, Kasper FK (2013) Scaffold/extracellular matrix hybrid constructs for bone-tissue engineering. Adv Healthc Mater 2(1):13–24

8. Parmaksiz M, Dogan A, Odabas S, Elçin AE, Elçin YM (2016) Clinical applications of decellularized extracellular matrices for tissue engineering and regenerative medicine. Biomed Mater 11(2):022003

9. Parmaksiz M, Elçin AE, Elçin YM (2020) Decellularized cell culture ECMs act as cell differentiation inducers. Stem Cell Rev Rep 16:569–584

10. Badylak SF, Lantz GC, Coffey A, Geddes LA (1989) Small intestinal submucosa as a large diameter vascular graft in the dog. J Surg Res 47(1):74–80

11. Badylak SF, Liang A, Record R, Tullius R, Hodde J (1999) Endothelial cell adherence to small intestinal submucosa: an acellular bioscaffold. Biomaterials 20(23–24):2257–2263

12. Parmaksiz M, Elçin AE, Elçin YM (2017) Decellularization of bovine small intestinal submucosa and its use for the healing of a critical-sized full-thickness skin defect, alone and in combination with stem cells, in a small rodent model. J Tissue Eng Regen M 11(6):1754–1765

13. Parmaksiz M, Elçin AE, Elçin YM (2019) Decellularized bovine small intestinal submucosa-PCL/hydroxyapatite-based multilayer composite scaffold for hard tissue repair. Mat Sci Eng C-Mater 94:788–797

14. Parmaksiz M, Elçin AE, Elçin YM (2018) Decellularized bSIS-ECM as a regenerative biomaterial for skin wound repair. In: Turksen K (ed) Skin stem cells, Methods in molecular biology, vol 1879. Humana Press, New York, pp 175–185

15. Bancroft GN, Sikavitsas VI, Mikos AG (2003) Design of a flow perfusion bioreactor system for bone tissue-engineering applications. Tissue Eng 9(3):549–554

16. Bancroft GN, Sikavitsas VI, van den Dolder J, Sheffield TL, Ambrose CG, Jansen JA, Mikos AG (2002) Fluid flow increases mincralized matrix deposition in 3D perfusion culture of marrow stromal osteoblasts in a dose-dependent manner. Proc Natl Acad Sci U S A 99(20):12600–12605

17. Gomes ME, Sikavitsas VI, Behravesh E, Reis RL, Mikos AG (2003) Effect of flow perfusion on the osteogenic differentiation of bone marrow stromal cells cultured on starch-based three-dimensional scaffolds. J Biomed Mater Res A 67(1):87–95

18. Baykan E, Koc A, Elçin AE, Elçin YM (2014) Evaluation of a biomimetic poly (ε-caprolactone)/β-tricalcium phosphate multispiral scaffold for bone tissue engineering: in vitro and in vivo studies. Biointerphases 9(2):029011

Chapter 18

Using the Air–Liquid Interface Approach to Foster Apical–Basal Polarization of Mammalian Female Reproductive Tract Epithelia In Vitro

Shuai Chen and Jennifer Schoen

Abstract

Oviduct and uterus are key female reproductive organs lined by ciliated simple columnar epithelia, which are the first line of maternal contact with gametes and the developing embryo during reproduction and which warrant the optimal developmental environment for the conceptus. A major challenge for modeling these epithelia in vitro is the preservation of apical–basal polarization and cilia formation. The air–liquid interface (ALI) culture approach is a technology originally invented for modeling epidermal and airway epithelia. It has recently been shown that it also allows the establishment of highly differentiated in vitro models of epithelia that do not have access to ambient air in vivo. In this chapter, we present a comprehensive ALI procedure to model female reproductive tract (FRT) epithelia of different mammalian species in vitro over extended time periods. As a working example, the protocol focuses on primary oviductal epithelial cells (OEC) isolated from domestic pig. Hints on protocol variations for the culture of OEC from other species are provided in the Subheading 4.

Key words Air–liquid interface, Apical–basal polarization, Cilia, Differentiation, Epithelium, Female reproductive tract, Oviduct, In vitro model

1 Introduction

Oviduct and uterus are key female reproductive organs hosting fertilization, transport and nourishment of zygotes and embryos, and (after the implantation process) fetal development until birth. The ciliated simple columnar epithelia of oviduct and uterus are the cells that first get in contact with gametes and the developing embryos. They represent an intricately regulated immune–endocrine interface ensuring the optimal developmental environment for the conceptus and at the same time an appropriate protection for the maternal organism [1]. Apical–basal polarization of the cells is warranted by specialized cellular junctions and essential for their

Tiziana A.L. Brevini et al. (eds.), *Next Generation Culture Platforms for Reliable In Vitro Models: Methods and Protocols*,
Methods in Molecular Biology, vol. 2273, https://doi.org/10.1007/978-1-0716-1246-0_18,

reproductive functions as well as for formation and maintenance of a proper biophysical barrier [2].

Therefore, a major demand and challenge for modeling these epithelia in vitro is to preserve the physiological and anatomical conditions of the native tissue, namely, apical–basal polarization and cilia formation. Routinely, epithelial cells are grown as adherent cultures on glass or plastic surfaces, which does, however, not appropriately mimic several basic in vivo settings, for example nutrition of the cells from the basolateral side, interactions with the extracellular matrix (ECM) or neighboring cells [3]. Therefore, epithelia cultured under adherent, submerged conditions often appear flat and stretched with no or loose cellular junctions, high proliferation speed, poor differentiation or dedifferentiation, and improper response to chemical or mechanical stimuli [4]. Recent progress in 3D organoid culture of both oviduct and endometrium epithelial cells allows to recapitulate the architecture and functional features of their tissue counterparts in vitro [5, 6]. Nevertheless, the inward-facing lumen of the organoid, caused by inward orientation of the epithelial cells, renders applications such as direct embryo coculture or apical stimulation experiments impossible. This, however, could be overcome by the compartmentalized air–liquid interface (ALI) culture approach, a technology originally invented for modeling epidermal and airway epithelia [7].

In this chapter, we present a comprehensive ALI procedure to model FRT epithelia of different mammalian species in vitro over extended time periods. As a working example, the protocol detailed here focuses on primary oviductal epithelial cells (OEC) isolated from domestic pig (Fig. 1). Hints on protocol variations for the culture of OEC from other species are provided in the Subheading 4. Initially, oviduct epithelial cells are isolated using a two-step enzymatic digestion in combination with two rounds of filtration, which yields a highly pure and vital cell population with vigorous cilia movement. The isolated porcine oviductal epithelial cells (POEC) are seeded on collagen-coated permeable polyethylene terephthalate (PET) inserts and are sequentially differentiated. First, cells are grown with medium supply from both the apical and basal compartment (liquid–liquid interface, LLI) followed by removal of the apical medium and exposure to ambient air (air–liquid interface, ALI), a condition better mimicking in vivo biochemical and biomechanical cues, such as composition and porosity of the ECM, and basal uptake of nutrients and signaling molecules. After approximately 3 weeks in culture, a polarized simple epithelium layer with a mixed population of ciliated and secretory cells is reconstituted (Fig. 2). The culture recapitulates key (ultra)-structural features of native oviduct epithelium tissue, including microvilli, motile cilia, and tight junctions (Fig. 3). Furthermore, moderately high transepithelial electrical resistance (TEER), an indicator of epithelial barrier formation, develops in fully differentiated ALI-POEC, which correlates with the extent of epithelial

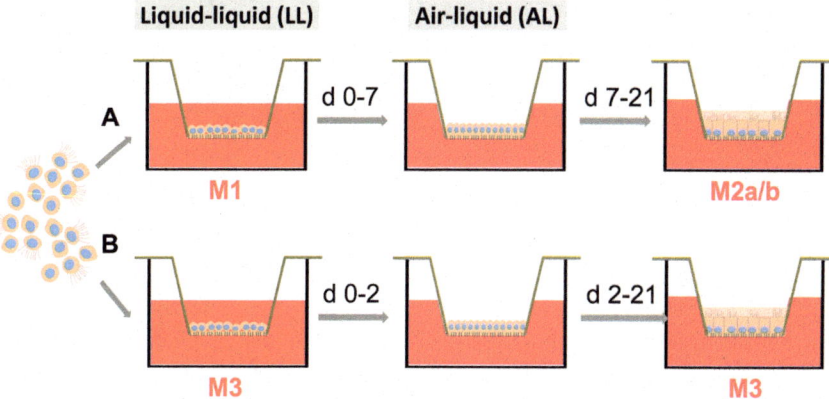

Fig. 1 Schematic illustration of the ALI-POEC culture procedure. (**a**) Two-step option: cells are first cultured in proliferation medium (M1) at the liquid–liquid interface (LLI) from d 0–7, then kept in the differentiation medium (M2a or M2b) at the air–liquid interface (ALI) until d21. (**b**) One-step option: Cells are first maintained at the liquid–liquid interface (LLI) until confluence on d2, and then further differentiated at the air–liquid interface (ALI) for 3 weeks, using the M3 culture medium during the entire process. Notice that an oviduct fluid surrogate (OFS) accumulates on the apical compartment of the insert

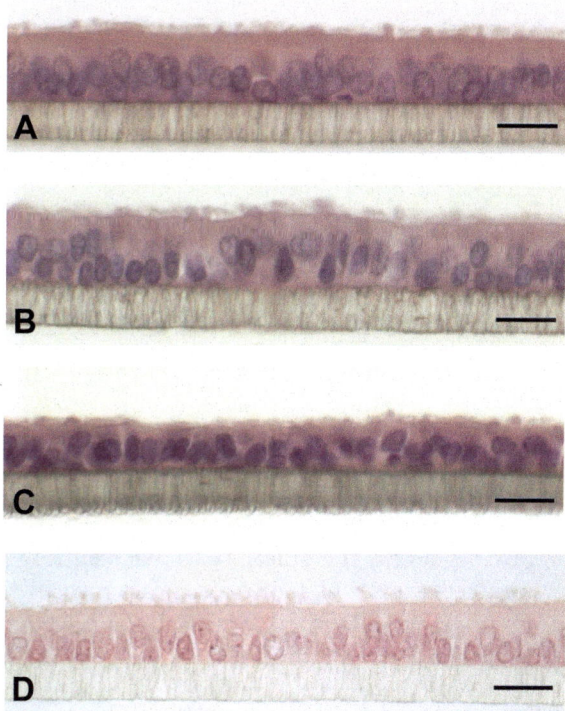

Fig. 2 Representative histological images of ALI-POEC gained via the two-step (**a**: medium M2a, **b**: medium M2b) and the one-step process (**c** and **d**: medium M3), respectively. (**a–c**) 3 weeks culture; (**d**) 12 weeks culture; Sections were stained with hematoxylin–eosin (HE), magnification ×400, scale bars = 20 μm

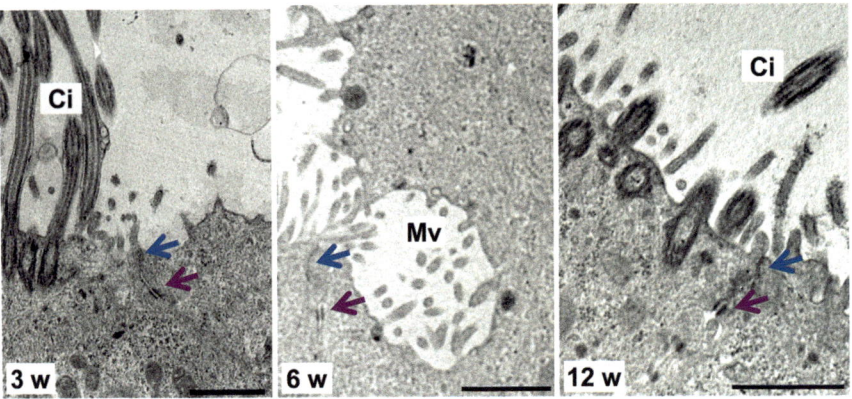

Fig. 3 Representative TEM pictures of ALI-POEC. The in vitro epithelium exhibits cilia (Ci), microvilli (Mv), and intercellular junctions after 3 weeks (left), 6 weeks (middle) and 12 weeks (right) of culture. Blue arrows: tight junctions, purple arrows: adherens junctions, scale bars = 1 μm. ALI-POEC: Air–liquid interface culture of porcine oviductal epithelial cells. TEM pictures were kindly provided by Dagmar Viertel, Leibniz-Institute for Zoo-and Wildlife Research (IZW) Berlin, Germany

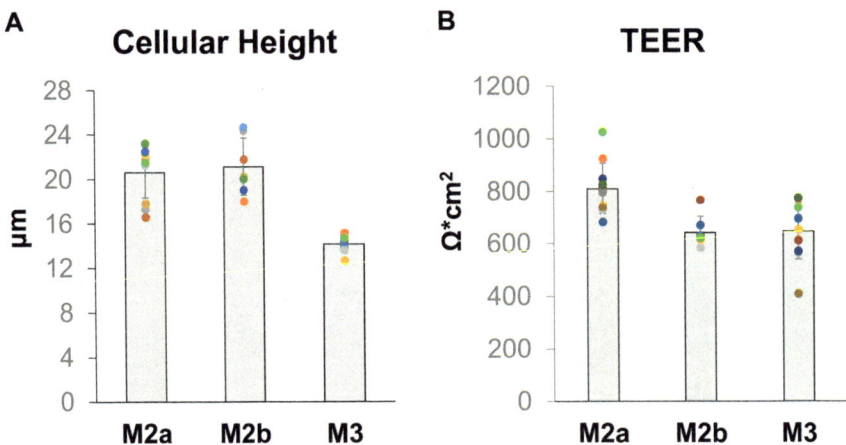

Fig. 4 Cellular height (**a**) and transepithelial electrical resistance (TEER, **b**) of ALI-POEC after 3 weeks cultivation in medium M2a, M2b, and M3, respectively. $N = 11$ animals in M2a, $N = 7$ animals in M2b, $N = 11$ animals in M3. ALI-POEC: Air–liquid interface culture of porcine oviductal epithelial cells

polarity along the apical–basal axis (Fig. 4) [8]. The ALI-POEC model possesses high physiological similarity to the in vivo tissue, as demonstrated by the swift responses to basolateral hormonal stimuli like 17b-estradiol, progesterone and cortisol [9–13]. Lastly, ALI-POEC secrete an oviduct fluid surrogate (OFS) on their apical surface at a steady rate that allows embryo coculture independent of any embryo culture medium [14, 15]. In sum, the ALI culture procedure described in this chapter may facilitate studies investigating the impact of maternal effectors on the early embryonic microenvironment, as well as on the mechanisms and relevance of early embryo–maternal interactions.

2   Materials

2.1   Cell Isolation and Dissociation

1. Instruments: lab trays, beakers, curved surgical scissors, tissue forceps, hemostatic forceps, petri dishes (100 mm), sterile cell strainers (40 μm), 15 ml conical tubes, 50 ml conical tubes, hemocytometer.

2. Trypan blue solution.

3. Accutase® cell dissociation reagent.

4. Collagenase from Clostridium histolyticum 1A (1 mg/ml, sterile filtered).

5. DPBS + antibiotics: Dulbecco's Phosphate Buffered Saline (DPBS, without Ca^{2+} & Mg^{2+}), penicillin/streptomycin (1%), gentamycin (0.05 mg/ml), amphotericin B (1 μg/ml).

6. Ham's F12 + antibiotics: Ham's F12 with 1 mM stable glutamine, penicillin/streptomycin (1%), gentamycin (0.05 mg/ml), amphotericin B (1 μg/ml), HEPES (15 mM).

2.2   ALI Culture

1. Instruments: 24 (or 12)-well plates, hanging cell culture inserts with PET membrane (see **Note 1**), ultra long pipette tips, humid chamber.

2. Dulbecco's Phosphate Buffered Saline (DPBS, without Ca^{2+} & Mg^{2+}).

3. Insert coating solution: human collagen from placenta Type IV (60 μg/ml), bovine collagen Type IV (60 μg/ml), or rat-tail collagen Type I (50 μg/ml).

4. OEC basic medium: DMEM/Ham's F12 with 4 mM stable glutamine, penicillin/streptomycin (1%), amphotericin B (0.25 μg/ml), HEPES (15 mM).

5. Proliferation medium (M1): OEC basic medium, fetal bovine serum (FBS, 5%), insulin (10 μg/ml), transferrin (5 μg/ml), cholera toxin (0.1 μg/ml), epidermal growth factor (25 ng/ml), bovine pituitary extract (30 μg/ml), retinoic acid (0.05 μM) (see **Note 2**).

6. Serum-free differentiation medium (M2a): OEC basic medium, bovine serum albumin (BSA, 1 mg/ml), insulin (5 μg/ml), transferrin (5 μg/ml), cholera toxin (0.025 μg/ml), epidermal growth factor (5 ng/ml), bovine pituitary extract (30 μg/ml), retinoic acid (0.05 μM).

7. Differentiation medium (M2b): OEC basic medium, FBS (3%), Corning® Nu-Serum™ Growth Medium Supplement (2%), retinoic acid (0.05 μM).

8. One-step medium (M3): Mix two volumes of Ham's F12 (with 1 mM stable glutamine) containing 10% FBS with one volume of the same medium conditioned by NIH/3T3 (ATCC®

CRL-1658™) mouse embryonic fibroblast cells [12]. Add penicillin/streptomycin (1%), gentamycin (0.05 mg/ml), amphotericin B (1 µg/ml), 10 µg/ml reduced glutathione, and 10 µg/ml ascorbic acid.

2.3 Sample Preparation for Histological Analysis

1. Instruments: scalpels, surgical scissors, tissue forceps.
2. Embedding solution: 2% agarose dissolved in distilled water.
3. Fixatives: 4% buffered formaldehyde solution; Bouin's solution [saturated picric acid, 35–37% formaldehyde, glacial acetic acid (volume ratio: 15:5:1)].
4. Infiltration and embedding: ethanol series (80–100%), intermedium (xylene or xylene substitute), paraffin.

2.4 TEER Measurement

1. Buffers: ethanol (70%), OEC basic medium.
2. Instruments: EVOM2 Epithelial Voltohmmeter, STX2 Electrode set.

2.5 Sample Preparation for Electron Microscopy

1. Fixative: 3% Glutaraldehyde.
2. Contrasting: 1% osmium tetroxide.
3. Infiltration and embedding: ethanol series, Epon 812, uranyl acetate and lead citrate.

3 Methods

3.1 Isolation and Dissociation of Porcine Oviductal Epithelial Cells

1. Collect oviduct with ovary and a piece of uterus from healthy pigs immediately after slaughter. Wash the sample twice with DPBS + antibiotics, then quickly transfer the tissue into a new tube and transport dry on ice to laboratory.
2. In the lab, first remove ovary and uterus, and then carefully clean off the connective tissue around the oviductal tube with curved surgical scissor. After cleaning, wash the sample twice in 5 ml DPBS + antibiotics.
3. Clamp one end of the tube with hemostatic forceps; fill the lumen with 1 mg/ml collagenase 1A (approx. 1 ml) from the other end, then use a second hemostatic forceps to close it up.
4. Transfer the filled tube into a container with preequilibrated Ham's F12+ antibiotics, incubate at 37 °C for 1 h in an incubator (*see* **Note 3**).
5. After incubation, open one end of the tube, gently and smoothly squeeze inner cell clusters into a petri dish with sterile forceps. Use the same petri dish to collect material from the other tube of the same animal if applicable (*see* **Note 4**).
6. Add 3 ml of Ham's F12 + antibiotics to the cell clusters, pipet up and down to dissociate big clusters. Transfer the suspension

onto a 40 μm cell strainer preloaded in a 50 ml falcon tube. Rinse the petri dish twice with 3 ml Ham's F12 + antibiotics, and collect cells using the same strainer (*see* **Note 5**).

7. The viable epithelial cell clusters should now be captured on the apical side of the strainer. Invert the strainer to flush back cell clusters into a new petri dish with 5 ml Ham's F12 + antibiotics.

8. Transfer the suspension into a falcon tube, add another 5 ml Ham's F12 + antibiotics, centrifuge at $250 \times g$ for 8 min.

9. Wash the pellet with 10 ml DPBS + antibiotics, centrifuge at $250 \times g$ for 8 min.

10. Carefully aspirate the supernatant, resuspend the pellet with 5 ml Accutase® and allow to digest at 37 °C for 20 min in water bath.

11. Stop the digestion process by adding 5 ml of Ham's F12 + antibiotics supplemented with 10% FBS, followed by 500 μl of pure FBS.

12. Mix well by gently pipetting up and down.

13. Filter the suspension through a fresh 40 μm strainer to remove cell clusters, collect the flow-through (single cells).

14. Centrifuge at $250 \times g$ for 8 min, resuspend the pellet in cell culture medium.

15. Count viable cells using Trypan Blue solution and hemocytometer chamber under microscope (*see* **Note 6**). Cells are ready for seeding or cryopreservation.

3.2 ALI Culture Procedure

3.2.1 Plate, Insert, and Humid Chamber Preparation

1. On the day before seeding, preload the desired number of inserts onto 24- or 12-well plates.

2. Warm chilled collagen solution to room temperature (*see* **Note 7**).

3. Add collagen solution onto the apical side of the insert (100 μl/24-well insert; 300 μl/12 well insert) to cover the membrane. Leave plate with lid slightly open overnight under the laminar flow hood.

4. On the day of seeding, wash the inserts three times with DPBS to remove excess collagen. Aspirate DPBS; dry the plate under laminar flow hood for 1 h.

5. Preparation of humid chamber: wash the humid chamber and lid with distilled water, disinfect with 70% ethanol, then cover the bottom of chamber with a film of autoclaved pure water to maintain humidity, add biocides that are suitable for water baths to prevent contamination, sterilize the chamber under UV light for 1 h. Afterward, return the humid chamber to the incubator at 37 °C with 20% O_2 and 5% CO_2.

3.2.2 Seeding and Maintenance of ALI-POEC

1. Two-step procedure.

    (a) Pipet proliferation medium M1 to the basolateral compartment (1 ml/24 well format; 1.5 ml/12 well format) of each well.

    (b) Seed cells at the density of 3.3–4.5×10^5 cells/cm^2 onto the apical side of the insert.

    (c) Fill the apical compartment to 200 µl in case of 24-well insert, or to 400 µl for 12 well insert with M1 culture medium.

    (d) Carefully transfer the plate into a humid chamber (optional), and maintain at 37 °C and 5% CO_2 (*see* **Note 8**).

    (e) Change medium every 2–3 days until day 7. This phase is termed LLI culture.

    (f) From day 7 onward, switch the culture to ALI conditions by carefully suctioning off the medium in the apical compartment and add differentiation medium (M2a or M2b) only to the basal compartment (1 ml/24 well; 1.5 ml/ 12 well, *see* **Note 9**).

    (g) Refresh culture medium twice per week.

    (h) Cells reach a differentiated status by day 21 (*see* **Note 10**), however, could be further maintained at least up to 3 months without any sign of senescence (Fig. 2D).

2. One-step procedure.

    (a) Follow the protocol described above using medium M3 in the basal compartment.

    (b) Switch from LLI culture to ALI as soon as cells are confluent (day 2 of culture).

3.3 ALI-POEC & Embryo Co-culture

In comparison to somatic cell culture, embryos of different species have their own requirements for culture conditions like oxygen concentration, temperature and humidity [16]. Therefore, it is necessary to adapt ALI cultures gradually to conditions that are suitable for in vitro embryo production before performing the co-culture experiment (*see* **Note 11**).

1. Sequential adaptation of culture conditions: gradually adjust ALI-POEC culture from 37 °C with 20% O_2 to 38.5 °C with 5% O_2 in three steps over at least 3 days.

2. Three to five days before performing the co-culture experiment, wash the apical compartment of ALI-POEC with warm OEC basic medium twice to get rid of dead cells and debris. Freshly secreted OFS accumulates in the apical side of the insert. When using a 24-well insert, a volume of ≥ 20 µl OFS is sufficient to cover the zygotes/embryos.

3. Transfer porcine zygotes onto the apical side of ALI-POEC cultures in groups of 10–30 zygotes/24 well insert, maintain the co-culture in an incubator at 38.5 °C with 5% O_2 and 5% CO_2 (*see* **Note 12**).

4. Preequilibrate cell culture medium in an incubator for at least 1 h before medium refreshment. Change medium for ALI-POEC every 3 days gently and swiftly, do not disturb embryos in the apical compartment.

5. By day 7, blastocyst stage embryos should be visible.

3.4 TEER Assessment

The TEER measurement can be routinely performed during the whole ALI culture period (Attention: not suitable during co-culture with embryos!) to monitor the growth and confluence of cells, or shortly before finalizing the culture to quantitatively assess the barrier function.

1. Preparation of blank insert: add 1000 μl complete culture medium to the 24-well basal compartment, fill the apical compartment of the insert with 200 μl medium, then equilibrate in the incubator for at least 1 h (*see* **Note 13**).

2. Equilibrate OEC basic medium, and cell culture medium in the incubator for at least 1 h to normalize the pH and temperature.

3. Setup the EVOM2 device following the manufactures' instruction, make sure the device is disconnected from charger during measurement.

4. Check the electrode surface. Disinfect the electrode with 70% ethanol for 10–15 min, allow to air dry for 15 s.

5. Place the electrode in OEC basic medium.

6. Pipet 200 μl preequilibrated cell culture medium onto the apical side of the insert just before the measurement.

7. First measure the resistance of the blank insert (R_{blank}), then move to the ALI-POEC samples ($R_{sample\ total}$).

8. Calculate the final unit area resistance using the below formula (*see* **Note 14**):

$$\text{Unit area resistance} = R_{true\ sample} \times \text{effective membrane area (cm}^2)$$
$$= (R_{sample\ total} -- R_{blank}) \times 0.33\ \text{cm}^2\,(\text{in case of a standard 24 well insert})$$

3.5 Sample Preparation for Histology

1. Remove cell culture medium from both the apical and basal compartment; rinse both compartments with warm PBS.

2. Pipet freshly prepared Bouin's solution to the apical compartment to immerse the membrane surface (200 μl/24 well; 400 μl/12 well), then quickly fill the bottom of the well with Bouin's solution (1 ml/24 well; 1.5 ml/12 well), and fix the inserts for 2 h.

3. Remove the membrane with cells from the insert by cutting around the insert edges with a clean scalpel. Cut the membrane into two equal pieces using a scissor.

4. Vertically embed the membrane halves at a 90-degree angle in 2% agarose (*see* **Note 15**).

5. Post-fix the agarose-embedded membrane in 4% formaldehyde for 1 h.

6. Dehydrate samples using a series of ethanol solutions (80–99%), clear with xylene or xylene substitute, then infiltrate and embed samples with paraffin at 60 °C.

3.6 Sample Preparation for Transmission Electron Microscopy (TEM)

1. Before harvesting, wash both the apical and basolateral sides of the inserts with DPBS.

2. Fix samples by adding 4 °C cold 3% Glutaraldehyde to both sides of insert (apical: 200 µl; basal: 1 ml per 24 well insert), fix for at least 1 h at 4 °C.

3. Cut membrane with cells off the insert, and then divide into two equal pieces.

4. Incubate samples in 1% osmium tetroxide, dehydrate in graded ethanol series, followed by embedding in Epon 812.

4 Notes

1. Choose a PET membrane with pore size of either 0.4 µm or 1.0 µm, which best supports the differentiation of the epithelial cell type. Due to the relatively larger pore size and lower pore density, the 1.0 µm-sized membrane products are often transparent; hence, they provide good optical properties for visualization of cell structures under the microscope.

2. Retinoic acid is sensitive to light. Keep it at −70 °C and add freshly during each medium preparation. The complete cell culture media should be prepared freshly every week.

3. Gentle shaking using a rocker in this step increases the cell yield.

4. We do not recommend pooling cells from different animals. In our experience, pooled cells do not properly differentiate, maybe due to immunological incompatibility.

5. To dissociate big clusters without shear stress for the cells, use wide-orifice pipette tips or cut a 1 ml tip end to widen it.

6. After isolation, viability of cells should be 80–95% to give good results after seeding. Normally the cell yield is $5–30 \times 10^6$ cells/animal.

7. Coating with human collagen Type IV suits most species (e.g. pig, mouse, and Callithrix). For bovine OEC culture, inserts could be coated with either bovine collagen Type IV or rat-tail collagen Type I.

8. It is important to keep the lid of the humid chamber slightly open during culture, to permit sufficient air ventilation. Once the culture is switched to ALI, meaning medium in the apical side of membrane is removed, usage of a humid chamber is indispensable.

9. Porcine cells differentiate well in both types of medium, while bovine OEC prefer M2b.

10. In case of bovine OEC, a 4 week culture period (1 week LL, 3 weeks AL) is compulsory.

11. In both mouse and pig, ALI-OEC maintained in serum free M2a medium are used for the embryo coculture experiment.

12. Pretrials demonstrate that placing the coculture plate in a humid chamber (with lid slightly open) better promotes embryo development.

13. Because a smaller membrane area permits more uniform delivery of current density, we prefer to use 24 well inserts for the TEER measurement to achieve stable and repeatable resistance readings.

14. Well-differentiated ALI-POEC cultures should exhibit moderate transepithelial resistance, which is between 400–1100 $\Omega*cm^2$ [8]. The proper TEER range may vary among different species.

15. Due to the property of membrane material, it is suggested to keep the temperature of the agarose gel below 60 °C. Higher temperature causes curling of the membrane.

Acknowledgments

We thank Lisa Speck, Caterina Poeppel, Petra Reckling, and Bianka Drawert, Leibniz Institute for Farm Animal Biology (FBN) Dummerstorf, Germany, for excellent technical assistance. We appreciate Dagmar Viertel, Leibniz Institute for Zoo and Wildlife research (IZW) Berlin, Germany, for providing electron microscopy pictures. The authors are active members of the COST Action CA16119 (In vitro 3D total cell guidance and fitness). The original work mentioned in this chapter was partially supported by grants from the German Research Foundation (DFG Schol231/7-1; CH2321/1-1). The Authors are members of the COST Action CA16119 In vitro 3-D total cell guidance and fitness (CellFit).

References

1. Wira CR, Grant-Tschudy KS, Crane-Godreau MA (2005) Epithelial cells in the female reproductive tract: a central role as sentinels of immune protection. Am J Reprod Immunol 53(2):65–76. https://doi.org/10.1111/j.1600-0897.2004.00248.x

2. Kurita T (2011) Normal and abnormal epithelial differentiation in the female reproductive tract. Differentiation 82(3):117–126. https://doi.org/10.1016/j.diff.2011.04.008

3. Haycock JW (2011) 3D cell culture: a review of current approaches and techniques. Methods Mol Biol 695:1–15. https://doi.org/10.1007/978-1-60761-984-0_1

4. Jensen C, Teng Y (2020) Is it time to start transitioning from 2D to 3D cell culture? Front Mol Biosci 7:33. https://doi.org/10.3389/fmolb.2020.000334

5. Kessler M, Hoffmann K, Brinkmann V, Thieck O, Jackisch S, Toelle B, Berger H, Mollenkopf HJ, Mangler M, Sehouli J, Fotopoulou C, Meyer TF (2015) The Notch and Wnt pathways regulate stemness and differentiation in human fallopian tube organoids. Nat Commun 6:8989. https://doi.org/10.1038/ncomms9989

6. Turco MY, Gardner L, Hughes J, Cindrova-Davies T, Gomez MJ, Farrell L, Hollinshead M, Marsh SGE, Brosens JJ, Critchley HO, Simons BD, Hemberger M, Koo BK, Moffett A, Burton GJ (2017) Long-term, hormone-responsive organoid cultures of human endometrium in a chemically defined medium. Nat Cell Biol 19(5):568–577. https://doi.org/10.1038/ncb3516

7. Chen S, Schoen J (2019) Air-liquid interface cell culture: from airway epithelium to the female reproductive tract. Reprod Domest Anim 54(Suppl 3):38–45. https://doi.org/10.1111/rda.13481

8. Chen S, Einspanier R, Schoen J (2015) Transepithelial electrical resistance (TEER): a functional parameter to monitor the quality of oviduct epithelial cells cultured on filter supports. Histochem Cell Biol 144(5):509–515. https://doi.org/10.1007/s00418-015-1351-1

9. Chen S, Einspanier R, Schoen J (2013) In vitro mimicking of estrous cycle stages in porcine oviduct epithelium cells: estradiol and progesterone regulate differentiation, gene expression, and cellular function. Biol Reprod 89(3):54. https://doi.org/10.1095/biolreprod.113.108829

10. Chen S, Palma-Vera SE, Kempisty B, Rucinski M, Vernunft A, Schoen J (2018) In vitro mimicking of estrous cycle stages: dissecting the impact of estradiol and progesterone on oviduct epithelium. Endocrinology 159 (9):3421–3432. https://doi.org/10.1210/en.2018-00567

11. Du S, Trakooljul N, Schoen J, Chen S (2020) Does maternal stress affect the early embryonic microenvironment? Impact of long-term cortisol stimulation on the oviduct epithelium. Int J Mol Sci 21(2):443. https://doi.org/10.3390/ijms21020443

12. Miessen K, Sharbati S, Einspanier R, Schoen J (2011) Modelling the porcine oviduct epithelium: a polarized in vitro system suitable for long-term cultivation. Theriogenology 76 (5):900–910. https://doi.org/10.1016/j.theriogenology.2011.04.021

13. Palma-Vera SE, Schoen J, Chen S (2017) Periovulatory follicular fluid levels of estradiol trigger inflammatory and DNA damage responses in oviduct epithelial cells. PLoS One 12(2): e0172192. https://doi.org/10.1371/journal.pone.0172192

14. Chen S, Palma-Vera SE, Langhammer M, Galuska SP, Braun BC, Krause E, Lucas-Hahn A, Schoen J (2017) An air-liquid interphase approach for modeling the early embryo-maternal contact zone. Sci Rep 7:42298. https://doi.org/10.1038/srep42298

15. van der Weijden VA, Chen S, Bauersachs S, Ulbrich SE, Schoen J (2017) Gene expression of bovine embryos developing at the air-liquid interface on oviductal epithelial cells (ALI-BOEC). Reprod Biol Endocrinol 15 (1):91. https://doi.org/10.1186/s12958-017-0310-1

16. Simopoulou M, Sfakianoudis K, Rapani A, Giannelou P, Anifandis G, Bolaris S, Pantou A, Lambropoulou M, Pappas A, Deligeoroglou E, Pantos K, Koutsilieris M (2018) Considerations regarding embryo culture conditions: from media to epigenetics. In Vivo 32(3):451–460. https://doi.org/10.21873/invivo.11261

Chapter 19

Preparation of Biological Scaffolds and Primary Intestinal Epithelial Cells to Efficiently 3D Model the Fish Intestinal Mucosa

Nicole Verdile, Anna Szabó, Rolando Pasquariello, Tiziana A.L. Brevini, Sandra Van Vlierberghe, and Fulvio Gandolfi

Abstract

Tissue engineering is an elegant tool to create organs in vitro, that can help obviate the lack of organ donors in transplantation medicine and provide the opportunity of studying complex biological systems in vitro, thereby reducing the need for animal experiments. Artificial intestine models are at the core of Fish-AI, an EU FET-Open research project dedicated to the development of a 3D in vitro platform that is intended to enable the aquaculture feed industry to predict the nutritional and health value of alternative feed sources accurately and efficiently.

At present, it is impossible to infer the health and nutrition value through the chemical characterization of any given feed. Therefore, each new feed must be tested through in vivo growth trials. The procedure is lengthy, expensive and requires the use of many animals. Furthermore, although this process allows for a precise evaluation of the final effect of each feed, it does not improve our basic knowledge of the cellular and molecular mechanisms determining such end-results. In turn, this lack of mechanistic knowledge severely limits the capacity to understand and predict the biological value of a single raw material and of their different combinations.

The protocol described herein allows to develop the two main components essential to produce a functional platform for the efficient and reliable screening of feeds that the feed industry is currently developing for improving their health and nutritional value. It is here applied to the Rainbow Trout, but it can be fruitfully used to many other fish species.

Key words Artificial intestine, In vitro model, Fish intestine, Polymer scaffold, Gelatin derivatives, Polymer synthesis, Primary cell line, Rainbow trout

---

Nicole Verdile and Anna Szabo contributed equally to this work.

Tiziana A.L. Brevini et al. (eds.), *Next Generation Culture Platforms for Reliable In Vitro Models: Methods and Protocols*, Methods in Molecular Biology, vol. 2273, https://doi.org/10.1007/978-1-0716-1246-0_19,
© The Author(s), under exclusive license to Springer Science+Business Media, LLC, part of Springer Nature 2021

1   Introduction

The gastrointestinal tract is the organ devoted to digestive and absorptive functions [1]. Specifically, the intestine is composed by an internal lumen and by a wall that in turn is organized in different layers [2]. The predominant layer is the mucosa, typically characterized by finger-like protrusions [1, 3] aimed to provide a wider absorptive surface [4] and lined by epithelium made up of a heterogeneous cell population [5].

So far, the in vitro models of the digestive system have been developed [6, 7] in the context of different applications including toxicology, drug testing, and tissue engineering [8, 9]. However, due to the structural and functional complexity of the organ, the challenge to develop an appropriate and predictive in vitro model that closely mimics the digestive intestinal physiology and its architecture is still open. Several models are being fabricated at different complexity levels. The most important variables used to generate reliable in vitro intestinal models include the option of growing cell lines in mono- or coculture conditions, the maintenance of the apical and basolateral side of the intestinal epithelium in static or advanced models as well as the option of continuous flow systems [10–12].

Irrespectively of the above, the majority of current in vitro models is based on the use of immortalized cell lines [13, 14] which lack a completely differentiated phenotype and therefore fail to reproduce the rich cell heterogeneity found in vivo [15]. As an alternative, 3D structures originating from pluripotent stem cells or from intestinal stem cells named organoids and enteroids respectively, are available [16]. These systems are characterized by more differentiated intestinal cells and by their progenitors but are not suitable for functional studies because of their enclosed lumen within a thick mass of hydrogel.

Several studies are trying to overcome this problem by applying well-tailored hydrogel supports [17]. To improve the suitability, their mechanical properties can be varied in a wide range, nonetheless, with increasing number of processing techniques, a structural variety of these hydrogels can be developed in order to achieve an advanced, life-like 3D in vitro intestinal model [18].

However, the keystone of a relevant and suitable tissue model is based on choosing the right combination of scaffold material and cell line. At present, an increasingly amount of data suggests that the physical and the mechanical properties of the cell culture surface play a crucial role in cell guidance and addressing cell differentiation [19–23] modulating mechano-sensing and mechano-transduction pathways [24].

In this protocol paper, we describe gelatin derivatives as versatile scaffold materials. Gelatin is a collagen derivative, which is the main component of the extracellular matrix (ECM), consisting of various amino acids. The RGD motif (arginine–glycine–aspartate triad) in the amino acid sequence of gelatin ensures the cell interactivity of these materials. In addition, the amino acid sequence of gelatin is highly versatile, with various side groups. These moieties can interact with each other, resulting in triple helix formation below a certain temperature, called the upper critical solution temperature (UCST, 30–35 °C), which leads to a physical gel formulation of gelatin at low temperatures, while having a liquid form above the UCST. The side group composition of gelatin gives a unique opportunity to modify these moieties, while introducing chemically cross-linkable side groups onto the macromolecule [25]. To this end, a cell-interactive, both physically and chemically crosslinked, hybrid polymer network is achievable at the utilized temperature (fish's body temperature is 20 °C, which is below the UCST of gelatin). The mechanical properties of the material are tailorable in a wide range via changing the modifying agents, the relative amount of the modifying agent compared to the number of modifiable moieties (degree of substitution, DS), etc.

We describe the synthesis of several gelatin derivatives and the procedure for the reliable derivation of Rainbow trout (*Oncorhynchus mykiss*) intestine primary cell lines. Their combination will lead to a faithful reproduction of the in vivo intestinal mucosa.

2 Materials

Prepare all solutions right before the use at room temperature (unless indicated otherwise).

2.1 Hydrogel Development

2.1.1 Gelatin-Methacrylamide (Gel-MA) Development

1. Unmodified gelatin type B.
2. Phosphate buffer, pH 7.8 (17.8 g Na_2HPO_4 and 6.8 g KH_2PO_4 dissolved in 1 L double distilled water).
3. Methacrylic anhydride.
4. Round bottom flask.
5. Mechanical stirrer.
6. Heating plate with water bath at 40 °C.
7. Thermostat.
8. Dialysis bath.
9. Distilled water for dialysis purposes.
10. Dialysis membrane (MWCO: 12–14 kDa).
11. Freeze-dryer.

2.1.2 Gelatin-Norbornene (Gel-NB) Synthesis

Activation

1. Stirring plate.
2. 3-neck flask.
3. Stirring bar.
4. Bunsen burner for flame drying.
5. Teflon sleeves.
6. 2 taps.
7. Stopper.
8. Ar balloon.
9. Vacuum pump.
10. Activation solution (2.21 g EDC, 1.99 g NHS, 1.77 mL 5-norbornene-2-carboxylic acid, dissolved in 75 mL dry DMSO).

Reaction

1. Heating plate with oil bath at 50 °C.
2. 3-neck flask.
3. Stirring bar.
4. Teflon sleeves.
5. 2 taps.
6. Reflux setup.
7. Stopper.
8. Ar balloon.
9. Vacuum pump.
10. Unmodified gelatin type B.
11. Activation solution.
12. Dry DMSO.

Purification

Precipitation

1. Buchner filter.
2. Filter paper (8–12 μm pore size).
3. RT acetone (tenfold excess).
4. Dropping funnel.
5. Distilled water for dialysis purposes.

Dialysis

1. Dialysis membrane (MWCO: 12–14 kDa).
2. Thermostat.
3. Dialysis bath.
4. Freeze dryer.

2.2 Film Casting of Gelatin Derivatives (Gel-X)

1. Double distilled water.
2. Gelatin solutions (unmodified gelatin type B or in-house synthetized gelatin derivatives, Gel-X, X = MA/NB): 1 g Gel-X in 10 mL double distilled water (sensitivity 18 mΩ at 25 °C).

3. Photoinitiator (2,4,6-trimethylbenzoyl) phenylphosphinate, Li-T-POL) solution: 10 mg Li-T-POL in 1 mL double distilled water (*see* **Note 1**).

4. Amber glass vials, 20 mL.

5. Heating plate.

6. Water bath at 40 °C.

7. UV-transparent glass plates.

8. UV-transparent Teflon foil.

9. 1 mm thick silicone spacer.

10. Tape, scissors, clamps.

11. 20 mL syringe.

12. Needles.

13. UV-A lamps.

2.3 Rheological Stiffness Characterization of Film-Casted Gelatin Films

1. Film-casted gelatin films.

2. 14 mm round puncher.

3. PBS solution.

4. 19 °C incubator.

5. Anton Paar MCR301 Physica rheometer.

6. 15 mm diameter spindle.

7. Torck paper.

8. Tweezers.

2.4 Scaffold Preparation from Film-Casted Gelatin Films

1. Film casted gelatin films.

2. 7 mm round puncher.

3. 4-well plates.

4. Spatula.

5. Tweezers.

2.5 Sample Collection

1. Rainbow trout weighing approximately 500 g from a fish farming.

2. Tricaine methane-sulfonate solution (MS-222): Dissolve 500 mg of MS-222 in 10 L tap water.

3. Dulbecco's Phosphate Buffered Saline (PBS): Weigh 8 g of NaCl (137 mM), 200 mg of KCl (2.7 mM), 1.44 g of Na_2HPO_4 (8 mM) 240 mg of KH_2PO_4 (2 mM) and dissolve in 800 mL of distilled water. Adjust pH to 7.4. Make up to 1 L adding distilled water Mix and sterilize the solution using autoclave.

4. Ice container.

5. Antibiotic–antimycotic solution.

6. 50 mL polypropylene tubes.

7. Surgical scissors.

8. Scalpel and surgical blades.

9. Tweezers.

Prepare all solutions in advance using ultrapure water and store all the reagents at +2–8 °C (unless indicated otherwise). Once the reagents are sterilized be sure to use them under a fume hood.

2.6 Primary Cell Culture Isolation and Maintenance

1. 25-cm^2 culture flasks.

2. 0.1% gelatin derived from pig skin: Weigh 0.1 g of gelatin derived from pig skin and dissolve it in 100 mL of water. Mix and sterilize the solution using autoclave.

3. Dulbecco's phosphate buffered saline (PBS): Weigh 8 g of NaCl (137 mM), 200 mg of KCl (2.7 mM), 1.44 g of Na$_2$HPO$_4$ (8 mM), 240 mg of KH$_2$PO$_4$ (2 mM) and dissolve in 800 mL of distilled water. Adjust pH to 7.4. Make up to 1 L adding distilled water. Mix and sterilize.

4. Refrigerated incubator.

5. Surgical scissors.

6. Scalpel and surgical blades.

7. Tweezers.

8. Glass Pasteur pipettes.

9. Micropipettes and tips.

10. Complete cell culture media: Prepare complete Leibovitz's L-15 medium without phenol red, adding 5% (v/v) fetal calf serum (FCS), 1% (v/v) antibiotic–antimycotic solution, and 200 mM L-glutamine (*see* **Note 2**).

11. Stereomicroscope.

12. 35 and 60-mm petri dishes.

13. Trypsin–EDTA solution: 0.5 g/L porcine trypsin, 0.2 g/L EDTA 4Na per liter of Hank's Balanced Salt Solution with phenol red.

2.7 Primary Cell Culture Seeding and Maintenance on Biological and Biocompatible Scaffolds

1. 70% ethanol in distilled water.

2. Dulbecco's Phosphate Buffered Saline (PBS): Weigh 8 g of NaCl (137 mM), 200 mg of KCl (2.7 mM), 1.44 g of Na$_2$HPO$_4$ (8 mM) 240 mg of KH$_2$PO$_4$ (2 mM) and dissolve in 800 mL of distilled water. Adjust pH to 7.4. Make up to 1 L adding distilled water. Mix and sterilize.

3. Complete cell culture media: Prepare complete Leibovitz's L-15 medium without phenol red, adding 5% (v/v) fetal calf serum (FCS), 1% (v/v) antibiotic–antimycotic solution, and 200 mM L-glutamine.

4. 4-wells multidish.

5. Refrigerated incubator.

6. 15 mL polystyrene tube.

7. Microscope.

8. Cell counting chamber.

9. Trypsin–EDTA solution: 0.5 g/L porcine trypsin, 0.2 g/L EDTA 4Na per liter of Hank's Balanced Salt Solution with phenol red.

3   Methods

Perform all procedures at room temperature, unless otherwise specified.

3.1   Hydrogel Development

3.1.1   Gelatin-Methacrylamide (Gel-MA) Development [25]

1. Dissolve 100 g gelatin type B in phosphate buffer (pH 7.8) at 40 °C.

2. Add 14.34 mL methacrylic anhydride to the gelatin solution while vigorously stirring with mechanical stirrer (*see* **Note 3**).

3. After 1 h reaction time, dilute the reaction mixture with the addition of 1 L double distilled water.

4. Dialyze the Gel-MA against distilled water (MWCO: 12–14 kDa) for 24 h at 40 °C.

5. Freeze dry the samples (Fig. 1).

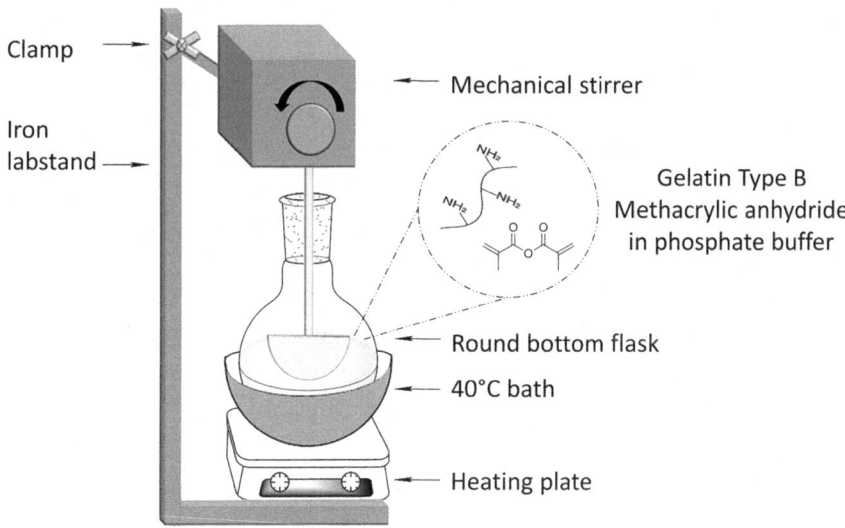

Fig. 1 Synthesis setup of Gelatin-methacrylamide (Gel-MA)

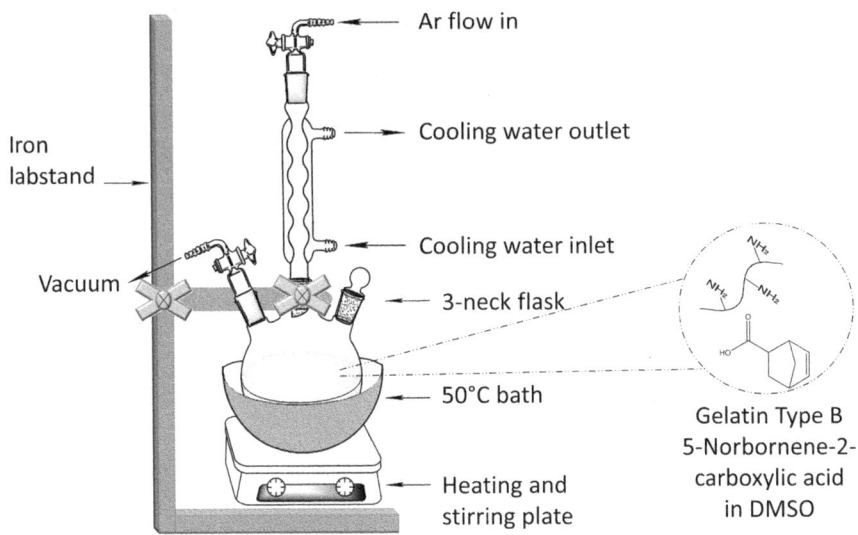

Ar flow in

Cooling water outlet

Iron labstand

Cooling water inlet

Vacuum

3-neck flask

50°C bath

Heating and stirring plate

Gelatin Type B
5-Norbornene-2-
carboxylic acid
in DMSO

Fig. 2 Synthesis setup of Gelatin-norbornene (Gel-NB)

3.1.2 Gelatin-Norbornene (Gel-NB) Synthesis [26]

Activation

1. Dissolve 1.77 mL 5-norbornene-2-carboxylic acid with 1.99 g N-hydroxysuccinimide (NHS) and 2.21 g N-(3-Dimethylaminopropyl)-N′-ethylcarbodiimide hydrochloride (EDC) in 75 mL dry DMSO at RT in a round bottom flask.

2. Apply Argon atmosphere.

3. The activation step should continue during 25 h.

Reaction

1. Dissolve 15 g gelatin type B in 450 mL dry DMSO at 50 °C under reflux conditions.

2. After full dissolution, add the activated solution.

3. Let the mixtures react overnight while vigorously stirring.

Purification

1. Precipitate the reaction solution in tenfold excess of acetone.

2. Dissolve the precipitate in double distilled water.

3. Dialyze the aqueous solution against distilled water (MWCO: 12–14 kDa) for 24 h at 40 °C.

4. Freeze dry the samples (Fig. 2).

3.2 Film Casting of Gelatins (Gel-X)

In order to test the stiffness and cell interactivity of the applicable scaffold materials, film casting of the gelatin derivatives was performed. Film casting provides the opportunity to create polymer sheets with a well-defined thickness. To this end, identical, flat polymer samples can be fabricated fast and efficiently in a reproducible manner.

1. Dissolve 1 g Gel-X in 10 mL double distilled water (sensitivity 18 mΩ at 25 °C) in amber glass vials in a 40 °C water bath.

2. Add 2 mol% photoinitiator (Li-T-POL, from stock solution, with respect to the amount of double bonds, *see* **Notes 4–7**) to the mixture and homogenize at 40 °C.

3. Prepare the glass plates: attach the Teflon foil on the glass plate with transparent tape.

4. Place the 1 mm thick silicone spacer on the top of the Teflon foil.

5. Transfer the gelatin solution into the spacer with the help of a syringe and a needle in order to prevent bubbles in the prepared film (*see* **Note 8**).

6. Place the second glass plate on the top of the filled spacer and clamp them together.

7. Place the samples in the fridge for 15 min (4 °C, *see* **Notes 9** and **10**).

8. Place the glass 2 UV-A lights for irradiation from top and bottom with total intensity of ± 10 mW/cm^2 for 30 min (Fig. 3).

3.3 Rheological Stiffness Characterization of Film-Casted Gelatin Films

To study the stiffness of the film-casted Gel-X films, oscillatory rheology was performed on the samples [27].

1. Punch out round samples from the film-casted polymer sheets with a puncher of 14 mm diameter.

2. Place them in well plates and let them swell for 24 h in PBS solution at 19 °C in an incubator.

3. The next day, start up the rheometer with a metal bottom plate and a 15 mm diameter spindle (*see* **Note 11**).

4. Apply a frequency sweep program between 0.5 and 5 Hz, with 0.1% strain applied on the samples (*see* **Note 12**).

3.4 Scaffold Preparation from Film-Casted Gelatin Films

1. Punch out round samples from the film-casted polymer sheets with a puncher of 7 mm diameter.

2. Place them in well plates before sterilization.

3.5 Sample Collection

1. Sacrifice the fish immersing it in the tank containing tricaine methane-sulfonate solution (*see* **Note 13**).

2. Immediately after euthanasia perform a longitudinal incision along the fish ventral line and gently remove the whole gastro-intestinal tract using surgical scissors and tweezers.

3. Identify starting and ending point of proximal and distal intestine and collect them into 50 mL tube stored in ice and containing freshly prepared PBS supplemented with 1% antibiotic–antimycotic solution.

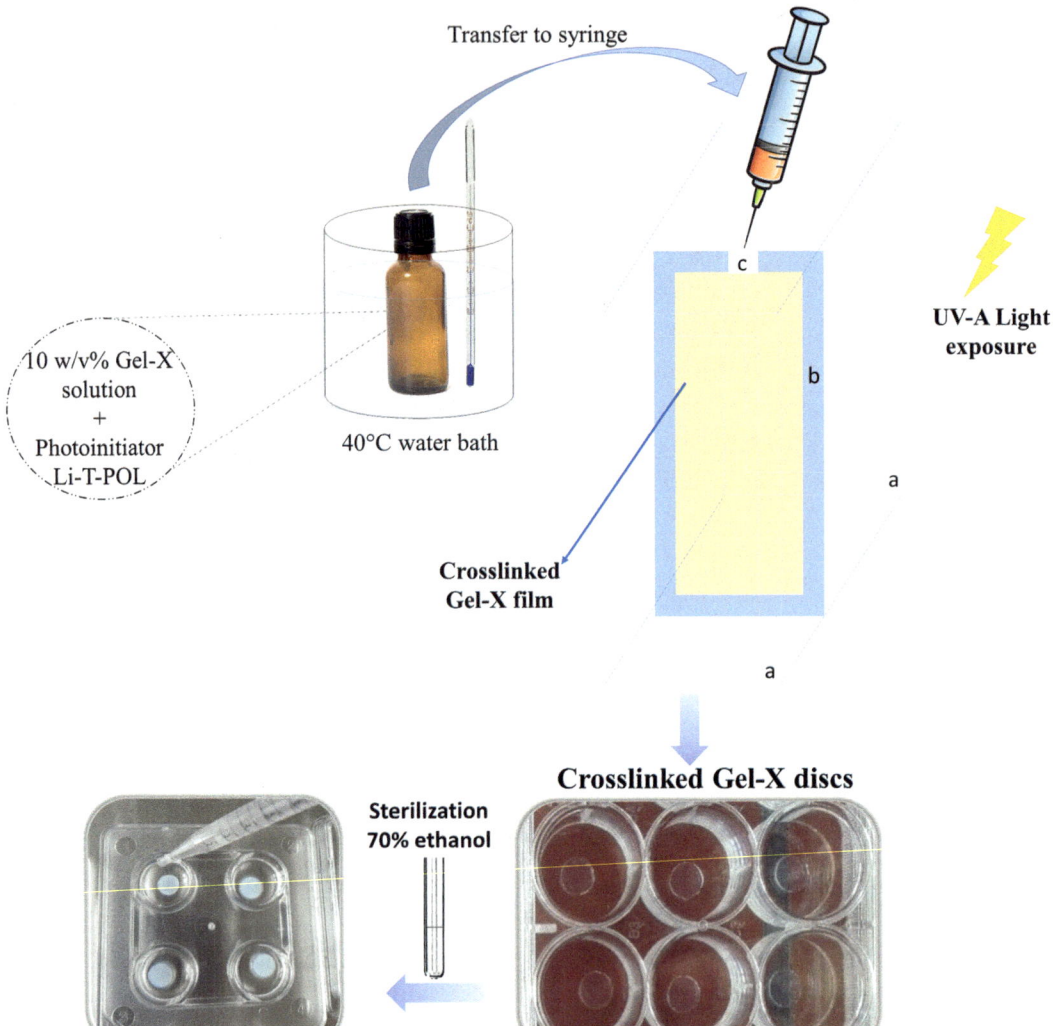

Fig. 3 Film casting and sterilization of Gel-X discs. Prepare a 10 w/v% Gel-X solution with photoinitiator (Li-T-POL) in an amber vial at 40 °C. Transfer the solution into a syringe for film casting. The film casting setup consists of (**a**) 2 UV-transparent glass plates, covered with Teflon foil, (**b**) a 1 mm thick spacer, (**c**) an inlet for the Gel-X solution. After transferring the Gel-X solution in the film casting setup, 30 min of UV-A light exposure is required. Punch out 7 mm diameter discs of the crosslinked Gel-X films and sterilize them in 70% ethanol

3.6 Primary Cell Culture Isolation and Maintenance

Perform all procedures at room temperature and under sterile conditions, unless otherwise specified.

1. Coat the 25-cm^2 culture flasks surface using 1.5 mL of 0.1% gelatin derived from pig skin and incubate for at least 20 min.

2. Place proximal and distal intestine into two different 60 mm petri dishes.

3. Carefully remove the intestinal content by slightly pressing the intestine from a side to another to let them get out using tweezers.

4. Gently, open the intestinal lumen longitudinally under the stereomicroscope using forceps and tweezers to expose the internal mucosa.

5. Move intestinal segment into two 35 mm petri dishes containing 2 mL PBS.

6. Vigorously, wash samples using a glass Pasteur pipette with fresh PBS to remove completely the mucus layer lining the mucosa (*see* **Note 14**).

7. Transfer tissues devoid of mucus into two new 35 mm petri dishes containing 2 mL PBS.

8. Cut samples into small diced (1 mm^2) using scalpel and tweezers.

9. Carefully, take each intestinal diced one by one and insert it into in 25-cm^2 culture flasks (3–4 pieces/flask) precoated with 0.1% gelatin using tweezers.

10. Very gently, add the sufficient complete L-15 medium to cover all explants (approximately 1.5 mL) making it percolate from a side of 25-cm^2 culture flasks.

11. Maintain cells in complete L-15 medium at 20 °C in incubator under ambient atmosphere (Fig. 4).

12. Refresh the medium once a week.

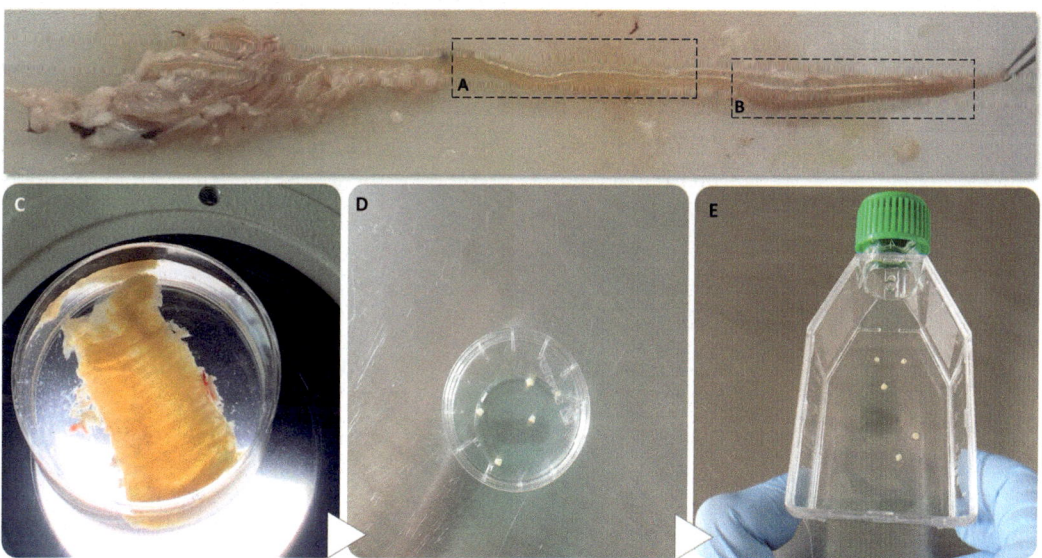

Fig. 4 Images showing selected regions for the samples collection (**a**: proximal; **b**:distal) and describing the main steps of primary cell culture isolation; **c**: distal intestine devoid of mucus after several vigorously wash in PBS, **d**: small intestinal pieces diced; **e**: explants inserted in a 25-cm^2 culture flask precoated with of 0.1% gelatin derived from pig skin

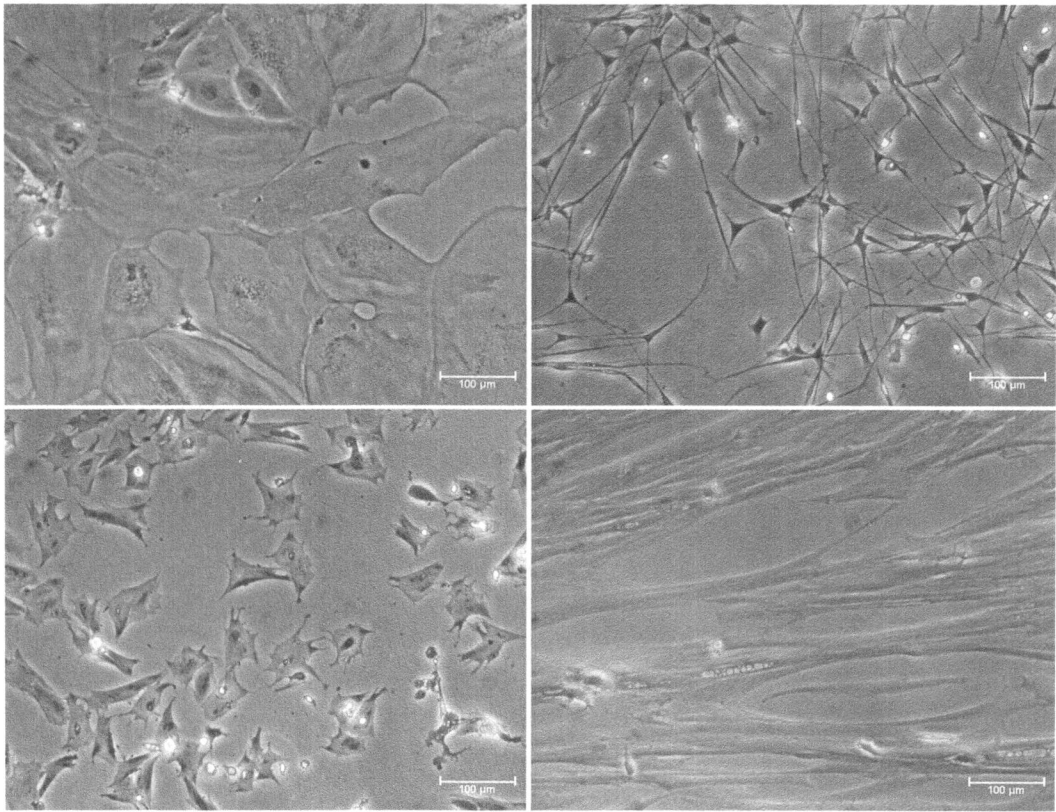

Fig. 5 Representative pictures showing culture morphology of different primary cells isolated from intestinal explants

13. After 5–10 days culture, cells will start to grow out from intestinal explants and will form colonies.

14. Maintain cells in 1.5 mL complete L-15 medium at 20 °C in incubator under ambient atmosphere at least for 3 months (Fig. 5) (*see* **Note 15**).

15. Once cells reach their confluence state, carefully remove the intestinal fragments from flasks surface gently aspirating them using a micropipette tip.

16. Maintain cells in 6 mL complete L-15 medium at 20 °C in incubator under ambient atmosphere.

17. Propagate cells refreshing medium once a week, splitting and passing adherent cells using a trypsin–EDTA solution.

3.7 Primary Cell Culture Seeding and Maintenance on Biological and Biocompatible Scaffolds

1. Gently, move freshly synthesized gelatin discs (previously punched out) into 4-wells multidish.

2. Incubate gelatin discs in 70% ethanol solution for 2 h to sterilize them.

3. Let ethanol evaporate for at least 20 min (*see* **Note 16**).

4. Take out from the incubator 25-cm^2 culture flasks and remove the maintenance medium.

5. Gently, wash 3 times adherent cells using around 2 mL PBS supplemented with 1% antibiotic–antimycotic solution.

6. Add 600 µL of trypsin–EDTA solution and let cells detach from the culture surface (*see* **Note 17**).

7. Add 5.4 mL of complete L-15 medium and vigorously mix to resuspend cells.

8. Collect the cell suspension solution in a 15 mL polystyrene tube.

9. Count cells within cell suspension under optical microscope using a cell counting chamber.

10. Based on the cell counting results, define the required cell suspension volume in order to seed each gelatin disc with 10^7 cells in a 20 µL complete culture media drop.

11. Culture cells in a humidified chamber at 20 °C normal atmosphere.

12. Add 10 µL of fresh complete medium 5 h after seeding.

13. The day after, add 10 µL of fresh complete medium to the drop.

14. The following day, add 40 µL of fresh complete medium to the drop (*see* **Note 18**).

15. After 24 h, wash once using complete fresh medium and cover gelatin discs completely adding 500 µL of fresh complete culture media.

16. Culture cells in complete L-15 medium at 20 °C in the incubator under ambient atmosphere.

4   Notes

1. Synthesis of the Li-T-POL in brief: Dissolve 9.45 g LiBr in 150 mL butanone at 65 °C and apply stirring. Add 8.6 g TPO-L to the solution and stir the reaction mixture for 24 h at 65 °C. Collect the precipitate by suction filtration and wash the precipitate with 400 mL petroleum ether. Dry the precipitate under vacuum at room temperature. Perform the reaction and drying step shielded from UV light.

2. All the medium should be at room temperature before starting the procedure.

3. In order to achieve different degree of substitutions (DS) for gelatin type B (to tailor the stiffness of the material), vary the amount of methacrylic anhydride added. Add 2.87 mL of methacrylic anhydride to achieve the DS of 30% (Gel-MA30),

5.74 mL methacrylic anhydride for the DS of 60% (Gel-MA60), 14.34 mL methacrylic anhydride for the DS of 90% (Gel-MA90).

4. No Li-T-POL was added to the unmodified Gelatin type B solution.

5. A 10 mg/mL stock solution was prepared from Li-T-POL, in order to overcome difficulties associated with the precise measurement of the Li-T-POL powder to the solutions, as only a small amount (some mgs) of it is needed to initiate the free radical polymerization.

6. Add the photoinitiator to the solution in the dark, and keep the solution shielded from UV light until use.

7. Bubbles on the surface of the polymer film are functioning as inhomogeneities, they have a weakening effect on the film.

8. According to the UCST behavior of gelatin, cooling the samples before UV crosslinking promotes the formulation of a physically crosslinked polymer network, and has a strengthening effect on the samples.

9. In case of the gelatin type B solution, no UV cross-linking is needed, the cooling of the sample is sufficient to maintain a stable, physically crosslinked network.

10. Wipe off the excess PBS from the surface of the samples before placing in the rheometer. If the samples are significantly swollen (diameter > 15 mm), repunch them with the 14 mm puncher.

11. In order to prevent the detachment of the spindle from the sample, if the sample would dry out during the process, apply a normal force on the sample (0.1–1 N).

12. Be sure to perform the procedure within 1–1.5 h after the sacrifice of the rainbow trout.

13. Be sure to completely remove the mucous layer during the cleaning procedure.

14. Three months represent the time frame required to obtain epithelial-like colonies able to be propagated.

15. Be sure to let ethanol completely evaporate.

16. Verify under the microscope cells detachment in order to avoid to let cells adherent on the plastic surface.

17. At this point, cells should be attached to gelatin discs.

18. Perform Hoechst staining and DNA quantification to evaluate cell growth and biocompatibility.

Acknowledgments

This work was supported by the European Union's Horizon 2020 research and innovation programme under grant agreement No 828835. The authors are members of the COST Action CA16119 In vitro 3-D total cell guidance and fitness (CellFit).

References

1. Barker N, Van Oudenaarden A, Clevers H (2012) Identifying the stem cell of the intestinal crypt: strategies and pitfalls. Cell Stem Cell 11:452–460. https://doi.org/10.1016/j.stem.2012.09.009

2. Randall KJ, Turton J, Foster JR (2011) Explant culture of gastrointestinal tissue: a review of methods and applications. Cell Biol Toxicol 27:267–284. https://doi.org/10.1007/s10565-011-9187-5

3. Fatehullah A, Tan SH, Barker N (2016) Organoids as an in vitro model of human development and disease. Nat Cell Biol 18:246–254. https://doi.org/10.1038/ncb3312

4. Ricci-Vitiani L, Fabrizi E, Palio E, De Maria R (2009) Colon cancer stem cells. J Mol Med 87:1097–1104. https://doi.org/10.1007/s00109-009-0518-4

5. Heo JM, Opapeju FO, Pluske JR et al (2013) Gastrointestinal health and function in weaned pigs: a review of feeding strategies to control post-weaning diarrhoea without using in-feed antimicrobial compounds. J Anim Physiol Anim Nutr 97:207 237. https://doi.org/10.1111/j.1439-0396.2012.01284.x

6. Verdile N, Mirmahmoudi R, Brevini TAL, Gandolfi F (2019) Evolution of pig intestinal stem cells from birth to weaning. Animal 13:2830–2839. https://doi.org/10.1017/S1751731119001319

7. Verdile N, Pasquariello R, Scolari M et al (2020) A detailed study of rainbow trout (onchorhynchus mykiss) intestine revealed that digestive and absorptive functions are not linearly distributed along its length. Animals 10 (4):745. https://doi.org/10.3390/ani10040745

8. Yu J, Carrier RL, March JC, Griffith LG (2014) Three dimensional human small intestine models for ADME-Tox studies. Drug Discov Today 19:1587–1594. https://doi.org/10.1016/j.drudis.2014.05.003

9. Dzobo K, Thomford NE, Senthebane DA et al (2018) Advances in regenerative medicine and tissue engineering: innovation and transformation of medicine. Stem Cells Int 2018:1–24. https://doi.org/10.1155/2018/2495848

10. Dosh RH, Essa A, Jordan-Mahy N et al (2017) Use of hydrogel scaffolds to develop an in vitro 3D culture model of human intestinal epithelium. Acta Biomater 62:128–143. https://doi.org/10.1016/j.actbio.2017.08.035

11. Fitzgerald KA, Malhotra M, Curtin CM et al (2015) Life in 3D is never flat: 3D models to optimise drug delivery. J Control Release 215:39–54. https://doi.org/10.1016/j.jconrel.2015.07.020

12. Huh D, Hamilton GA, Ingber DE (2011) From 3D cell culture to organs-on-chips. Trends Cell Biol 21:745–754. https://doi.org/10.1016/j.tcb.2011.09.005

13. Vázquez M, Vélez D, Devesa V (2014) In vitro characterization of the intestinal absorption of methylmercury using a caco-2 cell model. Chem Res Toxicol 27:254–264. https://doi.org/10.1021/tx4003758

14. Keemink J, Bergström CAS (2018) Caco-2 cell conditions enabling studies of drug absorption from digestible lipid-based formulations. Pharm Res 35:74. https://doi.org/10.1007/s11095-017-2327-8

15. Le Ferrec E, Chesne C, Artusson P et al (2001) In vitro models of the intestinal barrier: the report and recommendations of ECVAM workshop 461,2. ATLA Altern to Lab Anim 29:649–668. https://doi.org/10.1177/026119290102900604

16. Date S, Sato T (2015) Mini-gut organoids: reconstitution of the stem cell niche. Annu Rev Cell Dev Biol 31:269–289. https://doi.org/10.1146/annurev-cellbio-100814-125218

17. Kim GA, Spence JR, Takayama S (2017) Bioengineering for intestinal organoid cultures. Curr Opin Biotechnol 47:51–58. https://doi.org/10.1016/j.copbio.2017.05.006

18. Wu J, Chen Q, Liu W et al (2017) Recent advances in microfluidic 3D cellular scaffolds for drug assays. Trends Anal Chem 87:19–31. https://doi.org/10.1016/j.trac.2016.11.009

19. Brevini TAL, Manzoni EFM, Sergio L, Gandolfi F (2017) Use of a super-hydrophobic microbioreactor to generate and boost pancreatic mini-organoids. Methods Mol Biol

1576:257–284. https://doi.org/10.1007/7651_2017_47

20. Gjorevski N, Sachs N, Manfrin A et al (2016) Designer matrices for intestinal stem cell and organoid culture. Nature 539:560–564. https://doi.org/10.1038/nature20168

21. Wang L, Sun B, Ziemer KS et al (2010) Chemical and physical modifications to poly (dimethylsiloxane) surfaces affect adhesion of Caco-2 cells. J Biomed Mater Res Part A 93:1260–1271. https://doi.org/10.1002/jbm.a.32621

22. Buxboim A, Ivanovska IL, Discher DE (2010) Matrix elasticity, cytoskeletal forces and physics of the nucleus: how deeply do cells "feel" outside and in? J Cell Sci 123:297–308. https://doi.org/10.1242/jcs.041186

23. DiMarco RL, Hunt DR, Dewi RE, Heilshorn SC (2017) Improvement of paracellular transport in the Caco-2 drug screening model using protein-engineered substrates. Biomaterials 129:152–162. https://doi.org/10.1016/j.biomaterials.2017.03.023

24. Fratzl P, Barth FG (2009) Biomaterial systems for mechanosensing and actuation. Nature 462:442–448. https://doi.org/10.1038/nature08603

25. Van Hoorick J, Tytgat L, Dobos A et al (2019) (Photo-)crosslinkable gelatin derivatives for biofabrication applications. Acta Biomater 97:46–73. https://doi.org/10.1016/j.actbio.2019.07.035

26. Van Hoorick J, Gruber P, Markovic M et al (2018) Highly reactive thiol-norbornene photo-click hydrogels: toward improved processability. Macromol Rapid Commun 39:1–7. https://doi.org/10.1002/marc.201800181

27. Meyvis TKL, Stubbe BG, Van Steenbergen MJ et al (2002) A comparison between the use of dynamic mechanical analysis and oscillatory shear. Int J Pharm 244:163–116. https://doi.org/10.1016/s0378-5173(02)00328-9.

Chapter 20

Use of Porous Polystyrene Scaffolds to Bioengineer Human Epithelial Tissues In Vitro

Lydia Costello, Nicole Darling, Matthew Freer, Steven Bradbury, Claire Mobbs, and Stefan Przyborski

Abstract

In vitro epithelial models are valuable tools for both academic and industrial laboratories to investigate tissue physiology and disease. Epithelial tissues comprise the surface epithelium, basement membrane, and underlying supporting stromal cells. There are various types of epithelial tissue and they have a diverse and intricate architecture in vivo, which cannot be successfully recapitulated using two-dimensional (2D) cell culture. Tissue engineering strategies can be applied to bioengineer the organized, multilayered, and multicellular structure of epithelial tissues in vitro. Alvetex® is a porous, polystyrene scaffold that enables fibroblasts to synthesize a complex network of endogenous, humanized extracellular matrix proteins. This creates a physiologically relevant three-dimensional (3D) subepithelial microenvironment, enriched with mechanical and chemical cues, which supports the organization and differentiation of epithelial cells. Such technology has been used to bioengineer different epithelial architectures in vitro, including the simple, columnar structure of the intestine and the stratified, squamous, and keratinized structure of skin. Epithelial tissue models provide a useful platform for fundamental and translational research, with multifaceted applications including disease modeling, drug discovery, and product development.

Key words Epithelium, Bioengineering, Tissue, In vitro models, 3D, Polystyrene scaffold, Alvetex®, Platform technology, Skin, Intestine

1 Introduction

1.1 Architecture of Epithelial Tissues

Epithelial tissues are composed of a contiguous organization of cells that line the surfaces of the body, and provide a physical barrier against physical, chemical, and pathogenic insults [1]. Epithelial tissues have a conserved structure, with a polarized cellular layer interspersed with complex intercellular junctions, which mediate resistance to mechanical stress and enable cellular communication [2, 3]. The epithelial cells are anchored to the subepithelial compartment via the basement membrane, which facilitates the regulated exchange of substances, and aids the polarity of the epithelium [4]. The underlying subepithelial compartment is composed of a

Tiziana A.L. Brevini et al. (eds.), *Next Generation Culture Platforms for Reliable In Vitro Models: Methods and Protocols*, Methods in Molecular Biology, vol. 2273, https://doi.org/10.1007/978-1-0716-1246-0_20,

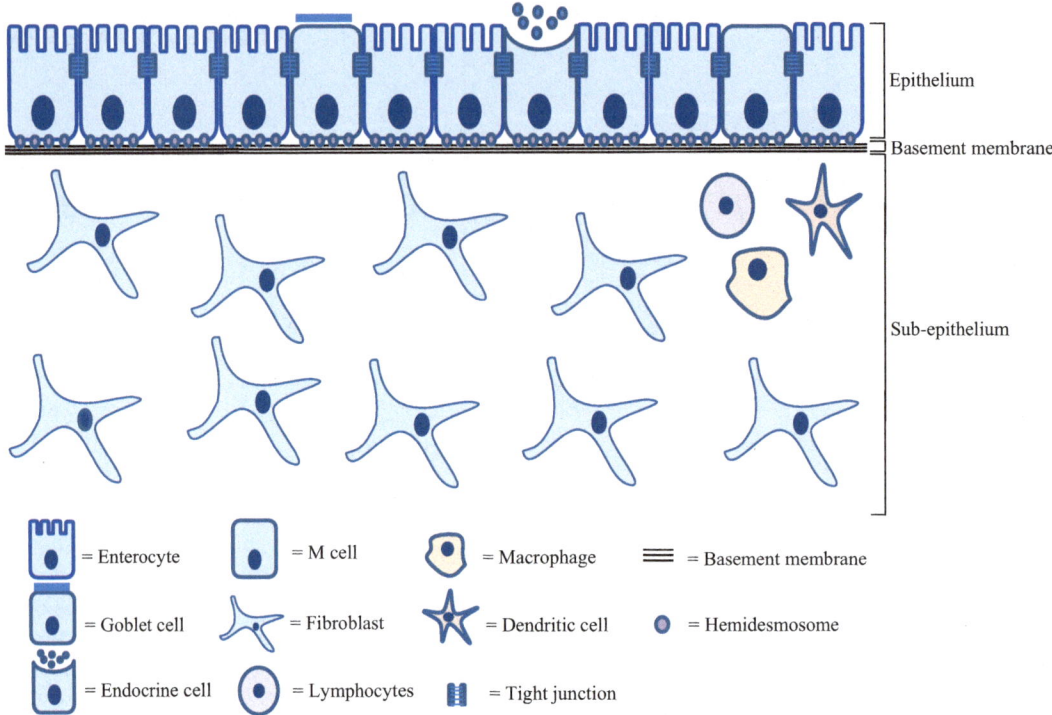

Fig. 1 Schematic of the human intestinal structure. The multicellular intestinal epithelium contains absorptive enterocytes, immunoregulatory M cells, mucus-producing goblet cells, and secretory endocrine cells. The epithelial cells are adhered to the subepithelium via the trilayered basement membrane. The subepithelial compartment contains a multitude of cell types including fibroblasts, which synthesize the extracellular matrix, and immune cells such as lymphocytes, macrophages, and dendritic cells

multitude of cell types and a complex network of extracellular matrix proteins, which provide mechanical and chemical cues to maintain epithelial homeostasis [5].

Despite this fundamental structure, epithelial tissues such as the intestine and skin, have unique architectures related to their anatomical location and specialized function. The intestine has a simple columnar epithelial structure, with microvilli on the apical surface, which increases the surface area and enables the efficient absorption of nutrients (Fig. 1).

In contrast, the skin has a stratified, squamous, and keratinized epithelium, with the primary function of providing a barrier against the external environment and protection against exogenous stressors [6] (Fig. 2).

Traditionally, epithelial research has been conducted using a reductionist approach, by culturing cells in a two-dimensional monolayer. There is a significant demand for more advanced epithelial models, which incorporate multiple cell types and successfully recreate the complex, intricate, and specialized tissue architectures in vitro.

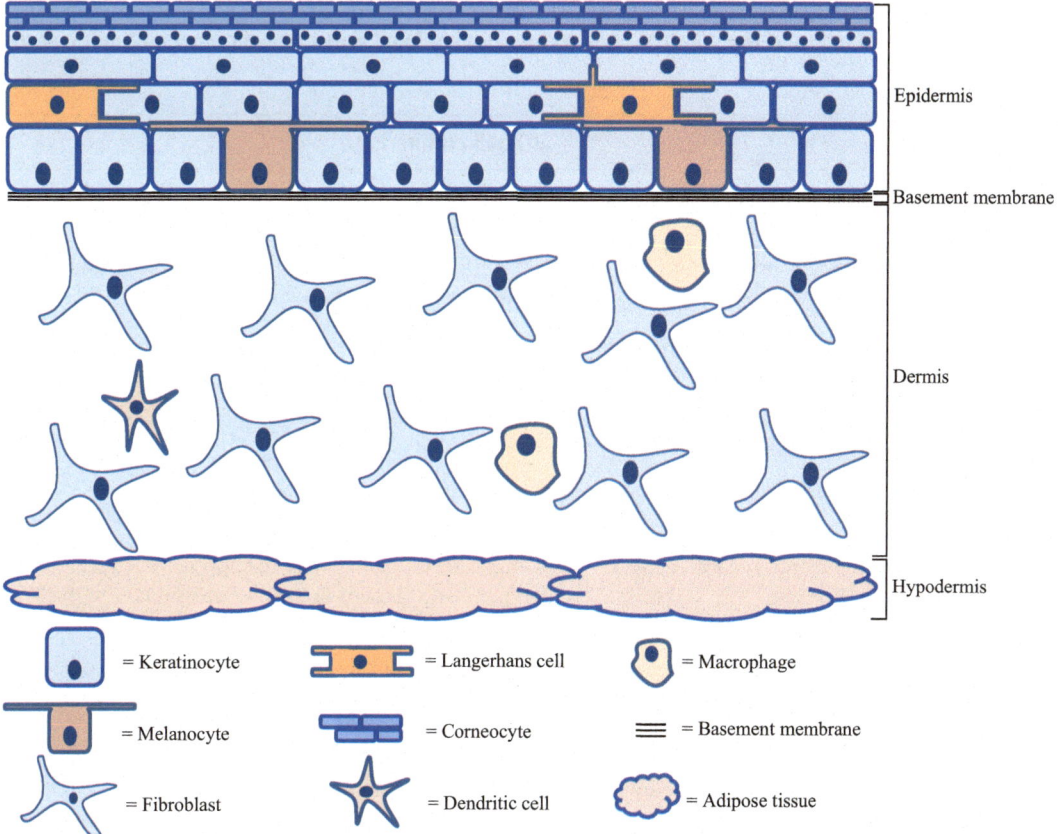

Fig. 2 Schematic of the human skin structure. The epithelial layer of skin, known as the epidermis, has a stratified, differentiated organization composed of the *stratum basale, stratum spinosum, stratum granulosum*, and *stratum corneum*. The epidermis is anchored to the subepithelial compartment via the basement membrane. The subepithelium, also known as the dermis, supports epidermal homeostasis through mechanical and secretory cues. The hypodermis is composed of adipose tissue, which provides insulation and additional support

1.2 The Importance of a Third Dimension to Bioengineer in Vitro Epithelial Models

Conventional two-dimensional cell culture involves homogeneous monocultures cultured on flat plastic substrates. This method has provided significant advances in the understanding of fundamental biology, and it enables the use of cost-effective, high-throughput assays. However, the nonphysiological culture conditions can induce unnatural phenotypic characteristics such as an aberrant, flattened structural morphology with significant differences in their cytoarchitecture, mechanotransduction, and polarity [7]. These disparate features ultimately influence gene and protein expression, and alter cellular responses and biological functions [8, 9].

Two-dimensional cell culture is also unable to capture the multicellular and multilayered topographical complexity of epithelial tissues. In vivo, epithelia have a three-dimensional geometry with multidirectional and spatiotemporal cellular communication,

Table 1
Differences in chemical and mechanical cues in two-dimensional and three-dimensional cell culture environments

2D cell culture	3D cell culture
Monolayer cultures do not mimic the native structures of tissues.	Possible to bioengineer complex, multilayered structures in vitro.
Sparse extracellular matrix production.	Organized extracellular matrix in transverse and longitudinal orientations.
Co-culture of multiple cell types cannot establish an in vivo-like niche.	Niche creation through the co-culture of multiple cell types in 3D.
No gradients of nutrients, oxygen, metabolites or signaling molecules.	Diffusion gradients of nutrients, oxygen, metabolites, and signaling molecules.
2D geometrical constraints induce a flattened cell morphology.	Maintenance of a 3D cell structure.
Forced apical–basal polarity, regardless of cell type.	Unrestricted polarity, able to support polarized epithelial cells and non-polarized subepithelial cells.
Limited cell–cell contacts at the edges of cells.	Physiological and spatiotemporal cell–cell contacts.
Cell adhesions restricted to the x–y plane.	Cell adhesions in x, y, and z planes.
Cell–substrate interactions dominate, as 50% of the cell is in contact with the plastic substrate.	Increased physiological relevance of cell–cell and cell–extracellular matrix interactions.
High, nonphysiological substrate stiffness.	Variable substrate stiffness.
Unconstrained migration and spreading in the x–y plane.	Migration and spreading through extracellular matrix in the x, y, and z planes is more physiological.
Altered mRNA splicing, gene expression, biochemistry and topology of cells.	mRNA splicing, gene expression, biochemistry and topology of cells are more similar to those in vivo.
Increased drug sensitivity.	Drug resistance closer to in vivo levels.

which creates a complex microenvironment of mechanical and chemical cues that influence cellular behavior. Due to the fundamental anatomical principle that structure is related to function, the successful recreation of epithelial tissue architecture in vitro is essential. Bioengineered three-dimensional tissue constructs circumvent many of the limitations of monolayer culture, and provide a more physiologically relevant microenvironment, for fundamental and translational research (Table 1).

To recreate the complex, three-dimensional architecture of epithelial tissues in vitro, tissue engineering approaches involve the formation of the subepithelium, with the sequential addition of epithelial cells and appropriate growth factors to enhance their differentiation and specialization [10, 11]. Scaffold-based approaches of engineering the subepithelial stratum in vitro can involve the use of hydrogel or porous scaffold matrices to mimic the dynamic reciprocity between fibroblasts and the extracellular matrix.

Hydrogel-based technologies encapsulate cells within a hydrophilic cross-linked matrix, composed of natural or synthetic polymers. Collagen hydrogels are often used to mimic the subepithelium due to the abundance of collagen in the extracellular matrix, and their tissue-like mechanical properties such as viscoelasticity, turgidity, and flexibility [12–16]. Nonetheless, limitations of hydrogels include batch-to-batch variation, and their often animal-derived origin such as bovine or rat-tail tendon. This may introduce interspecies effects, and the presence of only one extracellular matrix component is unable to successfully represent the complex architecture of the subepithelium. Furthermore, hydrogels, such as collagen gels, are known to contract during an experiment, leading to inconsistencies and variability. Hydrogel technologies have been demonstrated to support the differentiation of specialized epithelial tissues, and bioengineer the organized structure of the intestine, oral mucosa and skin in vitro [12–16].

Alternative matrices for tissue engineering include porous scaffolds such as Alvetex®, which enable the bioengineering of a self-assembled subepithelial compartment. Alvetex® Scaffold is a synthetic and inert polystyrene scaffold, which is manufactured using emulsion templating to produce a homogeneous network of 36–40 μm pores, linked by 13 μm interconnects [17] (Fig. 3). 90% of the volume of the material is space in which the cells grow in 3D, only 10% is inert plastic. This high porosity provides a physical space for fibroblasts to infiltrate the scaffold, retain their three-dimensional geometry and secrete a dynamic network of endogenous extracellular matrix proteins. The synthesized

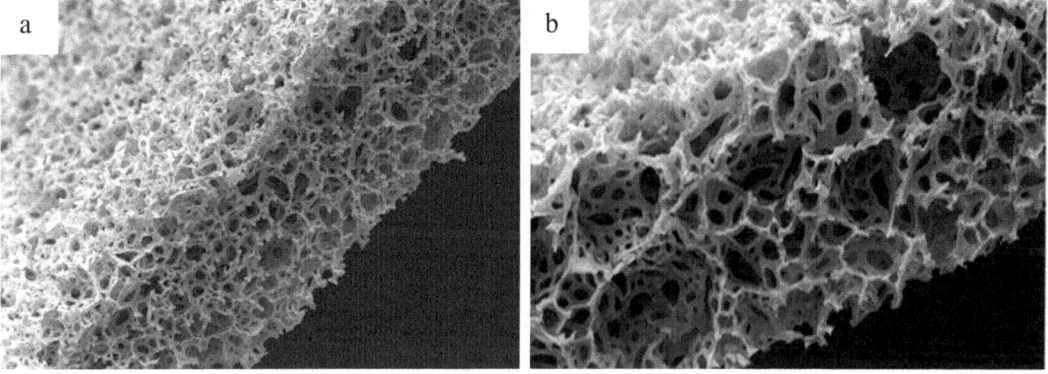

Fig. 3 Structure of porous polystyrene membrane used for 3D cell culture. Scanning electron microscopy micrographs of Alvetex® Scaffold porous polystyrene membrane in transverse section (low and high magnification, (**a**) and (**b**) respectively) demonstrating the high internal porosity which supports cellular geometry. Membrane thickness: 200 μm

extracellular matrix is more complex and physiologically relevant compared with existing hydrogel technologies. In addition to its porous architecture, Alvetex® is engineered into a membrane with a thickness of 200 µm, which enables the efficient mass transfer of gases, nutrients, and waste to maintain cellular viability [18]. Unlike many hydrogel technologies, which are associated with batch-to-batch variation, Alvetex® is commercially available, and it undergoes rigorous quality control procedures to enhance reproducibility [18]. Alvetex® Scaffold technology has been used to bioengineer epithelial tissues in vitro, including the intestine, buccal mucosa, endometrium, respiratory epithelia and skin [11, 19].

1.3 The Use of Polystyrene Scaffold Technology to Bioengineer Reproducible Epithelial Tissue Models In Vitro

Porous polystyrene scaffolds have been utilized to bioengineer the highly organized and polarized architecture of human epithelial tissues, which demonstrate a similar morphology to their native counterparts. These human tissue models are highly reproducible and transferable due to the exclusive use of commercially available materials such as cells, media, growth factors, plasticware, and Alvetex® Scaffold. To generate the subepithelial compartment, fibroblasts synthesize humanized extracellular matrix proteins within the scaffold, which alleviates the requirement for variable exogenous collagen matrices. Moreover, the plastic membrane provides a backbone to the subepithelial compartment, and contributes to mechanical integrity when suspended in a well insert, which prevents shrinkage during long-term tissue growth. The robust subepithelial layer is able to support the polarization, differentiation, and specialization of different epithelial structures, including the intestine and skin.

The human intestinal equivalent produced using this platform has a simple, columnar epithelium, with a distinctly organized and uniform cell monolayer (Fig. 4a). The epithelial cells are terminally differentiated and highly polarized in the apical–basal direction, with a similar structure to their in vivo counterparts (Fig. 4a, b). The intestinal epithelium is highly specialized due to the presence of microvilli which increase the surface area, mature junctional complexes which provide a permeability barrier, and specific carrier proteins that enable the efficient absorption of nutrients and pharmaceuticals. Such intestinal equivalents provide a platform for multifaceted downstream applications including drug permeability through the paracellular transport of compounds, investigating epithelial–subepithelial interactions, and modeling inflammatory diseases.

The human skin equivalent has a differentiated, stratified, and keratinized structure, similarly to in vivo skin (Fig. 4c, d). The organized keratinocytes within the *stratum basale* have a polarized, columnar phenotype, and they undergo characteristic sequential differentiation to form the suprabasal *stratum spinosum* and *stratum granulosum* layers. During terminal differentiation, the keratinocytes undergo cornification into anuclear, flattened corneocytes

In vitro epithelial tissue models | In vivo epithelial tissue

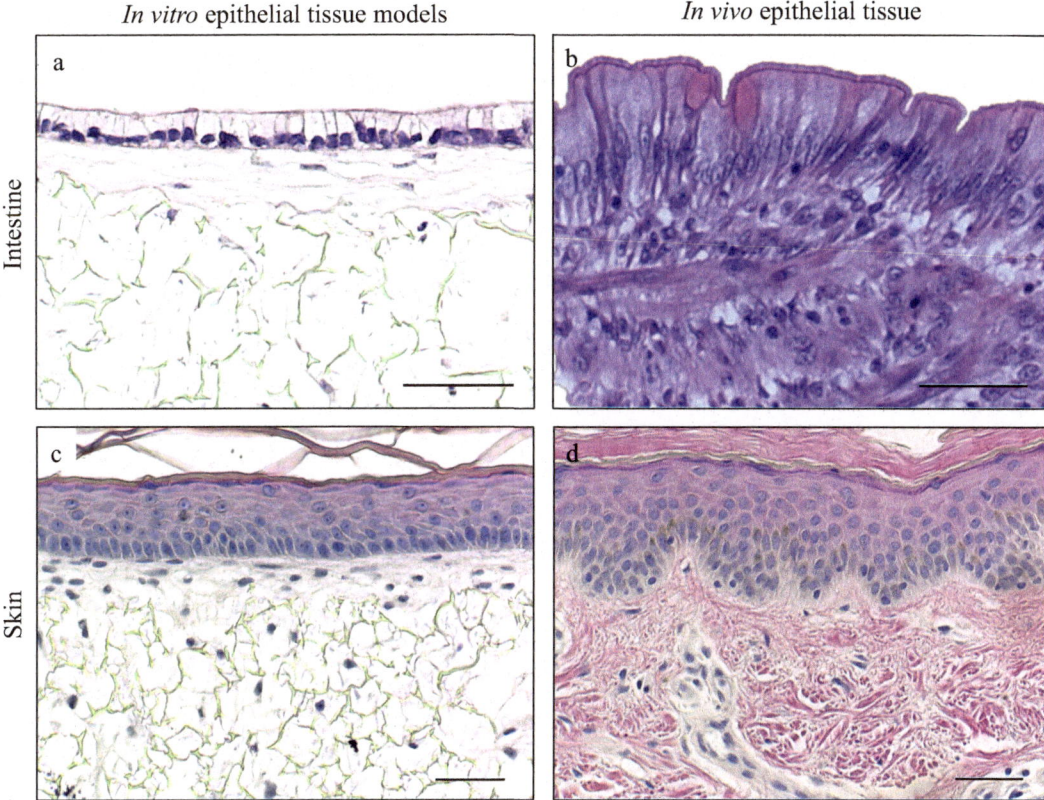

Fig. 4 Epithelial tissue models generated using porous polystyrene scaffold membranes. Representative histological images of bioengineered models (**a**, **c**) in comparison to their native tissue counterparts (**b**, **d**). Two tissue examples are shown: (**a**, **b**) human intestine; (**c**, **d**) human skin. The tissue sections are counterstained with H&E. Scale bars: 50 μm

in the *stratum corneum*. Skin equivalents can be used as a predictive platform for the screening of cosmetic actives, penetration profiling and investigating the impact of exogenous stressors such as ultraviolet radiation.

2 Materials

2.1 General

Alvetex® Scaffold porous polystyrene membrane housed in 12-well inserts (Reprocell).

6-well culture plate (Greiner Bio-One).

0.25% Trypsin–EDTA (Thermo Fisher Scientific).

Trypan blue (Sigma-Aldrich).

Sterile Dulbecco's phosphate-buffered saline (DPBS, Lonza).

Sterile forceps.

70% ethanol solution.

2.2 Intestinal Equivalent

Human neonatal dermal fibroblasts (HDFn, Thermo Fisher Scientific).

2.2.1 Subepithelial Compartment

Dulbecco's Modified Eagle Medium (DMEM, Thermo Fisher Scientific).

Fetal bovine serum (FBS, Thermo Fisher Scientific).

L-glutamine (Thermo Fisher Scientific).

Non-essential amino acids (NEAA, Thermo Fisher Scientific).

Penicillin–streptomycin solution (Thermo Fisher Scientific).

Transforming growth factor β1 (TGFβ1, PeproTech).

Ascorbic acid (Sigma-Aldrich).

2.2.2 Epithelial Compartment

Human epithelial colorectal adenocarcinoma cells (Caco-2, The European Collection for Authenticated Cell Cultures (ECACC)).

Dulbecco's Modified Eagle Medium (DMEM, Thermo Fisher Scientific).

Fetal bovine serum (FBS, Thermo Fisher Scientific).

L-glutamine (Thermo Fisher Scientific).

Non-essential amino acids (NEAA, Thermo Fisher Scientific).

Penicillin–streptomycin solution (Thermo Fisher Scientific).

2.3 Skin Equivalent

Human neonatal dermal fibroblasts (HDFn, Thermo Fisher Scientific).

2.3.1 Subepithelial Compartment

Medium 106 (Thermo Fisher Scientific).

Low-serum growth supplement (LSGS, Thermo Fisher Scientific).

Transforming growth factor β1 (TGFβ1, PeproTech).

Ascorbic acid (Sigma-Aldrich).

Gentamicin (Thermo Fisher Scientific).

Amphotericin B (Thermo Fisher Scientific).

Trypsin neutralizer (Thermo Fisher Scientific).

2.3.2 Epithelial Compartment

Human neonatal epidermal keratinocytes (HEKn, Thermo Fisher Scientific).

EpiLife® media (Thermo Fisher Scientific).

Human keratinocyte growth supplement (HKGS, Thermo Fisher Scientific).

Gentamicin (Thermo Fisher Scientific).

Amphotericin B (Thermo Fisher Scientific).

Keratinocyte growth factor (KGF, Thermo Fisher Scientific).

Calcium chloride (CaCl$_2$, Sigma-Aldrich).

Ascorbic acid (Sigma-Aldrich).

Trypsin neutralizer (Thermo Fisher Scientific).

3 Methods

The skin and intestinal equivalents are constructed using a biphasic step-by-step protocol (Fig. 5). Firstly, a robust subepithelial compartment is generated by the proliferation of fibroblasts within the porous polystyrene membrane, and the deposition of endogenous extracellular matrix proteins to establish a humanized extracellular matrix microenvironment. Epithelial cells are cultured on the apical surface of the subepithelial compartment and stimulated to undergo differentiation and specialization to form a well-organized, multilayered tissue equivalent. The intestinal and skin equivalents have been optimized for 12-well Alvetex® formats; however, 24-well inserts can also be used with appropriate scaling (*see* **Note 1**). All steps are to be carried out in a sterile, category 2 laminar flow hood at room temperature (*see* **Note 2**).

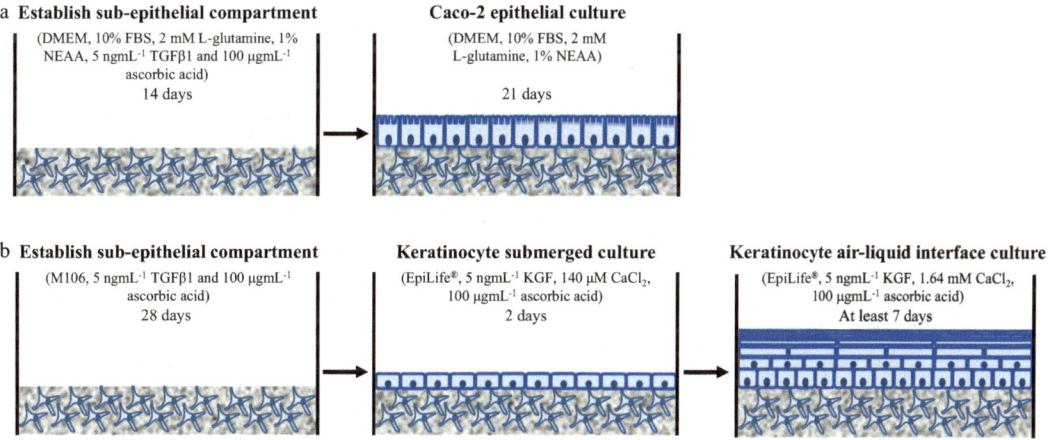

Fig. 5 Bioengineering intestinal and skin tissue equivalents in vitro. Epithelial tissue equivalents such as (**a**) intestinal equivalents and (**b**) skin equivalents are generated using detailed protocols. In the first phase, the robust subepithelial compartment is self-assembled by culturing fibroblasts within the porous polystyrene membrane and enabling them to secrete endogenous extracellular matrix proteins. In the second phase, epithelial cells such as Caco-2 cells or keratinocytes are seeded onto the apical surface of the subepithelium, and stimulated to proliferate, differentiate and undergo specialization

3.1 Subepithelial Compartment

3.1.1 Revival of Fibroblasts for the Intestinal Equivalent

1. Prepare a bottle of complete DMEM, by supplementing with 10% FBS, 1% NEAA and 2 mM L-glutamine. Additional supplementation with antimicrobial agents (100 U/mL penicillin and 100 µg/mL streptomycin) is also recommended.

2. Human neonatal dermal fibroblasts should be revived from long-term storage and cultured in complete DMEM in a standard culture flask. Cells should be seeded at the manufacturer's recommended density.

3.1.2 Revival of Fibroblasts for the Skin Equivalent

1. Prepare a bottle of complete Medium 106, by supplementing with 10 mL LSGS. Additional supplementation with antimicrobial agents (10 µg/mL gentamicin and 0.25 µg/mL amphotericin B) is also recommended.

2. Human neonatal dermal fibroblasts should be revived from long-term storage and cultured in complete Medium 106 in a standard culture flask. Cells should be seeded at the manufacturer's recommended density.

3.1.3 Preparation of the Porous Polystyrene Scaffold

The Alvetex® porous polystyrene membrane must be treated with ethanol before use, to render the scaffold hydrophilic (see **Note 3**).

1. Remove the Alvetex® Scaffold well inserts from their packaging using sterile forceps and place into a sterile petri dish or 6-well plate.

2. Submerge the well inserts in 70% ethanol for a minimum of 5 min.

3. Submerge the well inserts in sterile DPBS to remove all ethanol.

4. Submerge the well inserts in the appropriate culture medium until required.

3.1.4 Generation of the Subepithelial Compartment of the Intestinal Equivalent

1. Culture the human neonatal dermal fibroblasts in culture flasks to an 80% confluence (see **Note 4**).

2. Aspirate the cell culture medium and wash once in DPBS.

3. Trypsinize fibroblasts using a 0.25% trypsin–EDTA solution until all cells have detached in a single cell suspension.

4. Neutralize the trypsin–EDTA with complete DMEM.

5. Centrifuge cells at $200 \times g$ for 5 min to form a cell pellet.

6. Count the fibroblasts and test for viability using the trypan blue exclusion assay.

7. Seed 0.5×10^6 cells per 12-well Alvetex® Scaffold (see **Note 5**).

8. Incubate at 37 °C and 5% CO_2, in a humidified environment, submerged in complete DMEM supplemented with 5 ng/mL TGF β1 and 100 µg/mL ascorbic acid.

9. Culture for 14 days to form a mature subepithelial compartment.

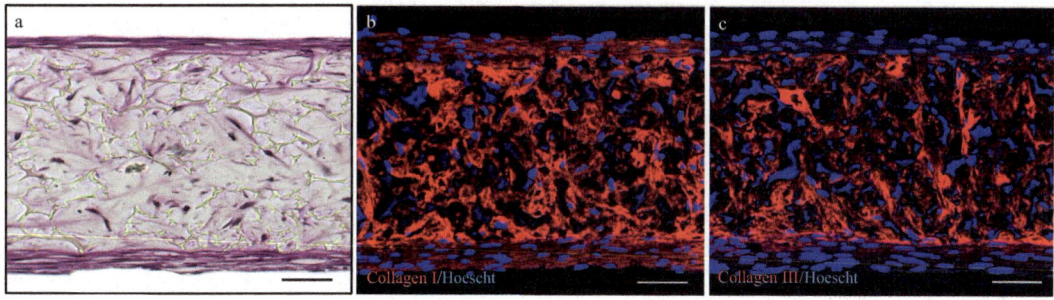

Fig. 6 The subepithelial compartment. (**a**) Representative H&E micrograph of fibroblasts cultured within the porous polystyrene scaffold to generate the subepithelial compartment. (**b**, **c**) Immunofluorescence analysis of collagen I and III synthesized de novo by fibroblasts within the subepithelium. Scale bars: 50 μm

3.1.5 Generation of the Subepithelial Compartment of the Skin Equivalent

1. Culture the human neonatal dermal fibroblasts in culture flasks to an 80% confluence.

2. Aspirate the cell culture medium and wash once in DPBS.

3. Trypsinize fibroblasts using a 0.25% trypsin–EDTA solution until all cells have detached in a single cell suspension.

4. Neutralize the trypsin–EDTA with trypsin neutralizer, according to the manufacturer's instructions.

5. Centrifuge cells at $200 \times g$ for 5 min to form a cell pellet.

6. Count the fibroblasts and test for viability using the trypan blue exclusion assay.

7. Seed 0.5×10^6 fibroblasts per 12-well Alvetex® Scaffold.

8. Incubate the subepithelial compartment at 37 °C and 5% CO_2 in a humidified environment, submerged in complete Medium 106 supplemented with 5 ng/mL TGF β1 and 100 μg/mL ascorbic acid.

9. Culture for up to 35 days to form a mature subepithelial compartment (Fig. 6).

3.2 Epithelial Compartment

3.2.1 Intestinal Equivalent

Revival of Caco-2 Cells for the Intestinal Equivalent

1. Prepare a bottle of complete DMEM, by supplementing with 10% FBS, 1% NEAA and 2 mM L-glutamine. Additional supplementation with antimicrobial agents (100 U/mL penicillin and 100 μg/mL streptomycin) is also recommended.

2. Caco-2 cells should be revived from long-term storage and cultured in complete DMEM, according to the manufacturer's instructions.

Seeding of Caco-2 Cells onto the Subepithelial Compartment

1. Culture the Caco-2 cells to a 70–90% confluence.

2. Aspirate the cell culture medium and wash once in DPBS.

3. Trypsinize the cells using a 0.25% trypsin–EDTA solution until all cells have detached in a single cell suspension.

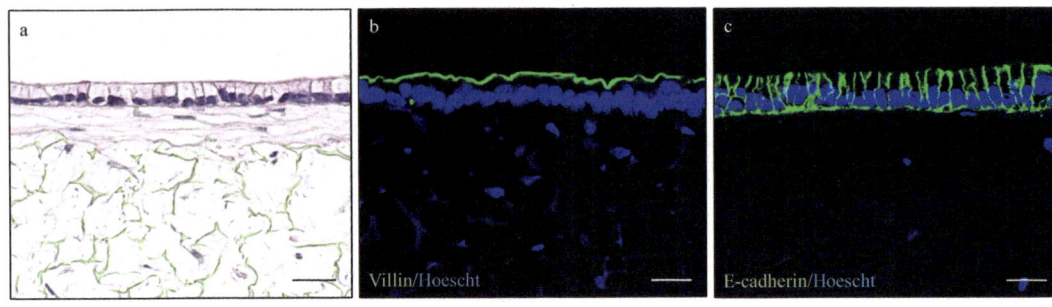

Fig. 7 Mature human intestinal equivalent. (**a**) Representative H&E micrograph of a mature intestinal equivalent, demonstrating the highly organized, polarized epithelial layer. Immunofluorescence analysis of (**b**) villin and (**c**) E-cadherin demonstrates the differentiation and specialization of the intestinal equivalent. Scale bars: 25 μm

4. Neutralize the trypsin–EDTA with complete DMEM.

5. Centrifuge the cells at 200 × g for 5 min to form a cell pellet.

6. Count the Caco-2 cells and test for viability using the trypan blue exclusion assay.

7. Seed 0.4×10^6 Caco-2 cells onto the mature subepithelial compartment.

8. Culture the intestinal equivalent at 37 °C and 5% CO_2 in a humidified environment in complete DMEM.

9. Culture for 21 days to form a mature intestinal equivalent with differentiated enterocytes (*see* **Note 6**) (Fig. 7).

3.2.2 Skin Equivalent

Revival of Keratinocytes for the Skin Equivalent

1. Prepare a bottle of complete EpiLife® medium, by supplementing with 5 mL HKGS. Additional supplementation with antimicrobial agents (10 μg/mL gentamicin and 0.25 μg/mL amphotericin B) is also recommended.

2. Human neonatal epidermal keratinocytes should be revived from long-term storage and cultured in complete EpiLife® medium, according to the manufacturer's instructions.

Seeding of Keratinocytes onto the Subepithelial Compartment

1. Culture the keratinocytes to an 80% confluence (*see* **Note 7**).

2. Aspirate the cell culture medium and wash once in DPBS.

3. Trypsinize the cells using a 0.25% trypsin–EDTA solution until all cells have detached in a single cell suspension.

4. Neutralize the trypsin–EDTA with trypsin neutralizer, according to the manufacturer's instructions.

5. Centrifuge the cells at 200 × g for 5 min to form a cell pellet.

6. Count the keratinocytes and test for viability using the trypan blue exclusion assay.

7. Seed 1.3×10^6 keratinocytes onto the mature subepithelial compartment.

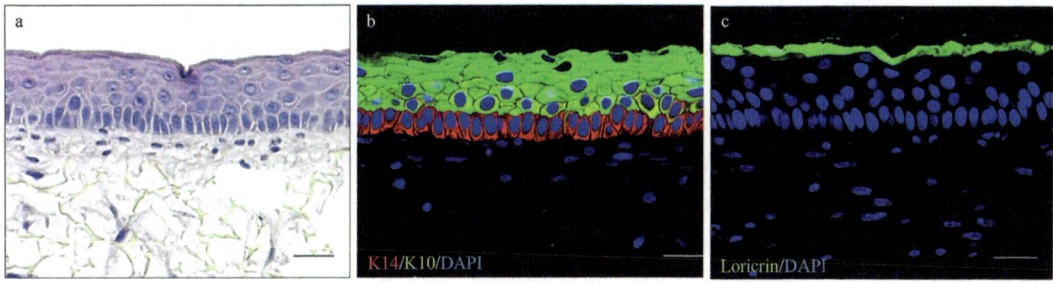

Fig. 8 Mature human skin equivalent. (**a**) Representative H&E micrograph of a mature skin equivalent, demonstrating the differentiated, stratified, and keratinized structure. Immunofluorescence of (**b**) cytokeratin 14 and cytokeratin 10 within the basal and suprabasal layers, respectively, and (**c**) loricrin within the *stratum granulosum*. These biomarkers demonstrate the organized, sequential differentiation to form the epidermis. Scale bars: 25 µm

8. Culture the skin equivalent for 48 h at 37 °C and 5% CO_2 in a humidified environment in complete EpiLife® media supplemented with 10 ng/mL KGF, 140 µM $CaCl_2$, and 100 µg/mL ascorbic acid.

Raising the Skin Equivalent to the Air–Liquid Interface

1. Carefully aspirate the media from inside the insert of the skin equivalent to raise it to the air–liquid interface (*see* **Note 8**).

2. Culture the skin equivalents at 37 °C and 5% CO_2 in a humidified environment in complete EpiLife® media supplemented with 10 ng/mL KGF, 1.64 mM $CaCl_2$ and 100 µg/mL ascorbic acid.

3. Culture for at least 7 days to form a mature skin equivalent with a differentiated, stratified, and keratinized epidermis (*see* **Note 9**) (Fig. 8).

3.3 Processing and Analyzing the Epithelial Models

In vitro epithelial tissue models must be characterized to demonstrate resemblance to their in vivo counterparts. The images included in this chapter were generated using the following protocols for wax embedding, histological, and immunofluorescence analysis.

3.3.1 Processing Samples for Paraffin Wax Embedding

1. Carefully unclip and remove the epithelial tissue models from the hanging insert using blunt tip forceps (*see* **Note 10**).

2. Wash the epithelial tissue models three times in DPBS to remove any residual culture media.

3. Fix in 4% paraformaldehyde solution for 2 h at room temperature, or overnight at 4 °C.

4. Wash the fixed epithelial tissue models three times in DPBS to remove residual paraformaldehyde.

5. Sequentially dehydrate the models using an ethanol gradient (30%, 50%, 70%, 80%, 90% and 95%), with 10-min incubations per ethanol concentration.

6. Incubate samples in 100% ethanol for 30 min to ensure the complete dehydration of the epithelial models.

7. Transfer the dehydrated samples into tissue-processing cassettes and submerge completely in Histoclear II for 30 min.

8. Incubate the epithelial tissue models in a 1:1 ratio of Histoclear II and molten paraffin wax for a further 30 min in a convection oven at 65 °C.

9. Incubate the samples in pure molten paraffin wax for a further 60 min at 65 °C.

10. Cut the tissue models in half across their diameter using a surgical scalpel, and embed in molten wax with the flat edge orientated to the base of the plastic embedding mould.

11. Allow the embedded models to completely set overnight at room temperature prior to sectioning and downstream analysis.

3.3.2 Histological Analysis

1. Section the paraffin embedded samples at a 5 μm thickness using a rotary manual microtome.

2. Place the transverse sections in a 37 °C water bath, and mount onto charged glass slides.

3. Allow the slides to completely dry on a 30 °C heated slide rack overnight.

4. Deparaffinize the slides in Histoclear I for 15 min.

5. Sequentially rehydrate the samples using an ethanol gradient (100%, 95%, 70%), for 1 min per ethanol concentration.

6. Incubate samples in distilled water for 1 min to ensure the complete hydration of the epithelial models.

7. Stain the nuclei using Mayer's hematoxylin for 5 min, which stains negatively charged components blue.

8. Wash the samples to remove any residual hematoxylin stain in distilled water for 30 s.

9. Place the samples in alkaline ethanol for 30 s, to ensure that the nuclei appear blue.

10. Sequentially dehydrate the samples using an ethanol gradient (70% and 95%), for 30 s each per ethanol concentration.

11. Incubate the tissue models in eosin solution for 30 s, which stains positively charged components pink.

12. Rapidly dehydrate the samples twice in 95% and 100% ethanol for 15–30 s.

13. Place the samples in Histoclear I twice for 5 min each.

14. Mount a coverslip on the slides using Omnimount inside a fume hood.

15. Leave to set overnight, prior to imaging with a light microscope.

3.3.3 Immuno-
fluorescence Analysis

1. Section the paraffin embedded samples at a 5 μm thickness using a rotary manual microtome, and place onto charged slides.

2. Allow the slides to completely dry on a 30 °C heated slide rack overnight.

3. Deparaffinize the slides in Histoclear I for 15 min.

4. Sequentially rehydrate the samples using an ethanol gradient (100% and 70%), for 5 min per ethanol concentration.

5. Incubate samples in phosphate buffered saline (PBS) for 5 min to ensure the complete hydration of the epithelial models.

6. Perform antigen retrieval for 20 min at 95 °C in pH 6 citrate buffer.

7. If permeabilization is required, incubate samples in 0.4% Triton X-100 in PBS for 5 min at room temperature.

8. Block the samples to prevent non-specific protein binding, by incubating in 20% neonatal calf serum in PBS for at least 1 h at room temperature in a humidified chamber.

9. Incubate the samples with the primary antibody diluted in a 20% neonatal calf serum in PBS solution, in a humidified chamber for 2 h at room temperature or at 4 °C overnight (Table 2).

Table 2
Primary antibodies used in immunofluorescence staining of epithelial models, including the supplier and working concentrations

Antibody	Supplier	Product code	Dilution
Collagen I	Abcam	ab34710	1:100
Collagen III	Abcam	ab7778	1:100
Villin	Abcam	ab130751	1:100
E-cadherin	BD Biosciences	610181	1:100
Cytokeratin 14	Abcam	ab7800	1:100
Cytokeratin 10	Abcam	ab76318	1:100
Loricrin	Abcam	ab85679	1:100

Table 3
Secondary antibodies used in immunofluorescence staining of epithelial models, including the supplier and working concentrations

Antibody	Supplier	Product code	Dilution
Alexa Fluor® 594 donkey anti-rabbit IgG (H + L)	ThermoFisher Scientific	A-21207	1:1000
Alexa Fluor® 488 donkey anti-rabbit IgG (H + L)	ThermoFisher Scientific	A-21206	1:1000
Alexa Fluor® 594 goat anti-mouse IgG (H + L)	ThermoFisher Scientific	A-11005	1:1000
Alexa Fluor® 488 goat anti-mouse IgG (H + L)	ThermoFisher Scientific	A-11001	1:1000

10. Wash the samples three times in PBS for 10 min each to remove any unbound non-specific primary antibody.

11. Incubate the samples with a species-specific, fluorescently tagged secondary antibody diluted in a 20% neonatal calf serum in PBS for 1 h at room temperature in a humidified chamber. Protect the samples from light to prevent photobleaching (Table 3).

12. Wash the samples three times in PBS for 10 min each to remove any unbound secondary antibody.

13. Stain the nuclei using either DAPI or Hoechst.

14. Mount a coverslip onto the slides using Hardset Vectashield mounting media.

15. Allow the Hardset Vectashield to set for 15 min at room temperature, prior to imaging with a confocal microscope (*see* **Note 11**).

4   Notes

1. *Alvetex® Scaffold formats.* Alvetex® Scaffold is available commercially in a 12-well and 24-well format. Epithelial tissues can be generated using these different formats, if cell seeding numbers are adjusted according to the growth area.

2. *Culturing long-term epithelial tissues in a sterile manner.* All reagents used to culture the epithelial tissues must be sterile and used within a Category 2 laminar flow hood. An aseptic technique must be used to prevent contamination.

3. *Preparation of the Alvetex® Scaffold.* The Alvetex® Scaffold porous polystyrene membrane is hydrophobic; therefore, it requires an ethanol solution to render it hydrophilic and amicable to cell attachment.

4. *The importance of fibroblast confluence.* Fibroblasts should be used in the exponential phase of growth, to ensure that a robust, mature subepithelial compartment is generated.

5. *Seeding cells onto the Alvetex® Scaffold.* Cell suspensions should be seeded onto the porous polystyrene membrane in a drop-wise manner, to ensure that they are evenly distributed across the model.

6. *Application of human intestinal equivalents.* The organized, differentiated intestinal equivalent can be cultured for an additional 9 days for downstream applications.

7. *The importance of keratinocyte confluence.* Keratinocytes must not be used at a higher confluence than 80%, as they begin to differentiate and lose their potential to generate a well-organized, stratified epidermis.

8. *Raising skin equivalents to the air–liquid interface.* Care should be taken when aspirating media from the surface of the skin equivalents, to ensure that the porous polystyrene membrane is not punctured.

9. *Application of human skin equivalents.* A well-organized differentiated, stratified, and keratinized epidermis is engineered after 7 days at the air–liquid interface; however, the skin equivalent can also be cultured for long-term experiments.

10. *Handling of porous polystyrene scaffold.* When unclipping the epithelial tissue models from the insert, take care to manipulate using forceps around the edge of the porous polystyrene membrane, to prevent damage to the models.

11. *Immunofluorescence staining of epithelial models.* After immunofluorescence staining, samples can be stored at 4 °C for up to 1 week. Long-term storage of immunofluorescence slides prior to imaging is not recommended.

Acknowledgments

This work was supported by Biotechnology and Biosciences Research Council (BBSRC BB/M015645; BB/K019260/1), NC3Rs (National Centre for the Replacement, Refinement and Reduction of Animals and Research (NC3Rs NC/N00289X/1), European Regional Development Fund Intensive Industrial Innovation and Reprocell Europe Ltd. (25R07P01847) and Procter & Gamble. The Authors are members of the COST Action CA16119 In vitro 3-D total cell guidance and fitness (CellFit)

References

1. Lowe J, Anderson P (2015) Epithelial cells. In: Lowes S (ed) Human histology, 4th edn. Mosby Elsevier, Maryland Heights

2. Farquhar M, Palade G (1963) Junctional complexes in various epithelia. J Cell Biol 17:375–412

3. Simons K, Fuller S (1985) Cell surface polarity in epithelia. Annu Rev Cell Biol 1(1):243–288

4. Yurchenco P (2011) Basement membranes: cell scaffoldings and signaling platforms. Cold Spring Harbour Perspect Biol 3(2):1–27

5. Frantz C, Stewart K, Weaver V (2010) The extracellular matrix at a glance. J Cell Sci 123 (24):4195–4200

6. Wickett R, Visscher M (1997) Structure and function of the epidermal barrier. Am J Infect Control 34(10):98–110

7. Baker B, Chen C (2012) Deconstructing the third dimension: how 3D culture microenvironments alter cellular cues. J Cell Sci 125 (3):3015–3024

8. Thomas C, Collier J, Sfeir C et al (2002) Engineering gene expression and protein synthesis by modulation of nuclear shape. Proc Natl Acad Sci 99(4):1972–1977

9. Vergani L, Grattarola M, Nicolini C (2004) Modifications of chromatin structure and gene expression following induced alterations of cellular shape. Int J Biochem Cell Biol 36 (8):1447–1461

10. Costello L, Fullard N, Roger M et al (2019) Engineering a multilayered skin equivalent: the importance of endogenous extracellular matrix maturation to provide robustness and reproducibility. In: Skin tissue engineering. Humana Press, New York

11. Roger M, Fullard N, Costello L et al (2019) Bioengineering the microanatomy of human skin. J Anat 234(4):438–455

12. Rossi A, Appelt-Menzel A, Kurdyn S et al (2015) Generation of a three-dimensional full thickness skin equivalent and automated wounding. J Visual Exp (96):e52576

13. Lotz C, Schmid F, Oechsle E et al (2017) Cross-linked collagen hydrogel matrix resisting contraction to facilitate full-thickness skin equivalents. ACS Appl Mater Interfaces 9 (24):20417–20425

14. Wang Y, Gunasekara D, Reed M et al (2017) A microengineered collagen scaffold for generating a polarized crypt-villus architecture of human small intestinal epithelium. Biomaterials 128:44–55

15. Loewa A, Vogt A, Kaessmeyer S et al (2018) Generation of full-thickness skin equivalents using hair follicle-derived primary human keratinocytes and fibroblasts. J Tissue Eng Regen Med 12(4):2134–2146

16. Mieremet A, Rietveld M, van Dijk R et al (2018) Recapitulation of native dermal tissue in a full-thickness human skin model using human collagens. Tissue Eng Part A 24 (11–12):873–881

17. Romo-Morales A, Knight E, Przyborski S (2017) Alvetex®, a highly porous polystyrene scaffold for routine three-dimensional cell culture. In: Technology platforms for 3D cell culture: a user's guide, 1st Edn. Wiley, New York

18. Knight E, Murray B, Carnachan R et al (2011) Alvetex®: polystyrene scaffold technology for routine three dimensional cell culture. In: 3D cell culture. Humana Press, New York

19. Marrazzo P, Maccari S, Taddei A et al (2016) 3D reconstruction of the human airway mucosa in vitro as an experimental model to study NTHi infections. PLoS One 11(4): e0153985

INDEX

A

Activin/Nodal/ERK signaling pathways 152
Additives 11
Adenosine 5′ Triphosphate (ATP) 21, 162
Adenylate kinase 43, 45, 46
Adipocyte derived stem cells (ADSCs)
    coculture 68
    culture 67
    culture condition 70
    immunocytochemistry analysis 68
    isolation 67
    neuronal differentiation 64
    neurons 67, 68
ADSC neuronal differentiation, small EVs
    cell differentiation inductor 70, 71
    characterization 69
    confocal microscopy analysis 71
    isolation, cell culture media 69
    TEM analysis 70
Aged uterine environment 152
AggreWell™ Anti-adherence 20
AggreWell™ plates 27–30
Air–liquid interface culture, FRT epithelia
    ALI-POEC
        embryo co-culture 258, 259
        maintenance 258
        seeding 258
    cell isolation and dissociation 255–257
    differentiation medium 255
    DPBS 255
    humid chamber preparation 257
    insert coating solution 255
    OEC basic medium 255
    one-step medium 255
    plates 255, 257
    proliferation medium 255
    sample preparation
        histological analysis 256, 259, 260
        TEM 256, 260
    serum-free differentiation medium 255
    TEER measurement 256, 259
ÄKTAprime plus chromatography system 191,
    192, 199

Albumin (ALB) 117
Alginate-encapsulated spheroids preparation 32, 33
Alginate-layered spheroids 33
ALI, see Air–liquid interface, FRT epithelia
ALI-OEC 261
ALI-POEC cultures 253, 254, 258, 259, 261
Aliquoting Q-gel™ 28
Aliquots 82
Alkali labile sites (ALS) 180
Alvetex® 282–285, 288, 294, 295
Amalgamation 202
Amino acid 265
Ammonium bicarbonate (ABC) 184
Amplex Red Cholesterol Assay 21, 49
Anesthesia/CO$_2$ chamber 66
Angiogenesis 112
Antibiotic/antimycotic solution (AAS) 78, 108
Anti-Biotin MicroBeads 93
Aovine serum albumin (BSA) 6
Apical–basal polarization
    cilia formation 252
    reproductive functions 251
    specialized cellular junctions 251
Apurinic/apyrimidinic (AP) 180
Arrhythmogenic ventricular cardiomyopathy 86
Artificial intestine
    biological/biocompatible scaffolds, primary cell
        culture seeding 268, 269, 273, 275
    cell lines growth, mono-or coculture
        conditions 264
    film-casted gelatin films
        rheological stiffness characterization 267,
        271
        sample collection 267, 271
        scaffold preparation 267, 271
    gelatin 265
    Gel-X 266, 267, 270–272
    hydrogel 264
        Gel-MA 265, 269
        Gel-NB synthesis 266, 270
    immortalized cell lines 264
    keystone 264
    mechanical properties 265
    primary cell culture isolation 268, 272–274

Tiziana A.L. Brevini et al. (eds.), *Next Generation Culture Platforms for Reliable In Vitro Models: Methods and Protocols*,
Methods in Molecular Biology, vol. 2273, https://doi.org/10.1007/978-1-0716-1246-0,
© The Author(s), under exclusive license to Springer Science+Business Media, LLC, part of Springer Nature 2021

Artificial intestine (*cont.*)
RGD motif 265
sample collection 268
ATP aliquots 59
ATP assay 179
ATP assay standard plate layout 44
ATP standard dilution series 43
Autoalignment 210
Autodigestion 185
AVI file 217

B

Benchtop columns 202
Benchtop SEC columns, EVs isolation
materials 202
packing with Sepharose 203, 204
prewash before use 204
sample preparation 203
samples analysis 204, 205
samples purification 204
washing column for reuse 205
Bicarbonate-based media 57
Bioactive ECM content 240, 248
Bioceramic-based bone scaffolds 240
Bioengineered three-dimensional tissue 282
Biogenesis 201
Bioinformatics analysis 233–236
Biomimetic bone scaffold 242, 243
Bioprosthetic ovary 140
BisBenzimide H 33258 reagent 37
BisBenzimide H 33258 solution 39
Bone morphogenetic protein 4
(BMP4) 152, 155
Bone tissue
auto/allografts 239
bioceramic-based 240
biomaterials fabrication technologies 240
decellularization 240
ECM 240
major defects/damages 239
organic-based living structure 239
synthetic bioceramics 239
tissue engineering 240
Bone tissue engineering 240
Bovine follicular fluid 203, 205
Bovine miRNAs 233
Bovine pituitary extract (BPE) 135
Bovine primary epithelial/stromal cells 104
Bovine serum albumin (BSA) 64, 104
Bright-field (BF) 97
Brownian motion 207, 210
BSA standard dilution series 41
Buried cells 18

C

C3A and cocultured spheroids 163
C3A and HMEC-1 cells 169
C3A spheroids preparation 29
Caco-2 cells 287, 289, 290
Cacodylate buffer 58
Cardiac differentiation 91, 99
Cardiac fibroblast 99
Cardiac fibroblast basal medium (CFBM) 92
Cardiac fibroblast differentiation 88, 89,
91, 92
Cardiac organoids 86
Cardiac patch engraftment 119, 120
Cardiac patch model 114
Cardiac patch transplantation 126, 127
Cardiomyocyte differentiation and culture 87, 88
Cardiomyocyte purification 88
Cardiomyocytes 86
Cassette 230
Cell–cell interactions 17
Cell clumping 27
CellCounter 99
Cell culture media 66, 67, 71, 110
Cell differentiation
cellular behavior 63
immunocytochemistry 68
in vitro 63
induction 63, 65
Cell labelling 146
Cell lines 25–27
Cell lysis 179
Cell proliferation 76
Cell quality 216
Cell removal techniques 112
Cell seeding 110, 248
Cell viability 23, 160
Cells cryopreservation 25
CellTiter-Glo reagents 179
CelVivo bioarray matrix (BAM) system 175
Cholesterol 49
concentrations 48
determination plate setup 49
dilution schema 47
Reaction Buffer 47
reference standard curve 47
standard sample series 48
Cholesterol esterase stock solution (CHES) 47, 48
Cholesterol oxidase stock solution (CHOX) 47, 48
Chorionic gonadotropin (CG) 151
Chromatin 3
Chromatography column 194
ClinoReactor™ bioreactor 19, 26
ClinoReactor™ spheroid cultures 27, 28

Clinostat 3D cell culture
 Aliquoting Q-gel™ 28
 assay
  DNA content 37, 38, 40
  protein content 38, 41, 42
 changing media 33–35
 ClinoReactor™ bioreactor 19, 26
 freezing media preparation 25
 functional assay
  adenylate kinase 43, 46
  cholesterol 45, 47–49
  glucose 47, 50
  intracellular ATP 41, 43, 44
  intracellular glutathione 50–53
  urea 51, 54
 functional assays
  equipment 20, 21
  reagents 21
 growth media preparation 22
 Lysis buffer B 38
 preparation
  equipments 19
  reagents 19
 spheroid culture 20
 spheroid/organoid culture splitting 34, 36
 spheroid/organoids visualization
  electron microscopy 37
  equipment 20
  immunomicroscopy 36, 37
  reagents 20
 spheroids preparation
  AggreWell™ plates 27–30
  alginate-encapsulated 32, 33
  cocultures 33
  Q-gel™ 31–32
  self-aggregated 26, 27
 stocks preparation, storage 25
 thawing cells 22, 23
 tissue construction reproducibility 18
 transition challenges 18
 2D cell sub-cultures 22, 24
CO₂ chamber 71
Coating material 86
COCs isolation
 EAFs
  L-IVCO 9
  ovarian cortex preparation 8
  stereomicroscope 9
 MAFs
  aspiration 7
  bovine ovaries collection 8
  IVM 8
  morphology 8
  pre-IVM 8
  retrieving..8

Co-culture cell suspension 94
Coculture methodology with inserts
  culture media 67
  EVs and cell label 67
  transwell coculture 68
Collagen 240
Collagen hydrogels 283
Collagenase 108, 146
Comet assay 180
Confocal microscopy analysis 71
Conventional two-dimensional (2D) culture
  systems 76
Coronaviridae 132
Corpus luteum (CL) 225, 234
Corrected fluorescence (CF) 54
Corrected sample fluorescence (CSF) 54
Count isolated cells 66
Cryomicrotome 20
Cryostat sectioning 127
Cryovials 25
Culture medium (CM) 78
Culture medium for the liver organoid
  (CMLO) 117
Cumulus cells
 COCs 8, 9
 compact layers 9
 GJs 3, 5
Cumulus oocyte complexes (COCs) 5
Customized cultural strategies 3
Cyclic guanosine monophosphate (cGMP) 4
Cytometer setup and tracking (CS&T) 163

D

Daily performance 210, 216
DAPI 105, 108, 143
Decellularization protocols 140, 145
Decellularization technology 240
Decellularized bovine SIS (bSIS) 241, 242, 246
Decellularized ovary 141
Decellularized scaffolds creation process 113
DeltaPMTV 169
Deoxyribonuclease (DNase) 78
Dermal fibroblasts 153
DIANA–miRPath 234
Diffusion gradients 18
Digestive system 264
Diluted trypsin solution 177
Disease modeling 87
Dislodge cells 90
Distilled water (DW) 117
DMEM-based medium 22, 55
DNA double-strand breaks (DSB) 180
DNA methylation 3
DNA quantification 276
DNA Quantitation Kit 21

DNA standard 58
DNase I activity 82
Doublet cells 170
Doubly negative cells 163
Drug screening 86
Dulbecco's Modified Eagle Media (DMEM) 64, 78
Dulbecco's Modified Eagle Medium/ Nutrient
    Mixture F-12 (DMEM/12) 117
Dulbecco's phosphate buffered saline (DPBS) 78
Dulbecco's Phosphate Buffered Saline with antibiotics
    (DPBS-Ab) 104
Dye reagent 42

E

Easy-to-use protocols 18
ECM-EVs enriched cardiac patch 116
Electrophoresis 180, 230, 231
Embryonic stem cells (ESCs) 152
Endometrial cell characterization 105
Endometrial cell culture 108
Endometrium 103
Engineered heart slices (EHSs) 112
Engineered livers slices (ELSs) 112, 127
Enzymatic digestion 175
Epidermis 281
Epigenetic signatures 3
Epithelial and stromal cells 110
Epithelial cells (EC) 134, 287
Epithelial tissue models, in vitro
    Alvetex® 283–285
    bioengineered three-dimensional tissue 282
    collagen hydrogels 283
    epithelial compartment
        intestinal equivalents 289, 290
        skin equivalents 290, 291
    histological analysis 292
    hydrogels 283
    immunofluorescence analysis 293, 294
    intestinal equivalent
        epithelial compartment 286
        subepithelial compartment 286, 287
    paraffin wax embedding 291, 292
    polystyrene scaffolds 282, 285
    porous scaffolds 283
    scaffold-based approaches 282
    skin equivalent
        epithelial compartment 286, 287
        subepithelial compartment 286, 287
    subepithelial compartment
        intestinal equivalents 287, 288
        polystyrene scaffolds 288
        skin equivalents 287–289
    three-dimensional geometry 281
    tissue engineering 282, 283
    two-dimensional cell culture 281

Epithelial tissues
    intestine 280
    skin 280
    structure 279
    subepithelial compartment 279
Epithelium 252
Ethylenediaminetetraacetic acid (EDTA) 117, 124
EVs miRNAs 234
Exosomes 201, 220
Extracellular matrix (ECM) 112, 240,
    252, 265
Extracellular vesicles (EVs)
    ability 63
    analytic approaches 222
    bioinformatics analysis 225
    biophysical properties 207
    characterization
        NTA 227, 228
        TEM 228, 229
        western blot analysis 229–231
    culturing protocols 113
    cytoplasmatic contents 114
    definition 63, 201
    EV isolation, OF and UF
        methods 221
        SEC 223, 226, 227
        ultracentrifugation 223, 226, 227
    exosomes 201, 220
    functional molecules transfer 220
    intercellular communication 201
    isolation 189, 220,
        (see also BenchtopSEC columns, EVs isolation)
    lipids 202
    membrane-confined biological nanoparticles 207
    microRNA isolation and analyses
        bioinformatics analysis 233–235
        microRNA profile 232
        miRNA expression analysis 232, 233
        RNA isolation 232
    microRNAs 220
    miRNA expression analysis 222, 225
    miRNA profile 225
    molecules 114
    MVs 220
    myocardial infarction 115
    NTA 223,
        (see also Nanoparticletracking analysis (NTA))
    OF 220
    oviduct/uterine flushing collection 223,
        225, 226
    paracrine communication 220
    purification 202, 221
    reproductive fluids 219
    RNA isolation 225
    SEC 221

SE-HPLC (*see* Size-exclusion high-performance liquid chromatography (SE-HPLC))
separation...202
size profile..213
technologies..222
TEM... 223, 224
trophectoderm cells 220
UF...220
ultracentrifugation 221
western blot analysis 224

F

Female germline stem cells (FGSCs)
bioscaffolds.. 140
culture medium 146
isolation 145, 146
MACS 140
Female reproductive tract (FRT), *see* Air–liquid interface, FRT epithelia
Fetal bovine serum (FBS)...............64, 78, 104
Fetal calf serum 182
FGSCs isolation...142
Fibroblast culture medium 154, 155
Fibroblasts 295
Filtration 72
Fish intestine 271
5-Azacytidine-CR (5-aza-CR)............ 155, 157
Flow cytometry 180, 181
data analysis 162, 170
dual strain 161
FACS Aria version 168
FSC A/SSC-A... 166
HepG2/C3A spheroid 167
measurement 161
PI-A... 168
Flow cytometry applications...........................181
FlowCytometryTools 215
Fluorescein diacetate (FDA).............160, 161, 164, 169
Fluorinated ethylene propylene (FEP) 78
copolymer 76
powder 77, 80
transfer preantral follicles 82
Fluorochrome(s) 180
Follicle reserve and categories 2
Follicle stimulating hormone (FSH)............... 4
Follicles
culture media... 76
flattened 76
in vitro growth 76
pipettor 83
porcine 78
primordial 76
secondary 82
Follicle-stimulating hormone (FSH) 79
Folliculogenesis 75

Forward scatter (FSC) 164
Freezing medium 25, 55
Fusion of liquid marble 97

G

Gap junctions (GJs) 3–5
Gastrointestinal tract...................................264
Gel filtration column 198
Gel polymerases...56
Gelatin 265, 276
Gelatin derivatives (Gel-X) 266, 267, 270–272
Gelatin-methacrylamide (Gel-MA)..................... 265, 269
Gelatin-norbornene (Gel-NB) synthesis
activation 266, 270
purification
dialysis... 266, 270
precipitation................................. 266, 270
reaction 266, 270
Gel-electrophoresis 186
Glucose 47, 50
Glutathione 52
Glutathione Lysis Buffer 50
Gold nanoparticles (AuNPs) 132
GSH standard curve...................................... 60
GV0 configuration 3

H

Haemocytometer 55
Hank's Balanced Salt Solution (HBSS) 19, 144, 163, 175
HAp microparticle suspension 243
Harvesting LM...94
Heat-treated medium dilution series 45
Hemocytometer 25
Hepatic Growth Factor.................................117
Hepatic progenitor cells (HPC)...................113
Hepatocytes 112
HepG2/C3A spheroids 174, 177
Heterogeneous population.............................201
Hexamethyldisilazane (nHMDS)..................86, 89, 94
hiPSC-derived cardiomyocytes.....................112
Histological analysis.....................................292
HNEpC and HTEpC 2D culture 134–136
Human induce pluripotent stem cells derived cardiomyocytes (hiPSCs-CM) 124
Human induced pluripotent stem cells (hiPSC).........112
Human nasal epithelial cells (HNEpCs).......134
Human pluripotent stem cells (hPSCs)......152
Human skin fibroblasts isolation..............154
Human skin fibroblasts seeding 154, 155
Human stem cell culture 87
Human tracheal epithelial cells (HTEpCs) 134
Hydrogel-based 3D culture system 134–136
Hydrogels 283

I

Immune–endocrine interface 251
Immunocytochemistry 68, 69
Immunofluorescence analysis 289, 293–295
Immunofluorescent staining 89, 98, 99
In vitro 2D monolayer models 103
In vitro culture (IVC) 1
   collection media 6
   IVM medium 7
   L-IVCO medium 7
   materials 5, 6
   methods (*see* COCs isolation)
   pre-IVM medium 7
In vitro embryo production (IVEP)
   embryonic development blastocyst stage 2
   embryos transfer 1
   follicle reserve and categories 2
   oocytes 2
   progressive chromatin compaction 3
   protocols 2, 5
In vitro fertilization (IVF) 1
In vitro maturation (IVM) 1
In vitro model 252
   artificial intestine (*see* Artificial intestine)
   digestive system 264
   immortalized cell lines 264
Incubation 186
Inferior vena cava (IVC) 120, 123
Infertility 139
Inhibitor 3-isobutyl-1-methyl-xanthine (IBMX) 9
Injection valve position 199
Intercellular communication 4, 201
International Society for Extracellular Vesicles
   (ISEV) 222
Intestinal epithelium 280, 282
Intestinal equivalent 295
   Caco-2 cells 289, 290
   epithelial compartment 286
   subepithelial compartment 286, 288
Intestinal equivalents 282
Intestine 264, 280
Intracellular ATP 41, 43, 44
Intracellular glutathione 50–53
Intraluminal vesicles 202
Isolated preantral follicles 76
IVM medium 7–9

J

JAr cell–derived EVs 208

K

Keratinocytes 282, 287, 290, 295
Keystone 264
KnockOut™ serum replacement 65

L

Large antral follicles (LAFs) 2
L-ascorbic acid stock solution 79
Least squares regression equation 39
Liberase ™ Thermolysin High (TH) Research
   Grade 78, 82
Lipid content 187
Lipidomics 178, 184
Liquid Chromatography System
   (ÄKTAprime plus) 190
Liquid chromatography-mass spectrometry
   (LC-MS) 184
Liquid–liquid interface (LLI) 252, 253
Liquid marbles (LM) 77, 80
   definition 86
   handling 96
   hydrophobicity 101
   organoid production 100
   outgrowth 100
   preparation 87, 94, 100
   stereomicroscope 101
   3D coculture approach 86
   3D structures 86
   ultimate capacity 101
Liquid nitrogen 20, 22, 25, 36
   safety guidelines 55
LM methods
   cardiac fibroblast differentiation 91, 92
   cardiomyocyte differentiation 90, 91
   cardiomyocyte purification 92, 93
   cardiospheres applications 98
   coalescence 97
   dissociating cardiac fibroblast 93
   dissociating cardiomyocytes 92
   ethics statement 90
   harvesting 94
   immunofluorescent staining 98
   microbioreactors 94, 95
   splitting 94, 97, 101
   stem cell culture 90
   waste/spent medium 94
LM technology
   cardiac fibroblast differentiation/culture 88, 89
   cardiomyocyte differentiation and culture 87, 88
   cardiomyocyte purification 88
   human stem cell culture 87
   immunofluorescent staining 89
   microbioreactor 89
   3D in vitro modeling 87
LOAD position 199
Long in vitro culture of oocytes (L-IVCO)
   BioCoat™ Collagen 6, 7
   COCs 9
   FSH 5
   GlutaMAX™ 7

in vitro 4
IVM 8
PVP 9, 11
Low melting point (LMP)................................180
Luciferin Generation Reagent 51
Lyophilized Luciferin Detection Reagent 51
Lysis buffer B 38, 59

M

Magnetic activated cell sorting (MACS) 92, 140
Mammalian cells..159
Mammalian ovary.. 2
Marfan syndrome 86
Masson's trichrome staining............................115
Mechanical cue 240
Meiotic arrest 4, 6, 9, 11
Mesenchymal stem cells (MSC)
    ADSCs 64, 66
    differentiation.. 63
    neuronal differentiation 72
    neurons medium 65
    protocols.. 63
Microbioreactors 77
Microenvironment factors 174
MicroRNA profile 232
MicroRNAs 220
Microscopy-based live/dead cell counting
        methods 160
Microvesicles (MVs)..220
Middle antral follicles (MAFs) 2
Mimetic tissues ,,..17
Mini-organoids..86
miRNA expression analysis 222, 232, 233
miRNA profile 225
miRNAs 233
miRTarBase database 234
Monolayer (2D) cultures....................................173
Monolayer cultures 160, 163, 165, 169
Mouse embryonic fibroblast (MEF) 152, 154
Multicellular spheroids 86
Multicolor staining..186

N

NaHCO$_3$-buffered media 11
Nanoparticle tracking analysis (NTA)................223, 227,
        228, 234
    Brownian motion 207
    EV analysis 207
    EVs measurement (see NTA measurement of EVs)
    ZetaView...208
    ZP 208
NanoSight 234
Natriuretic peptide precursor C (NPPC) ,,,,,,,,,,4, 11

Natriuretic peptide receptor type-2 (NPR2)................... 4
Natural polymers..240
Necrotic core 174
NTA measurement of EVs
    analytical grade reagents 208, 209
    cell chamber cleaning................................214
    size and concentration 210, 212, 213
    ZetaView instrument preparation 209–211
    ZetaView result analysis 215
    ZP 212, 214
NTA-ZetaView
    EVs concentration determination 208
    JAr cell–derived EVs 208
    PBS..210
    performance check 209–211
    PMX-110 V3.0 instrument 209
    result analysis 215
    sample measurements 210, 212
    user interface 210, 216
    ZP values 215

O

Omics analysis 182
    aim 182
    lipidomics..184
    proteomics 183, 184
    samples washing and freezing...............................182
    transcriptomic.................................... 182, 183
Oocytes
    bovine growing 5
    follicle growth 4
    FSH ,,,,,... 5
    GJ coupling ....,,,,, 5
    IVM 5
    L-IVCO 4
    MAFs 3, 4
    molecular machinery................................ 4
    synchronized... 4
Optimal cutting temperature (OCT)................. 116, 121
Ovarian cortex... 2
Ovarian follicles..75
Ovarian functions restoring.............................140
Ovary 76, 77, 80
Ovary bioengineering and repopulation
    decellularization protocols.....................................140
    fertility restoration 140
    FGSC isolation 143–147,
        (see also Female germline stem cells (FGSCs))
    in vitro ovarian tissue reconstruction....................141
    ovary collection 141, 144
    3D-scaffold 140
    whole-ovary decellularization.............. 142, 143, 145
Ovary dysfunction...139

Oviduct 251
Oviduct fluid surrogate (OFS) 254, 258
Oviductal epithelial cells (OEC) 252, 258
Oviductal fluid (OF) 220
Ovum pick up (OPU) 4
Oxidized Glutathione Lysis Buffer 51

P

Parafilm® M 31, 33, 56
Paraformaldehyde (PFA) 89
Particle Metrix' ZetaView 72
Permeabilization 98
PET membrane 260
Phosphate buffered saline (PBS) 64, 65, 134, 293
Phosphodiesterase-3 (PDE3) 4
Photobleaching 58, 169
PI intensity 170
Pig 80
Pipetting 71
Placental dysfunction 151
Polystyrene (PS) 216
Polystyrene scaffolds 282, 285, 288, 295
Polytetrafluoroethylene (PTFE) 77
Polyvinylpyrrolidone (PVP) 7
Polyvinylpyrrolidone-coated gold nanoparticle
        (PVP-AuNPs) 135–137
Porcine cells 261
Porcine follicular fluid (pFF) 79
Porcine oviductal epithelial cells (POEC) 252
Portion trypsin 55
Preantral follicle isolation
        DNase 78, 81, 82
        DPBS 78
        FEP microbioreactors 80
        HM 78
        micromanipulation 78
        porcine ovaries 79
Pre-IVM medium 4, 5, 7, 8
Prematuration 4, 5
Premature ovarian insufficiency (POI) 139
Primary antibodies 293
Primary cell line 268, 269
Primordial follicles 2
Principles of Laboratory Animals 125
Propidium iodide (PI) 160
Propidium iodide staining 179
Prostain™ Kit 38, 41, 42
ProStain™ Protein Quantification Kit 21
Proteomics 178, 183, 184

Q

Q-gel polymerization 56
Q-gel™ aliquots 56

Q-gel™ spheroids 31, 32, 56, 57
QuantGenious 186

R

Rabbit cells allotransplantation 122
Rainbow trout 265, 267, 276
Rat bone marrow-derived mesenchymal stem cells
        (rBM-MSCs) 243
Real-time PCR reaction 233
Real-time quantitative reverse transcription PCR
        (qRT-PCR) 232
Recombinant human FSH (r-hFSH) 7, 11
Regenerative medicine 111, 112, 240
Reproducibility 18, 50
Resuspensions 185
Retinoic acid 260
Retroperitoneal fat tissue 71
Reverse–transcription reaction components 233
RNA isolation 225, 232
RNA stabilization 182

S

Scaffold recellularization 114
Secondary antibodies 294
Self-aggregation 55
Self-assembled C3A spheroids 56
Sepharose 202
Sepharose beads 203, 205
Sepharose resin 206
Serdolyte resin 59
Severe acute respiratory syndrome coronavirus
        2 (SARS-CoV-2) 131, 132
Single cell suspensions 169, 175
        automated cell counters 178
        flow cytometer 179
        homogenous 180
        intact spheroid 185
        LMP 180
        spheroid fractions 179
        spheroids disassembly 174, 177, 178
        staining solution 181, 185
        trypsin incubation 185
        viability 180
Single-strand breaks (SSB) 180
Singlet cells 170
Size chromatography 199
Size exclusion chromatography (SEC) 189, 221,
                                        223, 226, 227
        benchtop columns 202
        EV yield 202
        EVs isoaltion (*see* Benchtop SEC columns, EVs
                isolation)
        molecular sieve chromatography 202

sepharose resin 202
Size-exclusion high-performance liquid chromatography
        (SE-HPLC)
    column disconnecting 198
    column washing 196, 198
    connecting systems 193, 194
    equilibration 194, 195
    EVs post-isolation handling 198
    flow path and SEC column 194
    liquid chromatography system
            preparation 191, 192
    reagents 190
    sample loading and EV isolation 195, 196
    sample preparation 190, 191
Skin 280, 281
Skin equivalent 282
    ALI 291, 295
    epithelial compartment 286, 287
    keratinocytes 290
    subepithelial compartment 286, 288, 289
Small intestinal submucosa (SIS) 241
Sodium dodecyl sulfate (SDS) 117
Sonication 186
Spheroid disassembly
    workflow steps 176
Spheroids core and rim methods
    cell viability (see Viability methods)
    comet assay 180
    disassembly (see Spheroids disassembly)
    flow cytometry 180, 181
    omics (see Omics analysis)
    seperation 174, 175
    single cell suspension 174
Spheroids disassembly
    core and rim 177, 178
    seperation 176, 177
    single-cell suspension 175, 176
Spheroids/organoids
    changing media 34
    ClinoReactor™ 30, 34, 43
    clumping 30
    cocultures 33
    Parafilm® covered PVC block 33
    protocols 18
    Q-gel™ 32
    self-aggregated 26, 27
    shape 18
    sizes 27
    splitting 34, 36
SSEA-4+ cells 146
Stacking gel solution 230
Standard operating procedure (SOP) 212, 214
Standard plate layout 42
Standards STOCK 1:2 dilutions 52

Stem cell differentiation, derived-EVs
    coculture methodology 67–68
    immunocytochemistry 68, 69
    isolation and characterization 65
    materials 64, 65
    MSC isolation and culture 66, 67
    multipotent cell type culture 64
    multipotent stem cell type 65
    neurons isolation and culture 66
Stem Cell Factor (SCF) 144
Steroid hormones secretion 76
Stokes–Einstein equation 208
Strand brakes 180
Subepithelium 281
Superhydrophobic microbioreactors 86
Synaptophysin 72

T

TaqMan probes 183
TargetScan 235
TEER measurement 256, 259, 261
Text documents 217
Therapy-induced ovarian failure 140
3D bioscaffold
    cellular components 140
    ECM-based 140
    whole-ovary 140
3D bone tissue model
    biochemical evaluation 246
    biomimetic bone scaffold 242, 243
    cell-devoid scaffolds 244, 246, 247
    cell seeding 243, 244
    cell suspension 243, 244
    equipment 241
    materials 242
    mechanical properties 244
    morphological characterization 246
    rBM-MSCs-laden construct 244, 246, 247
    reagents 242
    spectrophotometric measurements 247
3D cell cultures 85, 86, 159, 160, 162, 170
    in vitro assays 174
    in vivo conditions 173
    microenvironment 174
    model 180
    zones 174
3D cell models 174
3D coculture approach 86
3D culture system
    follicle encapsulation 80, 82
    follicles growth 77
    human oocytes in vitro 77
    isolated follicles 76
    isolated preantral follicle recovery 80

3D culture system (*cont.*)
LM 77
microbioreactor preparation 78
porcine ovary handling 78
preantral follicle culture 78, 79
preantral follicle isolation.............. 78, 79, 81
preantral ovarian follicles 76
3D follicle culture 76
3D in vitro models, infection platform
AuNPs... 132
bioengineered model 132
cell lines 134
HNEpC and HTEpC 2D culture 134–136
host–pathogen early interaction 132
hydrogel-based 3D culture system........ 132, 134–136
NPs cellular uptakes...................... 135–137
protocols 133
upper respiratory tract 132
3D in vitro tissue models
cell–cell interactions 103
cell–extracellular matrix interactions....... 103
endometrial cell characterization 105
endometrial cell isolation................ 104, 105
endometrial cells 104
epithelial/stromal cells isolation/primary
culture 105–107
mammalian endometrium 104
vimentin/cytokeratin double immunofluorescence
staining.................... 106, 108, 109
3D matrix and hypoxic environment 152
3D spheroids 160, 162
3D-ViaFlow method
ATP 162
calibrate flow cytometer......................... 164
cell collection.. 165
cellular debris/aggregates exclusion 166
define live/dead cell populations 167
dual stain and evaluation 162
dual viability 160, 165
exclude doublet cells...................... 166, 167
export high resolution plots 167, 168
FDA and PI 160, 161
flow cytometer preparation 163
live/dead cells isolation 162
monolayer cultures......................... 160, 163
multicellular spheroids 160, 162
protocol 162
set up flow cytometer............................. 164
spheroid disassembling 162, 165
spheroids viability 162
staining and trypsinization 163
start flow cytometer 163, 164

trypsin solution 165
viability stain 164
Thermolysin 82
Thiourea 59
Tissue characterization................................. 114
Tissue engineering 240, 283
cell-seeded scaffolds 112
disadvantages.. 112
heart decellularization
cardiac patch 118, 123, 124
EHS 118, 123
tissue retrieval 118, 123
heart recellularization
cardiac patch 119, 124, 125
EHS 118, 124
EVs enrichment, cardiac patch 119
patch EVs enriched 125
in vivo transplantation
animal protocols 125, 126
cardiac patch engraftment 119, 120
cardiac patch transplantation 126, 127
liver patch 119
liver decellularization
ELS 116, 117, 121
liver patch 117, 122
tissue retrieval 116, 120
liver recellularization
ELS 117, 122
enriched medium 117, 123
liver patch 117, 122, 123
multidisciplinary field............................ 112
regenerative medicine 112
Toxicological screenings 86
ToxiLight™ BioAssay Kit 21
Transcriptional silencing 3
Transcriptomics 178, 182, 183
Transepithelial electrical resistance (TEER) 252, 254, 256, 259, 261
Transmembrane proteins 222
Transmission electron microscopy (TEM) 70, 223, 224, 229, 234, 236, 256, 260
Trophectoderm 151
Trophoblast induction 155
Trophoblast-like cells
advantages.. 152
chemical induction 153
epigenetic erasing................................. 152
generation 153
5-aza-CR treatment 155, 157
human skin fibroblasts isolation 154, 156
human skin fibroblasts seeding 154, 155, 157
MEF thawing and inactivation 154–156
trophoblastic induction 155, 157

Trypan blue method 162
Trypan blue staining 176, 178, 179
Trypsin aliquots...169
Trypsin-EDTA solution 66, 110, 144,
        175, 177
Trypsinization...24, 32
Tumor spheroids 86
Two-dimensional cell culture 281
2D to 3D transition challenges
    amenability 19
    diffusion 18
    far-reaching structural transformation 18
    metabolic reprogramming 18
    reproducibility 18
    shape 18
    skill-set requirements 18

U

Ultracentrifugation 72, 221,
        223, 226, 227
Urea 51, 54, 59
Urea standard dilution series...53
Uterine fluid (UF) 220
Uterus...251

V

Variability 17
Versene/EDTA 91
Viability methods
    ATP assay 179
    flow cytometry 179
    strength and benefits..178
    trypan blue staining 178, 179
Vimentin/cytokeratin double immunofluorescence
        staining...106, 108, 109
Virus-mimicking nanoparticle uptake..........................133

W

WASTE 192
Western blot analysis....................................224, 229–231
Whole-ovary bioscaffold 140
Whole-ovary decellularization 141, 145

Z

Zeta potential (ZP) 208, 212, 214, 215, 217
ZetaView
    EVs concentration determination 208
Zona pellucida 11

Printed by Printforce, the Netherlands